Fundamentals of
Geotechnical Engineering

Fundamentals of Geotechnical Engineering

Braja M. Das

California State University, Sacramento

Brooks/Cole
Thomson Learning™

Australia • Canada • Mexico • Singapore • Spain • United Kingdom • United States

Sponsoring Editors: *Suzanne Jeans, Bill Stenquist*
Marketing Team: *Nathan Wilbur, Christina DeVeto*
Editorial Assistant: *Shelley Gesicki*
Production Editor: *Janet Hill*
Production Service: *Helen Walden*
Interior Design: *Helen Walden*

Cover Design: *Denise Davidson*
Art/Photo Editor: *Helen Walden*
Interior Illustration: *The PRD Group, Inc.*
Print Buyer: *Vena Dyer*
Typesetting: *The PRD Group, Inc.*
Cover Printing: *Phoenix Color Corporation*
Printing and Binding: *R. R. Donnelley & Sons*

For more information, contact:
BROOKS/COLE
511 Forest Lodge Road
Pacific Grove, CA 93950 USA
www.brookscole.com

Printed in United States of America

10 9 8 7 6 5 4 3 2 1

Library of Congress Cataloging-in-Publication Data
Das, Braja M.
    Fundamentals of geotechnical engineering / Braja M. Das.
        p. cm.
    Includes bibliographical references and index.
    ISBN 0-534-37114-0
    1. Foundations. 2. Soil mechanics. 3. Engineering geology. I. Title.

TA775.D2264   1999
624.1′5—dc21   99-047729

To Janice and Valerie

About the Author

Dr. Braja Das received his M.S. in Civil Engineering from the University of Iowa, Iowa City, and his Ph.D. in Geotechnical Engineering from the University of Wisconsin, Madison. He is the author of several geotechnical engineering texts and reference books including *Principles of Geotechnical Engineering, Principles of Foundation Engineering,* and *Principles of Soil Dynamics,* all published by Thomson Learning. He has authored more than 200 technical papers in the area of geotechnical engineering. His primary areas of research include shallow foundations, earth anchors, and geosynthetics.

Professor Das has previously served on ASCE's Shallow Foundations Committee, Deep Foundations Committee, and Grouting Committee. He was also a member of ASCE's editorial board for the *Journal of Geotechnical Engineering.* He is the founder of the Geotechnical Engineering Division of the International Society of Offshore and Polar Engineers and has served as an associate editor of ISOPE's *International Journal of Offshore and Polar Engineering.* Recently he served on the editorial board of the journal *Lowland Technology International,* which is published in Japan. He is presently the Chair of the Committee on Chemical and Mechanical Stabilization of the Transportation Research Board of the National Research Council of the United States.

Dr. Das has received numerous awards for teaching excellence, including the AMOCO Foundation Award, AT&T Award for Teaching Excellence from the American Society for Engineering Education, the Ralph Teetor Award from the Society of Automotive Engineers, and the Distinguished Achievement Award for Teaching Excellence from the University of Texas at El Paso.

Since 1994, Professor Das has served as Dean of the College of Engineering and Computer Science at California State University, Sacramento.

Preface

Principles of Foundation Engineering and *Principles of Geotechnical Engineering* were originally published in 1984 and 1985, respectively. These texts were well received by instructors, students, and practitioners alike. Depending on the needs of the users, the texts were revised and are presently in their fourth editions.

More recently, there have been several requests to prepare a single volume that is concise in nature but combines the essential components of *Principles of Foundation Engineering* and *Principles of Geotechnical Engineering*. This text is the product of those requests. It consists of 13 chapters and includes the fundamental concepts of soil mechanics as well as foundation engineering, including bearing capacity and settlement of shallow foundations (spread footings and mats), retaining walls, braced cuts, piles, and drilled shafts.

Research into the development of the fundamental principles of geotechnical engineering—that is, soil mechanics and rock mechanics—and their application to foundation analysis and design has been extensive during the last five decades. All authors are tempted to include all of the recent developments in a textbook; however, since this text is intended as an introductory text, it stresses the fundamental principles without becoming cluttered with too many details and alternatives.

Instructors must emphasize the difference between soil mechanics and foundation engineering in the classroom. Soil mechanics is the branch of engineering that involves the study of the properties of soils and their behavior under stress and strain in idealized conditions. Foundation engineering is the application of the principles of soil mechanics and geology in the planning, design, and construction of foundations for buildings, highways, dams, and so forth. Approximations and deviations from idealized conditions of soil mechanics become necessary for proper foundation design because natural soil deposits are not homogeneous in most cases. However, if a structure is to function properly, these approximations can be made only by an engineer who has a good background in soil mechanics. This book provides that background.

Fundamentals of Geotechnical Engineering is liberally illustrated to help students understand the material. Only SI units are used in this text. Several examples are presented in each chapter. Problems are provided at the end of each chapter for homework assignment, and they are all in SI units.

Acknowledgments

My wife, Janice, has been a continuing source of inspiration and help in completing the project.

I wish to acknowledge the following for their helpful reviews and comments on the manuscript:

Rob O. Davis, University of Canterbury
Jeffrey C. Evans, Bucknell University
Mark B. Jaksa, The University of Adelaide
C. Hsein Juang, Clemson University
Dilip K. Nag, Monash University—Gippsland Campus
Jean H. Prevost, Princeton University
Charles W. Schwartz, University of Maryland, College Park
Roly J. Salvas, Ryerson Polytechnic University
Nagaratnam Sivakugan, Purdue University
John Stormont, University of New Mexico
Dobroslav Znidarcic, University of Colorado
Manoochehr Zoghi, University of Dayton

I am grateful to Bill Stenquist and Suzanne Jeans of Brooks/Cole Publishing Company for their encouragement and understanding throughout the preparation and publication of the manuscript.

Braja M. Das
Sacramento, California

Contents

CHAPTER 11 SHALLOW FOUNDATIONS—BEARING CAPACITY AND SETTLEMENT 389

CHAPTER 12 RETAINING WALLS AND BRACED CUTS 445

1

Soil Deposits and Grain-Size Analysis

1.1 Introduction

In the general engineering sense, *soil* is defined as the uncemented aggregate of mineral grains and decayed organic matter (solid particles) along with the liquid and gas that occupy the empty spaces between the solid particles. Soil is used as a construction material in various civil engineering projects, and it supports structural foundations. Thus, civil engineers must study the properties of soil, such as its origin, grain-size distribution, ability to drain water, compressibility, shear strength, load-bearing capacity, and so on.

The record of the first use of soil as a construction material is lost in antiquity. For years, the art of soils engineering was based on only past experience. With the growth of science and technology, however, the need for better and more economical structural design and construction became critical. This need led to a detailed study of the nature and properties of soil as they relate to engineering. The publication of *Erdbaumechanik* by Karl Terzaghi in 1925 gave birth to modern soil mechanics.

Soil mechanics is the branch of science that deals with the study of the physical properties of soil and the behavior of soil masses subjected to various types of forces. *Soils engineering* is the application of the principles of soil mechanics to practical problems. *Geotechnical engineering* is the science and practice of that part of civil engineering that involves natural materials found close to the surface of the earth. In a general sense, it includes the application of the fundamental principles of soil mechanics and rock mechanics to foundation design problems.

1.2 Natural Soil Deposits

Soil is produced by weathering—the breaking down of various types of rocks into smaller pieces through mechanical and chemical processes. Some soils stay where they are formed and cover the rock surface from which they are derived. These soils are referred to as *residual soils*. In contrast, some weathered products are transported by physical processes to other places and deposited. These soils are called *transported soils*. Based on the *transporting agent*, transported soils can be subdivided into three major categories:

1. *Alluvial or fluvial:* deposited by running water
2. *Glacial:* deposited by glacier action
3. *Aeolian:* deposited by wind action

In addition to transported and residual soils, *peats* are derived from the decomposition of organic materials usually found in low-lying areas where the water table is near or above the ground surface. The presence of a high water table helps support the growth of aquatic plants that, when decomposed, form peat. This type of deposit is usually encountered in coastal areas and glaciated regions. When a relatively large percentage of peat is mixed with inorganic soil, it is referred to as *organic soil*. Organic soils characteristically have a natural moisture content ranging from 200% to 300%, and they are highly compressible. Laboratory tests have shown that, under loads, a large amount of settlement results from the secondary consolidation of organic soils (see Chapter 6).

During the planning, design, and construction of foundations, embankments, and earth-retaining structures, engineers find it helpful to know the origin of the soil deposit over which the foundation is to be constructed because each soil deposit has it own unique physical attributes.

1.3   Soil-Particle Size

Irrespective of the origin of soil, the sizes of particles, in general, that make up soil vary over a wide range. Soils are generally called *gravel*, *sand*, *silt*, or *clay*, depending on the predominant size of particles within the soil. To describe soils by their particle size, several organizations have developed *soil-separate-size limits*. Table 1.1 shows the soil-separate-size limits developed by the Massachusetts Institute of Technology, the U.S. Department of Agriculture, the American Association of State Highway and Transportation Officials, and the U.S. Army Corps of Engineers and U.S. Bureau of Reclamation. In this table, the MIT system is presented for illustration purposes only because it plays an important role in the history of the development of soil-separate-size limits. Presently, however, the Unified System is almost universally accepted. The Unified Soil Classification System has now been adopted by the American Society for Testing and Materials.

Gravels are pieces of rocks with occasional particles of quartz, feldspar, and other minerals.

Sand particles are made of mostly quartz and feldspar. Other mineral grains may also be present at times.

Silts are the microscopic soil fractions that consist of very fine quartz grains and some flake-shaped particles that are fragments of micaceous minerals.

Clays are mostly flake-shaped microscopic and submicroscopic particles of mica, clay minerals, and other minerals. As shown in Table 1.1, clays are generally

Table 1.1 Soil-separate-size limits

Name of organization	Grain size (mm)			
	Gravel	Sand	Silt	Clay
Massachusetts Institute of Technology (MIT)	> 2	2 to 0.06	0.06 to 0.002	< 0.002
U.S. Department of Agriculture (USDA)	> 2	2 to 0.05	0.05 to 0.002	< 0.002
American Association of State Highway and Transportation Officials (AASHTO)	76.2 to 2	2 to 0.075	0.075 to 0.002	< 0.002
Unified Soil Classification System (U.S. Army Corps of Engineers; U.S. Bureau of Reclamation; American Society for Testing and Materials)	76.2 to 4.75	4.75 to 0.075	Fines (i.e., silts and clays) < 0.075	

defined as particles smaller than 0.002 mm. In some cases, particles between 0.002 and 0.005 mm in size are also referred to as clay. Particles are classified as *clay* on the basis of their size; they may not necessarily contain clay minerals. Clays are defined as those particles "which develop plasticity when mixed with a limited amount of water" (Grim, 1953). (Plasticity is the puttylike property of clays when they contain a certain amount of water.) Nonclay soils can contain particles of quartz, feldspar, or mica that are small enough to be within the clay size classification. Hence, it is appropriate for soil particles smaller than 2 μ, or 5 μ as defined under different systems, to be called clay-sized particles rather than clay. Clay particles are mostly of colloidal size range (< 1 μ), and 2 μ appears to be the upper limit.

1.4   Clay Minerals

Clay minerals are complex aluminum silicates composed of one of two basic units: (1) *silica tetrahedron* and (2) *alumina octahedron*. Each tetrahedron unit consists of four oxygen atoms surrounding a silicon atom (Figure 1.1a). The combination of tetrahedral silica units gives a *silica sheet* (Figure 1.1b). Three oxygen atoms at the base of each tetrahedron are shared by neighboring tetrahedra. The octahedral units consist of six hydroxyls surrounding an aluminum atom (Figure 1.1c), and the combination of the octahedral aluminum hydroxyl units gives an *octahedral sheet*. (This is also called a *gibbsite sheet;* Figure 1.1d.) Sometimes magnesium replaces the aluminum atoms in the octahedral units; in that case, the octahedral sheet is called a *brucite sheet*.

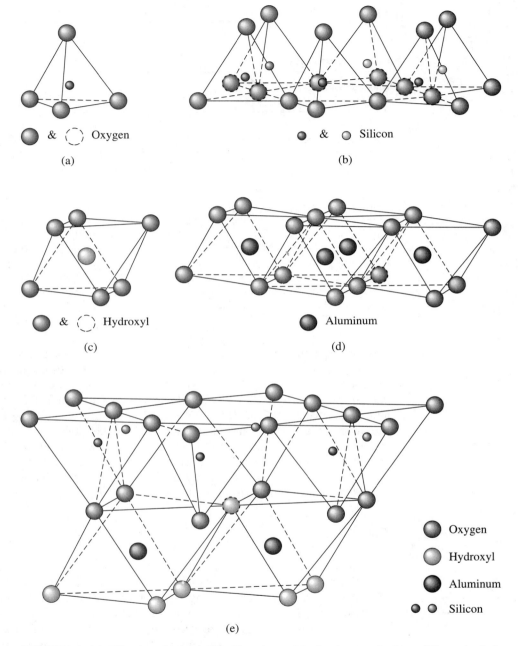

FIGURE 1.1 (a) Silica tetrahedron; (b) silica sheet; (c) alumina octahedron; (d) octahedral (gibbsite) sheet; (e) elemental silica-gibbsite sheet (After Grim, 1959)

In a silica sheet, each silicon atom with a positive valence of four is linked to four oxygen atoms with a total negative valence of eight. But each oxygen atom at the base of the tetrahedron is linked to two silicon atoms. This means that the top oxygen atom of each tetrahedral unit has a negative valence charge of one to be counterbalanced. When the silica sheet is stacked over the octahedral sheet, as shown in Figure 1.1e, these oxygen atoms replace the hydroxyls to satisfy their valence bonds.

Kaolinite consists of repeating layers of elemental silica-gibbsite sheets, as shown in Figure 1.2a. Each layer is about 7.2 Å thick. The layers are held together by hydrogen bonding. Kaolinite occurs as platelets, each with a lateral dimension of 1000 to 20,000 Å and a thickness of 100 to 1000 Å. The surface area of the kaolinite particles per unit mass is about 15 m²/g. The surface area per unit mass is defined as *specific surface*.

Illite consists of a gibbsite sheet bonded to two silica sheets—one at the top and another at the bottom (Figure 1.2b). It is sometimes called *clay mica*. The illite layers are bonded together by potassium ions. The negative charge to balance the potassium ions comes from the substitution of aluminum for some silicon in the tetrahedral sheets. Substitution of one element for another with no change in the crystalline form is known as *isomorphous substitution*. Illite particles generally have lateral dimensions ranging from 1000 to 5000 Å and thicknesses from 50 to 500 Å. The specific surface of the particles is about 80 m²/g.

Montmorillonite has a similar structure as illite—that is, one gibbsite sheet sandwiched between two silica sheets (Figure 1.2c). In montmorillonite, there is isomorphous substitution of magnesium and iron for aluminum in the octahedral sheets. Potassium ions are not present here as in the case of illite, and a large amount of water is attracted into the space between the layers. Particles of montmorillonite have lateral dimensions of 1000 to 5000 Å and thicknesses of 10 to 50 Å. The specific surface is about 800 m²/g.

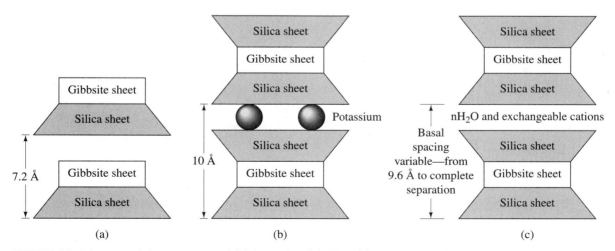

FIGURE 1.2 Diagram of the structures of (a) kaolinite; (b) illite; (c) montmorillonite

Besides kaolinite, illite, and montmorillonite, other common clay minerals generally found are chlorite, halloysite, vermiculite, and attapulgite.

The clay particles carry a net negative charge on their surfaces. This is the result both of isomorphous substitution and of a break in continuity of the structure at its edges. Larger negative charges are derived from larger specific surfaces. Some positively charged sites also occur at the edges of the particles.

In dry clay, the negative charge is balanced by exchangeable cations, like Ca^{++}, Mg^{++}, Na^+, and K^+, surrounding the particles being held by electrostatic attraction. When water is added to clay, these cations and a small number of anions float around the clay particles. This is referred to as *diffuse double layer* (Figure 1.3a). The cation concentration decreases with distance from the surface of the particle (Figure 1.3b).

Water molecules are polar. Hydrogen atoms are not arranged in a symmetric manner around an oxygen atom; instead, they occur at a bonded angle of 105°. As a result, a water molecule acts like a small rod with a positive charge at one end and a negative charge at the other end. It is known as a *dipole*.

The dipolar water is attracted both by the negatively charged surface of the clay particles and by the cations in the double layer. The cations, in turn, are attracted to the soil particles. A third mechanism by which water is attracted to clay particles is *hydrogen bonding,* in which hydrogen atoms in the water molecules are shared with oxygen atoms on the surface of the clay. Some partially hydrated cations in the pore water are also attracted to the surface of clay particles. These cations attract dipolar water molecules. The force of attraction between water and clay decreases with distance from the surface of the particles. All of the water held to clay particles by force of attraction is known as *double-layer water*. The innermost layer of double-layer water, which is held very strongly by clay, is known as *adsorbed water*. This water is more viscous than is free water. The orientation of water around the clay particles gives clay soils their plastic properties.

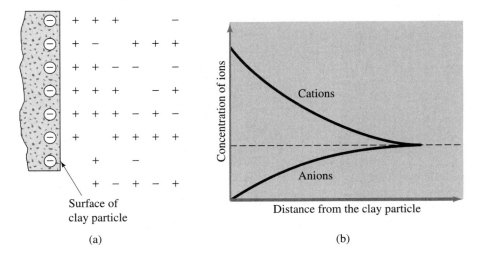

(a)   (b)

FIGURE 1.3 Diffuse double layer

Table 1.2 Specific gravity of important minerals

Mineral	Specific gravity, G_s
Quartz	2.65
Kaolinite	2.6
Illite	2.8
Montmorillonite	2.65–2.80
Halloysite	2.0–2.55
Potassium feldspar	2.57
Sodium and calcium feldspar	2.62–2.76
Chlorite	2.6–2.9
Biotite	2.8–3.2
Muscovite	2.76–3.1
Hornblende	3.0–3.47
Limonite	3.6–4.0
Olivine	3.27–3.37

1.5 Specific Gravity (G_s)

The specific gravity of the soil solids is used in various calculations in soil mechanics. The specific gravity can be determined accurately in the laboratory. Table 1.2 shows the specific gravity of some common minerals found in soils. Most of the minerals have a specific gravity that falls within a general range of 2.6 to 2.9. The specific gravity of solids of light-colored sand, which is made of mostly quartz, may be estimated to be about 2.65; for clayey and silty soils, it may vary from 2.6 to 2.9.

1.6 Mechanical Analysis of Soil

Mechanical analysis is the determination of the size range of particles present in a soil, expressed as a percentage of the total dry weight (or mass). Two methods are generally used to find the particle-size distribution of soil: (1) *sieve analysis*—for particle sizes larger than 0.075 mm in diameter, and (2) *hydrometer analysis*—for particle sizes smaller than 0.075 mm in diameter. The basic principles of sieve analysis and hydrometer analysis are described next.

Sieve Analysis

Sieve analysis consists of shaking the soil sample through a set of sieves that have progressively smaller openings. U.S. standard sieve numbers and the sizes of openings are given in Table 1.3.

First the soil is oven dried, and then all lumps are broken into small particles before they are passed through the sieves. Figure 1.4 shows a set of sieves in a sieve shaker used for conducting the test in the laboratory. After the shaking period

Table 1.3 U.S. standard
sieve sizes

Sieve no.	Opening (mm)
4	4.750
6	3.350
8	2.360
10	2.000
16	1.180
20	0.850
30	0.600
40	0.425
50	0.300
60	0.250
80	0.180
100	0.150
140	0.106
170	0.088
200	0.075
270	0.053

FIGURE 1.4 A set of sieves for a test in the laboratory

Table 1.4 Sieve analysis (mass of dry soil sample = 450 g)

Sieve no. (1)	Diameter (mm) (2)	Mass of soil retained on each sieve (g) (3)	Percent of soil retained on each sieve* (4)	Percent passing† (5)
10	2.000	0	0	100.00
16	1.180	9.90	2.20	97.80
30	0.600	24.66	5.48	92.32
40	0.425	17.60	3.91	88.41
60	0.250	23.90	5.31	83.10
100	0.150	35.10	7.80	75.30
200	0.075	59.85	13.30	62.00
Pan	—	278.99	62.00	0

*Column 4 = (column 3)/(total mass of soil) × 100
†This is also referred to as *percent finer*.

is over, the mass of soil retained on each sieve is determined. When cohesive soils are analyzed, it may be difficult to break lumps into individual particles. In that case, the soil is mixed with water to make a slurry and then washed through the sieves. The portions retained on each sieve are collected separately and oven dried before the mass retained on each sieve is measured.

The results of sieve analysis are generally expressed as the percentage of the total weight of soil that passed through different sieves. Table 1.4 shows an example of the calculations performed in a sieve analysis.

Hydrometer Analysis

Hydrometer analysis is based on the principle of sedimentation of soil grains in water. When a soil specimen is dispersed in water, the particles settle at different velocities, depending on their shape, size, and weight. For simplicity, it is assumed that all the soil particles are spheres and the velocity of soil particles can be expressed by *Stokes' law*, according to which

$$v = \frac{\rho_s - \rho_w}{18\eta} D^2 \tag{1.1}$$

where v = velocity
   ρ_s = density of soil particles
   ρ_w = density of water
   η = viscosity of water
   D = diameter of soil particles

Thus, from Eq. (1.1),

$$D = \sqrt{\frac{18\eta v}{\rho_s - \rho_w}} = \sqrt{\frac{18\eta}{\rho_s - \rho_w}} \sqrt{\frac{L}{t}} \tag{1.2}$$

where $v = \dfrac{\text{distance}}{\text{time}} = \dfrac{L}{t}$

Note that·

$$\rho_s = G_s \rho_w \qquad (1.3)$$

Thus, combining Eqs. (1.2) and (1.3) gives

$$D = \sqrt{\frac{18\eta}{(G_s - 1)\rho_w}} \sqrt{\frac{L}{t}} \qquad (1.4)$$

If the units of η are (g · sec)/cm², ρ_w is in g/cm³, L is in cm, t is in min, and D is in mm, then

$$\frac{D\,(\text{mm})}{10} = \sqrt{\frac{18\eta[(\text{g} \cdot \text{sec})/\text{cm}^2]}{(G_s - 1)\rho_w(\text{g/cm}^3)}} \sqrt{\frac{L\,(\text{cm})}{t\,(\text{min}) \times 60}}$$

or

$$D = \sqrt{\frac{30\eta}{(G_s - 1)\rho_w}} \sqrt{\frac{L}{t}}$$

Assuming ρ_w to be approximately equal to 1 g/cm³, we have

$$D\,(\text{mm}) = K \sqrt{\frac{L\,(\text{cm})}{t\,(\text{min})}} \qquad (1.5)$$

where $K = \sqrt{\dfrac{30\eta}{(G_s - 1)}}$ $\qquad (1.6)$

Note that the value of K is a function of G_s and η, which are dependent on the temperature of the test.

In the laboratory, the hydrometer test is conducted in a sedimentation cylinder with 50 g of oven-dry sample. The sedimentation cylinder is 457 mm high and 63.5 mm in diameter. It is marked for a volume of 1000 ml. Sodium hexametaphosphate is generally used as the *dispersing agent*. The volume of the dispersed soil suspension is brought up to 1000 ml by adding distilled water.

When an ASTM 152H type of hydrometer is placed in the soil suspension (Figure 1.5) at a time t, measured from the start of sedimentation, it measures the specific gravity in the vicinity of its bulb at a depth L. The specific gravity is a function of the amount of soil particles present per unit volume of suspension at that depth. Also, at a time t, the soil particles in suspension at a depth L will have a diameter smaller than D as calculated in Eq. (1.5). The larger particles would have settled beyond the zone of measurement. Hydrometers are designed to give the amount of soil, in grams, that is still in suspension. Hydrometers are calibrated for soils that have a specific gravity (G_s) of 2.65; for soils of other specific gravity, it is necessary to make corrections.

By knowing the amount of soil in suspension, L, and t, we can calculate the percentage of soil by weight finer than a given diameter. Note that L is the depth measured from the surface of the water to the center of gravity of the hydrometer bulb at which the density of the suspension is measured. The value of L will change

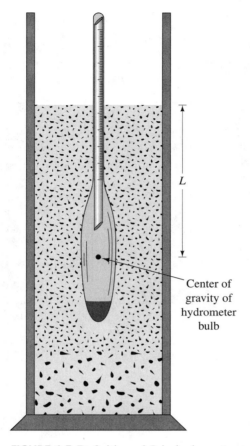

FIGURE 1.5 Definition of L in hydrometer test

with time t; its variation with the hydrometer readings is given in the *ASTM Book of Standards* (1998—see Test Designation D-422). Hydrometer analysis is effective for separating soil fractions down to a size of about 0.5 μ.

Particle-Size Distribution Curve

The results of mechanical analysis (sieve and hydrometer analyses) are generally presented on semilogarithmic plots known as *particle-size* (or grain-size) *distribution curves*. The particle diameters are plotted in log scale and the corresponding percent finer in arithmetic scale. As an example, the particle-size distribution curves for two soils are shown in Figure 1.6. The particle-size distribution curve for soil A is the combination of the sieve analysis results presented in Table 1.4 and the results of the hydrometer analysis for the finer fraction. When the results of sieve analysis and hydrometer analysis are combined, a discontinuity generally occurs in the range where they overlap. The reason for the discontinuity is that soil particles are generally irregular in shape. Sieve analysis gives the intermediate dimension of a

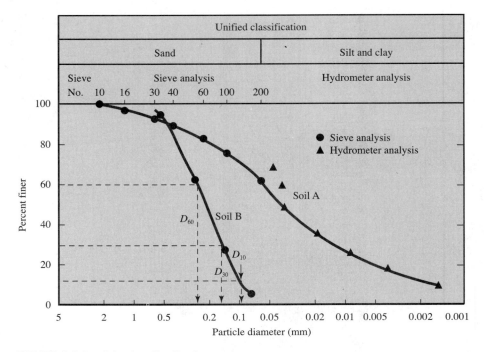

FIGURE 1.6 Particle-size distribution curves

particle; hydrometer analysis gives the diameter of a sphere that would settle at the same rate as the soil particle.

The percentages of gravel, sand, silt, and clay-size particles present in a soil can be obtained from the particle-size distribution curve. According to the Unified Soil Classification System, soil A in Figure 1.6 has these percentages:

Gravel (size limits—greater than 4.75 mm) = 0%
Sand (size limits—4.75 to 0.075 mm) = percent finer than 4.75 mm diameter − percent finer than 0.075 mm diameter = 100 − 62 = 38%
Silt and clay (size limits—less than 0.075 mm) = 38%

1.7 *Effective Size, Uniformity Coefficient, and Coefficient of Gradation*

The particle-size distribution curves can be used to compare different soils. Also, three basic soil parameters can be determined from these curves, and they can be used to classify granular soils. The three soil parameters are:

1. Effective size
2. Uniformity coefficient
3. Coefficient of gradation

The diameter in the particle-size distribution curve corresponding to 10% finer is defined as the *effective size,* or D_{10}. The *uniformity coefficient* is given by the relation

$$C_u = \frac{D_{60}}{D_{10}} \qquad (1.7)$$

where C_u = uniformity coefficient
   D_{60} = the diameter corresponding to 60% finer in the particle-size distribution curve

The *coefficient of gradation* may be expressed as

$$C_z = \frac{D_{30}^2}{D_{60} \times D_{10}} \qquad (1.8)$$

where C_z = coefficient of gradation
   D_{30} = diameter corresponding to 30% finer

The particle-size distribution curve shows not only the range of particle sizes present in a soil but also the distribution of various size particles. Three curves are shown in Figure 1.7. Curve I represents a type of soil in which most of the soil

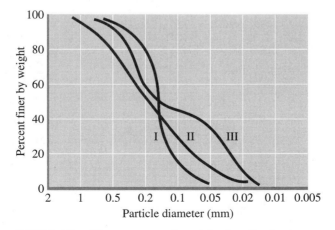

FIGURE 1.7 Different types of particle-size distribution curves

grains are the same size. This is called *poorly graded* soil. Curve II represents a soil in which the particle sizes are distributed over a wide range and is termed *well graded*. A well-graded soil has a uniformity coefficient greater than about 4 for gravels and 6 for sands, and a coefficient of gradation between 1 and 3 (for gravels and sands). A soil might have a combination of two or more uniformly graded fractions. Curve III represents such a soil, termed *gap graded*.

Problems

1.1   The table gives the results of a sieve analysis:

U.S. sieve no.	Mass of soil retained on each sieve (g)
4	0
10	21.6
20	49.5
40	102.6
60	89.1
100	95.6
200	60.4
Pan	31.2

   a. Determine the percent finer than each sieve size and plot a particle-size distribution curve.
   b. Determine D_{10}, D_{30}, and D_{60} from the particle-size distribution curve.
   c. Calculate the uniformity coefficient, C_u.
   d. Calculate the coefficient of gradation, C_z.

1.2   A soil has these values:

$$D_{10} = 0.1 \text{ mm}$$
$$D_{30} = 0.41 \text{ mm}$$
$$D_{60} = 0.62 \text{ mm}$$

Calculate the uniformity coefficient and the coefficient of gradation of the soil.

1.3   Repeat Problem 1.2 for a soil with the following values:

$$D_{10} = 0.082 \text{ mm}$$
$$D_{30} = 0.29 \text{ mm}$$
$$D_{60} = 0.51 \text{ mm}$$

1.4   Repeat Problem 1.1 with these results of a sieve analysis:

U.S. sieve no.	Mass of soil retained on each sieve (g)
4	0
6	30
10	48.7
20	127.3
40	96.8
60	76.6
100	55.2
200	43.4
Pan	22

1.5 Repeat Problem 1.1 with the sieve analysis results given in the table:

U.S. sieve no.	Mass of soil retained on each sieve (g)
4	0
6	0
10	0
20	9.1
40	249.4
60	179.8
100	22.7
200	15.5
Pan	23.5

1.6 The particle characteristics of a soil are given in the table. Draw the particle-size distribution curve and find the percentages of gravel, sand, silt, and clay according to the MIT system (Table 1.1).

Size (mm)	Percent finer by weight
0.850	100.0
0.425	92.1
0.250	85.8
0.150	77.3
0.075	62.0
0.040	50.8
0.020	41.0
0.010	34.3
0.006	29.0
0.002	23.0

1.7   Redo Problem 1.6 according to the USDA system (Table 1.1).

1.8   Redo Problem 1.6 according to the AASHTO system (Table 1.1).

1.9   The particle-size characteristics of a soil are given in the table. Find the percentages of gravel, sand, silt, and clay according to the MIT system (Table 1.1).

Size (mm)	Percent finer by weight
0.850	100.0
0.425	100.0
0.250	94.1
0.150	79.3
0.075	34.1
0.040	28.0
0.020	25.2
0.010	21.8
0.006	18.9
0.002	14.0

1.10   Redo Problem 1.9 according to the USDA system (Table 1.1).

1.11   Redo Problem 1.9 according to the AASHTO system (Table 1.1).

References

American Society for Testing and Materials (1998). *ASTM Book of Standards,* Vol. 04.08, West Conshohocken, PA.

Grim, R. E. (1953). *Clay Mineralogy,* McGraw-Hill, New York.

Grim, R. E. (1959). "Physico-Chemical Properties of Soils: Clay Minerals," *Journal of the Soil Mechanics and Foundations Division,* ASCE, Vol. 85, No. SM2, 1–17.

Terzaghi, K. (1925). *Erdbaumechanik auf Bodenphysikalischer Grundlage,* Deuticke, Vienna.

Supplementary References for Further Study

Mitchell, J. K. (1993). *Fundamentals of Soil Behavior,* 2nd ed., Wiley, New York.

Van Olphen, H. (1963). *An Introduction to Clay Colloid Chemistry,* Wiley Interscience, New York.

2

Weight–Volume Relationships, Plasticity, and Soil Classification

The preceding chapter presented the geological processes by which soils are formed, the description of the soil-particle size limits, and the mechanical analysis of soils. In natural occurrence, soils are three-phase systems consisting of soil solids, water, and air. This chapter discusses the weight–volume relationships of soil aggregates, their structures and plasticity, and their engineering classification.

2.1 Weight–Volume Relationships

Figure 2.1a shows an element of soil of volume V and weight W as it would exist in a natural state. To develop the weight–volume relationships, we separate the three phases (that is, solid, water, and air) as shown in Figure 2.1b. Thus, the total volume of a given soil sample can be expressed as

$$V = V_s + V_v = V_s + V_w + V_a \tag{2.1}$$

where V_s = volume of soil solids
V_v = volume of voids
V_w = volume of water in the voids
V_a = volume of air in the voids

Assuming the weight of the air to be negligible, we can give the total weight of the sample as

$$W = W_s + W_w \tag{2.2}$$

where W_s = weight of soil solids
W_w = weight of water

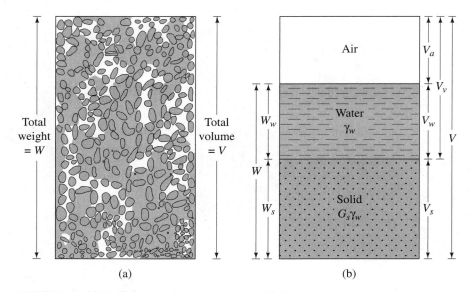

FIGURE 2.1 (a) Soil element in natural state; (b) three phases of the soil element

The volume relationships commonly used for the three phases in a soil element are *void ratio, porosity,* and *degree of saturation. Void ratio* (e) is defined as the ratio of the volume of voids to the volume of solids, or

$$e = \frac{V_v}{V_s} \tag{2.3}$$

Porosity (n) is defined as the ratio of the volume of voids to the total volume, or

$$n = \frac{V_v}{V} \tag{2.4}$$

Degree of saturation (S) is defined as the ratio of the volume of water to the volume of voids, or

$$S = \frac{V_w}{V_v} \tag{2.5}$$

The degree of saturation is commonly expressed as a percentage.

The relationship between void ratio and porosity can be derived from Eqs. (2.1), (2.3), and (2.4), as follows:

$$e = \frac{V_v}{V_s} = \frac{V_v}{V - V_v} = \frac{\left(\dfrac{V_v}{V}\right)}{1 - \left(\dfrac{V_v}{V}\right)} = \frac{n}{1 - n} \qquad (2.6)$$

Also, from Eq. (2.6), we have

$$n = \frac{e}{1 + e} \qquad (2.7)$$

The common weight relationships are *moisture content* and *unit weight*. *Moisture content* (w) is also referred to as *water content* and is defined as the ratio of the weight of water to the weight of solids in a given volume of soil, or

$$w = \frac{W_w}{W_s} \qquad (2.8)$$

Unit weight (γ) is the weight of soil per unit volume:

$$\gamma = \frac{W}{V} \qquad (2.9)$$

The unit weight can also be expressed in terms of weight of soil solids, moisture content, and total volume. From Eqs. (2.2), (2.8), and (2.9), we have

$$\gamma = \frac{W}{V} = \frac{W_s + W_w}{V} = \frac{W_s \left[1 + \left(\dfrac{W_w}{W_s}\right)\right]}{V} = \frac{W_s (1 + w)}{V} \qquad (2.10)$$

Soils engineers sometimes refer to the unit weight defined by Eq. (2.9) as the *moist unit weight*.

It is sometimes necessary to know the weight per unit volume of soil excluding water. This is referred to as the *dry unit weight*, γ_d. Thus,

$$\gamma_d = \frac{W_s}{V} \qquad (2.11)$$

From Eqs. (2.10) and (2.11), we can give the relationship among unit weight, dry unit weight, and moisture content as

$$\gamma_d = \frac{\gamma}{1 + w} \qquad (2.12)$$

Unit weight is expressed in kilonewtons per cubic meter (kN/m³). Since the newton is a derived unit, it may sometimes be convenient to work with densities

(ρ) of soil. The SI unit of density is kilograms per cubic meter (kg/m³). We can write the density equations [similar to Eqs. (2.9) and (2.11)] as

$$\rho = \frac{m}{V} \tag{2.13a}$$

and

$$\rho_d = \frac{m_s}{V} \tag{2.13b}$$

where ρ = density of soil (kg/m³)
ρ_d = dry density of soil (kg/m³)
m = total mass of the soil sample (kg)
m_s = mass of soil solids in the sample (kg)

The unit of total volume, V, is m³.

The unit weights of soil in N/m³ can be obtained from densities in kg/m³ as

$$\gamma = \rho \cdot g = 9.81\rho \tag{2.14a}$$

and

$$\gamma_d = \rho_d \cdot g = 9.81\rho_d \tag{2.14b}$$

where g = acceleration due to gravity = 9.81 m/sec².

2.2   Relationships Among Unit Weight, Void Ratio, Moisture Content, and Specific Gravity

To obtain a relationship among unit weight (or density), void ratio, and moisture content, consider a volume of soil in which the volume of the soil solids is 1, as shown in Figure 2.2. If the volume of the soil solids is 1, then the volume of voids is numerically equal to the void ratio, e [from Eq. (2.3)]. The weights of soil solids and water can be given as

$$W_s = G_s\gamma_w$$
$$W_w = wW_s = wG_s\gamma_w$$

where G_s = specific gravity of soil solids
w = moisture content
γ_w = unit weight of water

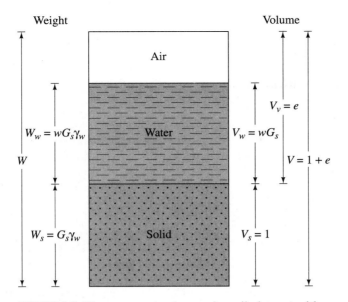

FIGURE 2.2 Three separate phases of a soil element with volume of soil solids equal to 1

The unit weight of water is 9.81 kN/m³. Now, using the definitions of unit weight and dry unit weight [Eqs. (2.9) and (2.11)], we can write

$$\gamma = \frac{W}{V} = \frac{W_s + W_w}{V} = \frac{G_s\gamma_w + wG_s\gamma_w}{1 + e} = \frac{(1 + w)\,G_s\gamma_w}{1 + e} \tag{2.15}$$

and

$$\gamma_d = \frac{W_s}{V} = \frac{G_s\gamma_w}{1 + e} \tag{2.16}$$

Since the weight of water in the soil element under consideration is $wG_s\gamma_w$, the volume occupied by it is

$$V_w = \frac{W_w}{\gamma_w} = \frac{wG_s\gamma_w}{\gamma_w} = wG_s$$

Hence, from the definition of degree of saturation [Eq. (2.5)], we have

$$S = \frac{V_w}{V_v} = \frac{wG_s}{e}$$

or

$$Se = wG_s \tag{2.17}$$

This is a very useful equation for solving problems involving three-phase relationships.

If the soil sample is *saturated*—that is, the void spaces are completely filled with water (Figure 2.3)—the relationship for saturated unit weight can be derived in a similar manner:

$$\gamma_{\text{sat}} = \frac{W}{V} = \frac{W_s + W_w}{V} = \frac{G_s \gamma_w + e\gamma_w}{1 + e} = \frac{(G_s + e)\gamma_w}{1 + e} \tag{2.18}$$

where γ_{sat} = saturated unit weight of soil.

As mentioned before, because it is convenient to work with densities, the following equations [similar to the unit–weight relationships given in Eqs. (2.15), (2.16), and (2.18)] are useful:

$$\text{Density} = \rho = \frac{(1 + w)G_s \rho_w}{1 + e} \tag{2.19a}$$

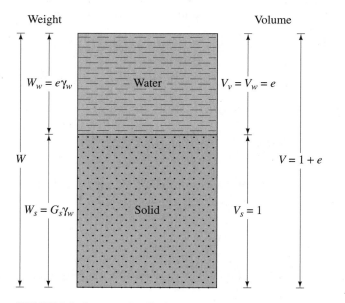

FIGURE 2.3 Saturated soil element with volume of soil solids equal to 1

$$\text{Dry density} = \rho_d = \frac{G_s \rho_w}{1 + e} \tag{2.19b}$$

$$\text{Saturated density} = \rho_{sat} = \frac{(G_s + e)\rho_w}{1 + e} \tag{2.19c}$$

where ρ_w = density of water = 1000 kg/m³.

The relationships among unit weight, porosity, and moisture content can also be developed by considering a soil specimen that has a total volume equal to 1.

2.3 *Relative Density*

The term *relative density* is commonly used to indicate the *in situ* denseness or looseness of granular soil. It is defined as

$$D_r = \frac{e_{max} - e}{e_{max} - e_{min}} \tag{2.20}$$

where D_r = relative density, usually given as a percentage
 e = *in situ* void ratio of the soil
 e_{max} = void ratio of the soil in the loosest condition
 e_{min} = void ratio of the soil in the densest condition

The values of D_r may vary from a minimum of 0 for very loose soil to a maximum of 1 for very dense soil. Soils engineers qualitatively describe the granular soil deposits according to their relative densities, as shown in Table 2.1. Some typical values of void ratio, moisture content in a saturated condition, and dry unit weight as encountered in a natural state are given in Table 2.2.

By using the definition of dry unit weight given in Eq. (2.16), we can also express relative density in terms of maximum and minimum possible dry unit weights. Thus,

Table 2.1 Qualitative description of granular soil deposits

Relative density (%)	Description of soil deposit
0–15	Very loose
15–50	Loose
50–70	Medium
70–85	Dense
85–100	Very dense

Table 2.2 Void ratio, moisture content, and dry unit weight for some typical soils in a natural state

Type of soil	Void ratio, e	Natural Moisture content in a saturated state (%)	Dry unit weight, γ_d (kN/m³)
Loose uniform sand	0.8	30	14.5
Dense uniform sand	0.45	16	18
Loose angular-grained silty sand	0.65	25	16
Dense angular-grained silty sand	0.4	15	19
Stiff clay	0.6	21	17
Soft clay	0.9–1.4	30–50	11.5–14.5
Loess	0.9	25	13.5
Soft organic clay	2.5–3.2	90–120	6–8
Glacial till	0.3	10	21

$$D_r = \frac{\left[\dfrac{1}{\gamma_{d(min)}}\right] - \left[\dfrac{1}{\gamma_d}\right]}{\left[\dfrac{1}{\gamma_{d(min)}}\right] - \left[\dfrac{1}{\gamma_{d(max)}}\right]} = \left[\frac{\gamma_d - \gamma_{d(min)}}{\gamma_{d(max)} - \gamma_{d(min)}}\right]\left[\frac{\gamma_{d(max)}}{\gamma_d}\right] \tag{2.21}$$

where $\gamma_{d(min)}$ = dry unit weight in the loosest condition (at a void ratio of e_{max})
γ_d = *in situ* dry unit weight (at a void ratio of e)
$\gamma_{d(max)}$ = dry unit weight in the densest condition (at a void ratio of e_{min})

EXAMPLE 2.1

In the natural state, a moist soil has a volume of 0.0093 m³ and weighs 177.6 N. The oven dry weight of the soil is 153.6 N. If G_s = 2.71, calculate the moisture content, moist unit weight, dry unit weight, void ratio, porosity, and degree of saturation.

Solution   Refer to Figure 2.4. The moisture content [Eq. (2.8)] is

$$w = \frac{W_w}{W_s} = \frac{W - W_s}{W_s} = \frac{177.6 - 153.6}{153.6} = \frac{24}{153.6} \times 100 = \textbf{15.6\%}$$

The moist unit weight [Eq. (2.9)] is

$$\gamma = \frac{W}{V} = \frac{177.6}{0.0093} = 19{,}906 \text{ N/m}^3 \approx \textbf{19.1 kN/m}^3$$

For dry unit weight [Eq. (2.11)], we have

$$\gamma_d = \frac{W_s}{V} = \frac{153.6}{0.0093} = 16{,}516 \text{ N/m}^3 \approx \textbf{16.52 kN/m}^3$$

The void ratio [Eq. (2.3)] is found as follows:

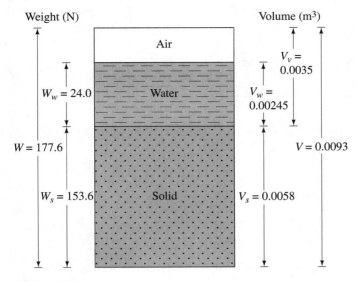

FIGURE 2.4

$$e = \frac{V_v}{V_s}$$

$$V_s = \frac{W_s}{G_s \gamma_w} = \frac{0.1536}{2.71 \times 9.81} = 0.0058 \text{ m}^3$$

$$V_v = V - V_s = 0.0093 - 0.0058 = 0.0035 \text{ m}^3$$

so

$$e = \frac{0.0035}{0.0058} \approx \mathbf{0.60}$$

For porosity [Eq. (2.7)], we have

$$n = \frac{e}{1 + e} = \frac{0.60}{1 + 0.60} = \mathbf{0.375}$$

We find the degree of saturation [Eq. (2.5)] as follows:

$$S = \frac{V_w}{V_v}$$

$$V_w = \frac{W_w}{\gamma_w} = \frac{0.024}{9.81} = 0.00245 \text{ m}^3$$

so

$$S = \frac{0.00245}{0.0035} \times 100 = \mathbf{70\%}$$

∎

**EXAMPLE
2.2**

For a given soil, $e = 0.75$, $w = 22\%$, and $G_s = 2.66$. Calculate the porosity, moist unit weight, dry unit weight, and degree of saturation.

Solution   The porosity [Eq. (2.7)] is

$$n = \frac{e}{1 + e} = \frac{0.75}{1 + 0.75} = \mathbf{0.43}$$

To find the moist unit weight, we use Eq. (2.19a) to calculate the moist density:

$$\rho = \frac{(1 + w)G_s\rho_w}{1 + e}$$

$$\rho_w = 1000 \text{ kg/m}^3$$

$$\rho = \frac{(1 + 0.22)2.66 \times 1000}{1 + 0.75} = 1854.4 \text{ kg/m}^3$$

Hence, the moist unit weight is γ (kN/m^3) $= \rho \cdot g = \dfrac{9.81 \times 1854.4}{1000} = \mathbf{18.19 \text{ kN/m}^3}$

To find the dry unit weight, we use Eq. (2.19b):

$$\rho_d = \frac{G_s\rho_w}{1 + e} = \frac{2.66 \times 1000}{1 + 0.75} = 1520 \text{ kg/m}^3$$

so

$$\gamma_d = \frac{9.81 \times 1520}{1000} = \mathbf{14.91 \text{ kN/m}^3}$$

The degree of saturation [Eq. (2.17)] is

$$S\,(\%) = \frac{wG_s}{e} \times 100 = \frac{0.22 \times 2.66}{0.75} \times 100 = \mathbf{78\%} \qquad \blacksquare$$

**EXAMPLE
2.3**

The following data are given for a soil: porosity $= 0.45$, specific gravity of the soil solids $= 2.68$, and moisture content $= 10\%$. Determine the mass of water to be added to 10 m^3 of soil for full saturation.

Solution   From Eq. (2.6), we have

$$e = \frac{n}{1 - n} = \frac{0.45}{1 - 0.45} = 0.82$$

The moist density of soil [Eq. (2.19a)] is

$$\rho = \frac{(1 + w)G_s\rho_w}{1 + e} = \frac{(1 + 0.1)2.68 \times 1000}{1 + 0.82} = 1619.8 \text{ kg/m}^3$$

The saturated density of soil [Eq. (2.19c)] is

$$\rho_{sat} = \frac{(G_s + e)\rho_w}{1 + e} = \frac{(2.68 + 0.82)1000}{1 + 0.82} = 1923 \text{ kg/m}^3$$

The mass of water needed per cubic meter equals

$$\rho_{sat} - \rho = 1923 - 1619.8 = 303.2 \text{ kg}$$

So, the total mass of water to be added is

$$303.2 \times 10 = \textbf{3032 kg}$$ ■

2.4 *Consistency of Soil*

When clay minerals are present in fine-grained soil, that soil can be remolded in the presence of some moisture without crumbling. This cohesive nature is because of the adsorbed water surrounding the clay particles. In the early 1900s, a Swedish scientist named Albert Mauritz Atterberg developed a method to describe the consistency of fine-grained soils with varying moisture contents. At a very low moisture content, soil behaves more like a brittle solid. When the moisture content is very high, the soil and water may flow like a liquid. Hence, on an arbitrary basis, depending on the moisture content, the nature of soil behavior can be broken down into four basic states. They are *solid, semisolid, plastic,* and *liquid,* as shown in Figure 2.5.

The moisture content, in percent, at which the transition from solid to semisolid state takes place is defined as the *shrinkage limit.* The moisture content at the point of transition from semisolid to plastic state is the *plastic limit,* and from plastic to liquid state is the *liquid limit.* These limits are also known as *Atterberg limits.*

Liquid Limit (LL)

A schematic diagram (side view) of a liquid limit device is shown in Figure 2.6a. This device consists of a brass cup and hard rubber base. The brass cup can be dropped on the base by a cam operated by a crank. For the liquid limit test, a soil

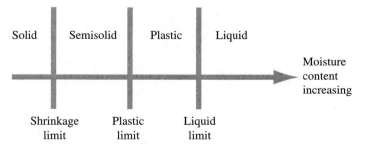

FIGURE 2.5 Atterberg limits

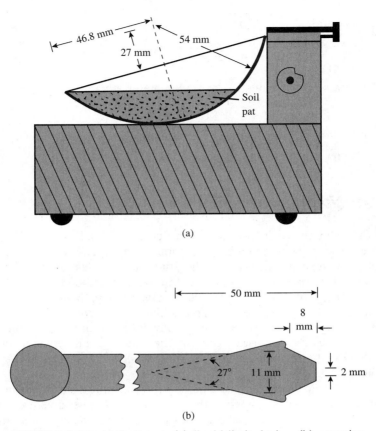

FIGURE 2.6 Liquid limit test: (a) liquid limit device; (b) grooving tool; (c) soil pat before test; (d) soil pat after test

paste is placed in the cup. A groove is cut at the center of the soil pat, using the standard grooving tool (Figure 2.6b). Then, with the crank-operated cam, the cup is lifted and dropped from a height of 10 mm. The moisture content, in percent, required to close a distance of 12.7 mm along the bottom of the groove (see Figures 2.6c and 2.6d) after 25 blows is defined as the *liquid limit*. The procedure for the liquid limit test is given in ASTM Test Designation D-4318.

Casagrande (1932) concluded that each blow in a standard liquid limit device corresponds to a soil shear strength of about 1 g/cm^2 ($\approx$0.1 kN/m^2). Hence, the liquid limit of a fine-grained soil gives the moisture content at which the shear strength of the soil is approximately 25 g/cm^2 ($\approx$2.5 kN/m^2).

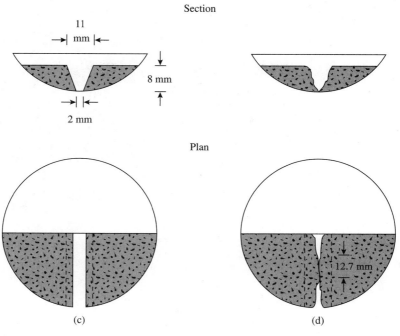

FIGURE 2.6 (Continued)

Plastic Limit (PL)

The *plastic limit* is defined as the moisture content, in percent, at which the soil, when rolled into threads of 3.2 mm in diameter, crumbles. The plastic limit is the lower limit of the plastic stage of soil. The test is simple and is performed by repeated rollings of an ellipsoidal size soil mass by hand on a ground glass plate (Figure 2.7).

The *plasticity index* (*PI*) is the difference between the liquid limit and plastic limit of a soil, or

$$PI = LL - PL \tag{2.22}$$

The procedure for the plastic limit test is given in ASTM Test Designation D-4318.

Shrinkage Limit (SL)

Soil mass shrinks as moisture is gradually lost from the soil. With continuous loss of moisture, a stage of equilibrium is reached at which more loss of moisture will

FIGURE 2.7 Plastic limit test

result in no further volume change (Figure 2.8). The moisture content, in percent, at which the volume change of the soil mass ceases is defined as the *shrinkage limit.*

Shrinkage limit tests (ASTM Test Designation D-427) are performed in the laboratory with a porcelain dish about 44 mm in diameter and about 13 mm in height. The inside of the dish is coated with petroleum jelly and is then filled completely with wet soil. Excess soil standing above the edge of the dish is struck off with a straightedge. The mass of the wet soil inside the dish is recorded. The soil pat in the dish is then oven dried. The volume of the oven-dried soil pat is determined by the displacement of mercury. Because handling mercury may be hazardous, ASTM Test Designation D-4943 describes a method of dipping the oven-dried soil pat in a pot of melted wax. The wax-coated soil pat is then cooled. Its volume is determined by submerging it in water.

Referring to Figure 2.8, we can determine the shrinkage limit in the following manner:

$$SL = w_i\,(\%) - \Delta w\,(\%) \tag{2.23}$$

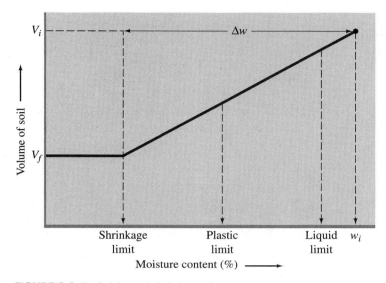

FIGURE 2.8 Definition of shrinkage limit

where w_i = initial moisture content when the soil is placed in the shrinkage limit dish

Δw = change in moisture content (that is, between the initial moisture content and the moisture content at shrinkage limit)

However,

$$w_i\,(\%) = \frac{m_1 - m_2}{m_2} \times 100 \tag{2.24}$$

where m_1 = mass of the wet soil pat in the dish at the beginning of the test (g)
m_2 = mass of the dry soil pat (g) (see Figure 2.9)

Also,

$$\Delta w\,(\%) = \frac{(V_i - V_f)\rho_w}{m_2} \times 100 \tag{2.25}$$

where V_i = initial volume of the wet soil pat (that is, inside volume of the dish, cm^3)
V_f = volume of the oven-dried soil pat (cm^3)
ρ_w = density of water (g/cm^3)

Now, combining Eqs. (2.23), (2.24), and (2.25), we have

$$SL = \left(\frac{m_1 - m_2}{m_2}\right)(100) - \left[\frac{(V_i - V_f)\,\rho_w}{m_2}\right)(100) \tag{2.26}$$

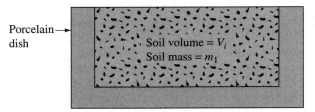

(a) Before drying

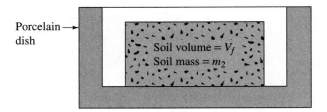

(b) After drying

FIGURE 2.9 Shrinkage limit test

 2.5   Activity

Since the plastic property of soil results from the adsorbed water that surrounds the clay particles, we can expect that the type of clay minerals and their proportional amounts in a soil will affect the liquid and plastic limits. Skempton (1953) observed that the plasticity index of a soil increases linearly with the percent of clay-size fraction (percent finer than $2\ \mu$ by weight) present in it. On the basis of these results, Skempton defined a quantity called *activity,* which is the slope of the line

Table 2.3 Activity of clay minerals

Mineral	Activity, A
Smectites	1–7
Illite	0.5–1
Kaolinite	0.5
Halloysite ($2H_2O$)	0.5
Holloysite ($4H_2O$)	0.1
Attapulgite	0.5–1.2
Allophane	0.5–1.2

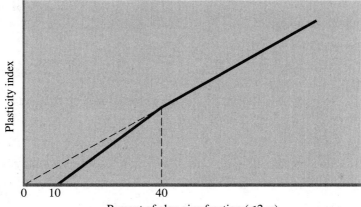

FIGURE 2.10 Simplified relationship between plasticity index and percent of clay-size fraction by weight

correlating *PI* and percent finer than 2 μ. This activity may be expressed as

$$A = \frac{PI}{\text{(percent of clay-size fraction, by weight)}} \tag{2.27}$$

where A = activity. Activity is used as an index for identifying the swelling potential of clay soils. Typical values of activities for various clay minerals are listed in Table 2.3 (Mitchell, 1976).

Seed, Woodward, and Lundgren (1964) studied the plastic property of several artificially prepared mixtures of sand and clay. They concluded that, although the relationship of the plasticity index to the percent of clay-size fraction is linear, as observed by Skempton, the line may not always pass through the origin. They showed that the relationship of the plasticity index to the percent of clay-size fraction present in a soil can be represented by two straight lines. This relationship is shown qualitatively in Figure 2.10. For clay-size fractions greater than 40%, the straight line passes through the origin when it is projected back.

2.6 Liquidity Index

The relative consistency of a cohesive soil in the natural state can be defined by a ratio called the *liquidity index* (*LI*):

$$LI = \frac{w - PL}{LL - PL} \tag{2.28}$$

where w = *in situ* moisture content of soil.

The *in situ* moisture content of a sensitive clay may be greater than the liquid limit. In that case,

$$LI > 1$$

These soils, when remolded, can be transformed into a viscous form to flow like a liquid.

Soil deposits that are heavily overconsolidated may have a natural moisture content less than the plastic limit. In that case,

$$LI < 1$$

The values of the liquidity index for some of these soils may be negative.

2.7   Plasticity Chart

Liquid and plastic limits are determined by relatively simple laboratory tests that provide information about the nature of cohesive soils. The tests have been used extensively by engineers for the correlation of several physical soil parameters as well as for soil identification. Casagrande (1932) studied the relationship of the plasticity index to the liquid limit of a wide variety of natural soils. On the basis of the test results, he proposed a plasticity chart as shown in Figure 2.11. The important feature of this chart is the empirical *A*-line that is given by the equation $PI = 0.73(LL - 20)$. The *A*-line separates the inorganic clays from the inorganic

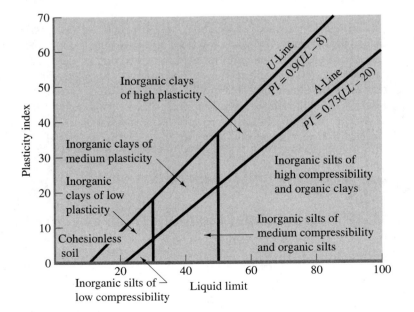

FIGURE 2.11 Plasticity chart

silts. Plots of plasticity indexes against liquid limits for inorganic clays lie above the *A*-line, and those for inorganic silts lie below the *A*-line. Organic silts plot in the same region (below the *A*-line and with *LL* ranging from 30 to 50) as the inorganic silts of medium compressibility. Organic clays plot in the same region as the inorganic silts of high compressibility (below the *A*-line and *LL* greater than 50). The information provided in the plasticity chart is of great value and is the basis for the classification of fine-grained soils in the Unified Soil Classification System.

Note that a line called the *U*-line lies above the *A*-line. The *U*-line is approximately the upper limit of the relationship of the plasticity index to the liquid limit for any soil found so far. The equation for the *U*-line can be given as

$$PI = 0.9(LL - 8) \tag{2.29}$$

2.8 *Soil Classification*

Soils with similar properties may be classified into groups and subgroups based on their engineering behavior. Classification systems provide a common language to express concisely the general characteristics of soils, which are infinitely varied, without a detailed description. At the present time, two elaborate classification systems that use the grain-size distribution and plasticity of soils are commonly used by soils engineers. They are the AASHTO classification system and the Unified Soil Classification System. The AASHTO system is used mostly by state and county highway departments, whereas geotechnical engineers usually prefer to use the Unified System.

AASHTO Classification System

This system of soil classification was developed in 1929 as the Public Road Administration Classification System. It has undergone several revisions, with the present version proposed by the Committee on Classification of Materials for Subgrades and Granular Type Roads of the Highway Research Board in 1945 (ASTM Test Designation D-3282; AASHTO method M145).

The AASHTO classification system in present use is given in Table 2.4. According to this system, soil is classified into seven major groups: A-1 through A-7. Soils classified into groups A-1, A-2, and A-3 are granular materials, where 35% or less of the particles pass through the No. 200 sieve. Soils where more than 35% pass through the No. 200 sieve are classified into groups A-4, A-5, A-6, and A-7. These are mostly silt and clay-type materials. The classification system is based on the following criteria:

1. *Grain size*
   Gravel: fraction passing the 75 mm sieve and retained on the No. 10 (2 mm) U.S. sieve
   Sand: fraction passing the No. 10 (2 mm) U.S. sieve and retained on the No. 200 (0.075 mm) U.S. sieve
   Silt and clay: fraction passing the No. 200 U.S. sieve

(cont. on p. 37)

Table 2.4 Classification of highway subgrade materials

General classification	Granular materials (35% or less of total sample passing No. 200)						
	A-1		A-3	A-2			
Group classification	A-1-a	A-1-b		A-2-4	A-2-5	A-2-6	A-2-7
Sieve analysis (percent passing)							
No. 10	50 max.						
No. 40	30 max.	50 max.	51 min.				
No. 200	15 max.	25 max.	10 max.	35 max.	35 max.	35 max.	35 max.
Characteristics of fraction passing No. 40							
Liquid limit				40 max.	41 min.	40 max.	41 min.
Plasticity index	6 max.		NP	10 max.	10 max.	11 min.	11 min.
Usual types of significant constituent materials	Stone fragments, gravel, and sand		Fine sand	Silty or clayey gravel and sand			
General subgrade rating	Excellent to good						

General classification	Silt-clay materials (more than 35% of total sample passing No. 200)			
Group classification	A-4	A-5	A-6	A-7 A-7-5* A-7-6[†]
Sieve analysis (percent passing)				
No. 10				
No. 40				
No. 200	36 min.	36 min.	36 min.	36 min.
Characteristics of fraction passing No. 40				
Liquid limit	40 max.	41 min.	40 max	41 min.
Plasticity index	10 max.	10 max.	11 min.	11 min.
Usual types of significant constituent materials	Silty soils		Clayey soils	
General subgrade rating	Fair to poor			

*For A-7-5, $PI \leq LL - 30$
[†]For A-7-6, $PI > LL - 30$

36

2. *Plasticity:* The term *silty* is applied when the fine fractions of the soil have a plasticity index of 10 or less. The term *clayey* is applied when the fine fractions have a plasticity index of 11 or more.
3. If cobbles and *boulders* (size larger than 75 mm) are encountered, they are excluded from the portion of the soil sample on which classification is made. However, the percentage of such material is recorded.

To classify a soil according to Table 2.4, the test data are applied from left to right. By process of elimination, the first group from the left into which the test data will fit is the correct classification.

For the evaluation of the quality of a soil as a highway subgrade material, a number called the *group index* (*GI*) is also incorporated with the groups and subgroups of the soil. This number is written in parentheses after the group or subgroup designation. The group index is given by the equation

$$GI = (F - 35)[0.2 + 0.005(LL - 40)] + 0.01(F - 15)(PI - 10) \qquad (2.30)$$

where F = percent passing the No. 200 sieve
   LL = liquid limit
   PI = plasticity index

The first term of Eq. (2.30)—that is, $(F - 35)[0.2 + 0.005(LL - 40)]$—is the partial group index determined from the liquid limit. The second term—that is, $0.01(F - 15)(PI - 10)$—is the partial group index determined from the plasticity index. Following are some rules for determining the group index:

1. If Eq. (2.30) yields a negative value for *GI*, it is taken as 0.
2. The group index calculated from Eq. (2.30) is rounded off to the nearest whole number (for example, $GI = 3.4$ is rounded off to 3; $GI = 3.5$ is rounded off to 4).
3. There is no upper limit for the group index.
4. The group index of soils belonging to groups A-1-a, A-1-b, A-2-4, A-2-5, and A-3 is always 0.
5. When calculating the group index for soils that belong to groups A-2-6 and A-2-7, use the partial group index for *PI*, or

$$GI = 0.01(F - 15)(PI - 10) \qquad (2.31)$$

In general, the quality of performance of a soil as a subgrade material is inversely proportional to the group index.

EXAMPLE 2.4

Classify the soils given in the table by the AASHTO classification system.

Soil no.	Sieve analysis; percent finer			Plasticity for the minus No. 40 fraction	
	No. 10 sieve	No. 40 sieve	No. 200 sieve	Liquid limit	Plasticity index
1	100	82	38	42	23
2	48	29	8	—	2
3	100	80	64	47	29
4	90	76	34	37	12

Solution   For soil 1, the percent passing the No. 200 sieve is 38%, which is greater than 35%, so it is a silty clay material. Proceeding from left to right in Table 2.4, we see that it falls under A-7. For this case, $PI = 23 > LL - 30$, so it is A-7-6. From Eq. (2.30), we have

$$GI = (F - 35)[0.2 + 0.005(LL - 40)] + 0.01(F - 15)(PI - 10)$$

For this soil, $F = 38$, $LL = 42$, and $PI = 23$, so

$$GI = (38 - 35)[0.2 + 0.005(42 - 40)] + 0.01(38 - 15)(23 - 10) = 3.88 \approx 4$$

Hence, the soil is **A-7-6(4)**.

For soil 2, the percent passing the No. 200 sieve is less than 35%, so it is a granular material. Proceeding from left to right in Table 2.4, we find that it is A-1-a. The group index is 0, so the soil is **A-1-a(0)**.

For soil 3, the percent passing the No. 200 sieve is greater than 35%, so this is a silty clay material. Proceeding from left to right in Table 2.4, we find it to be A-7-6.

$$GI = (F - 35)[0.2 + 0.005(LL - 40)] + 0.01(F - 15)(PI - 10)$$

Given $F = 64$, $LL = 47$, and $PI = 29$, we have

$$GI = (64 - 35)[0.2 + 0.005(47 - 40)] + 0.01(64 - 15)(29 - 10) = 16.1 \approx 16$$

Hence, the soil is **A-7-6(16)**.

For soil 4, the percent passing the No. 200 sieve is less than 35%, so it is a granular material. According to Table 2.4, it is A-2-6.

$$GI = 0.01(F - 15)(PI - 10)$$

Now, $F = 34$ and $PI = 12$, so

$$GI = 0.01(34 - 15)(12 - 10) = 0.38 \approx 0$$

Thus, the soil is **A-2-6(0)**.   ∎

Table 2.5 Unified classification system—group symbols for gravelly soil

Group symbol	Criteria
GW	Less than 5% passing No. 200 sieve; $C_u = D_{60}/D_{10}$ greater than or equal to 4; $C_z = (D_{30})^2/(D_{10} \times D_{60})$ between 1 and 3
GP	Less than 5% passing No. 200 sieve; not meeting both criteria for GW
GM	More than 12% passing No. 200 sieve; Atterberg's limits plot below A-line (Figure 2.12) or plasticity index less than 4
GC	More than 12% passing No. 200 sieve; Atterberg's limits plot above A-line (Figure 2.12); plasticity index greater than 7
GC-GM	More than 12% passing No. 200 sieve; Atterberg's limits fall in hatched area marked CL-ML in Figure 2.12
GW-GM	Percent passing No. 200 sieve is 5 to 12; meets the criteria for GW and GM
GW-GC	Percent passing No. 200 sieve is 5 to 12; meets the criteria for GW and GC
GP-GM	Percent passing No. 200 sieve is 5 to 12; meets the criteria for GP and GM
GP-GC	Percent passing No. 200 sieve is 5 to 12; meets the criteria for GP and GC

Unified Soil Classification System

The original form of this system was proposed by Casagrande in 1942 for use in the airfield construction works undertaken by the Army Corps of Engineers during World War II. In cooperation with the U.S. Bureau of Reclamation, this system was revised in 1952. At present, it is widely used by engineers (ASTM Test Designation D-2487). The Unified Classification System is presented in Tables 2.5, 2.6, and 2.7. This system classifies soils into two broad categories:

1. Coarse-grained soils that are gravelly and sandy in nature with less than 50% passing through the No. 200 sieve. The group symbols start with a prefix of either G or S. G stands for gravel or gravelly soil, and S for sand or sandy soil.
2. Fine-grained soils with 50% or more passing through the No. 200 sieve. The group symbols start with a prefix of M, which stands for inorganic silt, C for inorganic clay, or O for organic silts and clays. The symbol Pt is used for peat, muck, and other highly organic soils.

Other symbols are also used for the classification:

- W—well graded
- P—poorly graded
- L—low plasticity (liquid limit less than 50)
- H—high plasticity (liquid limit more than 50)

Table 2.6 Unified classification system—group symbols for sandy soil

Group symbol	Criteria
SW	Less than 5% passing No. 200 sieve; $C_u = D_{60}/D_{10}$ greater than or equal to 6; $C_z = (D_{30})^2/(D_{10} \times D_{60})$ between 1 and 3
SP	Less than 5% passing No. 200 sieve; not meeting both criteria for SW
SM	More than 12% passing No. 200 sieve; Atterberg's limits plot below A-line (Figure 2.12) or plasticity index less than 4
SC	More than 12% passing No. 200 sieve; Atterberg's limits plot above A-line (Figure 2.12); plasticity index greater than 7
SC-SM	More than 12% passing No. 200 sieve; Atterberg's limits fall in hatched area marked CL-ML in Figure 2.12
SW-SM	Percent passing No. 200 sieve is 5 to 12; meets the criteria for SW and SM
SW-SC	Percent passing No. 200 sieve is 5 to 12; meets the criteria for SW and SC
SP-SM	Percent passing No. 200 sieve is 5 to 12; meets the criteria for SP and SM
SP-SC	Percent passing No. 200 sieve is 5 to 12; meets the criteria for SP and SC

Table 2.7 Unified classification system—group symbols for silty and clayey soils

Group symbol	Criteria
CL	Inorganic; $LL < 50$; $PI > 7$; plots on or above A-line (see CL zone in Figure 2.12)
ML	Inorganic: $LL < 50$; $PI < 4$ or plots below A-line (see ML zone in Figure 2.12)
OL	Organic; $(LL - \text{oven-dried})/(LL - \text{not dried}) < 0.75$; $LL < 50$ (see OL zone in Figure 2.12)
CH	Inorganic; $LL \geq 50$; PI plots on or above A-line (see CH zone in Figure 2.12)
MH	Inorganic; $LL \geq 50$; PI plots below A-line (see MH zone in Figure 2.12)
OH	Organic; $(LL - \text{oven-dried})/(LL - \text{not dried}) < 0.75$; $LL \geq 50$ (see OH zone in Figure 2.12)
CL-ML	Inorganic; plot in the hatched zone in Figure 2.12
Pt	Peat, muck, and other highly organic soils

For proper classification according to this system, some or all of the following information must be known:

1. Percent of gravel—that is, the fraction passing the 76.2-mm sieve and retained on the No. 4 sieve (4.75-mm opening)
2. Percent of sand—that is, the fraction passing the No. 4 sieve (4.75-mm opening) and retained on the No. 200 sieve (0.075-mm opening)
3. Percent of silt and clay—that is, the fraction finer than the No. 200 sieve (0.075-mm opening)
4. Uniformity coefficient (C_u) and the coefficient of gradation (C_z)
5. Liquid limit and plasticity index of the portion of soil passing the No. 40 sieve

The group symbols for coarse-grained gravelly soils are GW, GP, GM, GC, GC-GM, GW-GM, GW-GC, GP-GM, and GP-GC. Similarly, the group symbols for fine-grained soils are CL, ML, OL, CH, MH, OH, CL-ML, and Pt. Following is a step-by-step procedure for the classification of soils:

Step 1: Determine the percent of soil passing the No. 200 sieve (F). If $F <$ 50%, it is a coarse-grained soil—that is, gravelly or sandy soil (where F = percent finer than No. 200 sieve). Go to Step 2. If $F \geq$ 50%, it is a fine-grained soil. Go to Step 3.

Step 2: For a coarse-grained soil, $(100 - F)$ is the coarse fraction in percent. Determine the percent of soil passing the No. 4 sieve and retained on the No. 200 sieve, F_1. If $F_1 < (100 - F)/2$, then it has more gravel than sand, so it is a gravelly soil. Go to Table 2.5 and Figure 2.12 for determination of the *group symbol,* and then proceed to Figure 2.13

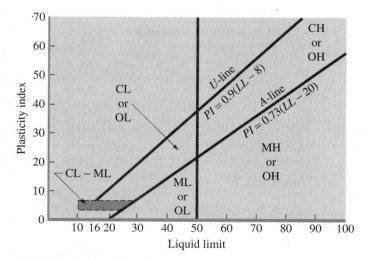

FIGURE 2.12 Plasticity chart

Group Symbol   **Group Name**

GW ———→ <15% sand ———→ Well-graded gravel
      ↘ ≥15% sand ———→ Well-graded gravel with sand
GP ———→ <15% sand ———→ Poorly graded gravel
      ↘ ≥15% sand ———→ Poorly graded gravel with sand

GW-GM ⤡ <15% sand ———→ Well-graded gravel with silt
        ≥15% sand ———→ Well-graded gravel with silt and sand
GW-GC ⤡ <15% sand ———→ Well-graded gravel with clay (or silty clay)
        ≥15% sand ———→ Well-graded gravel with clay and sand (or silty clay and sand)

GP-GM ⤡ <15% sand ———→ Poorly graded gravel with silt
        ≥15% sand ———→ Poorly graded gravel with silt and sand
GP-GC ⤡ <15% sand ———→ Poorly graded gravel with clay (or silty clay)
        ≥15% sand ———→ Poorly graded gravel with clay and sand (or silty clay and sand)

GM ———→ <15% sand ———→ Silty gravel
     ↘ ≥15% sand ———→ Silty gravel with sand
GC ———→ <15% sand ———→ Clayey gravel
     ↘ ≥15% sand ———→ Clayey gravel with sand
GC-GM ⤡ <15% sand ———→ Silty clayey gravel
        ≥15% sand ———→ Silty clayey gravel with sand

SW ———→ <15% gravel ———→ Well-graded sand
     ↘ ≥15% gravel ———→ Well-graded sand with gravel
SP ———→ <15% gravel ———→ Poorly graded sand
     ↘ ≥15% gravel ———→ Poorly graded sand with gravel

SW-SM ⤡ <15% gravel ———→ Well-graded sand with silt
        ≥15% gravel ———→ Well-graded sand with silt and gravel
SW-SC ⤡ <15% gravel ———→ Well-graded sand with clay (or silty clay)
        ≥15% gravel ———→ Well-graded sand with clay and gravel (or silty clay and gravel)

SP-SM ⤡ <15% gravel ———→ Poorly graded sand with silt
        ≥15% gravel ———→ Poorly graded sand with silt and gravel
SP-SC ⤡ <15% gravel ———→ Poorly graded sand with clay (or silty clay)
        ≥15% gravel ———→ Poorly graded sand with clay and gravel (or silty clay and gravel)

SM ———→ <15% gravel ———→ Silty sand
     ↘ ≥15% gravel ———→ Silty sand with gravel
SC ———→ <15% gravel ———→ Clayey sand
     ↘ ≥15% gravel ———→ Clayey sand with gravel
SC-SM ⤡ <15% gravel ———→ Silty clayey sand
        ≥15% gravel ———→ Silty clayey sand with gravel

FIGURE 2.13 Flowchart group names for gravelly and sandy soil (After ASTM, 1998)

for the proper *group name* of the soil. If $F_1 \geq (100 - F)/2$, then it is a sandy soil. Go to Table 2.6 and Figure 2.12 to determine the group symbol, and to Figure 2.13 for the group name of the soil.

Step 3: For a fine-grained soil, go to Table 2.7 and Figure 2.12 to obtain the *group symbol*. If it is an inorganic soil, go to Figure 2.14 to obtain the *group name*. If it is an organic soil, go to Figure 2.15 to get the group name.

Note that Figure 2.12 is the plasticity chart originally developed by Casagrande (1948) and modified to some extent here.

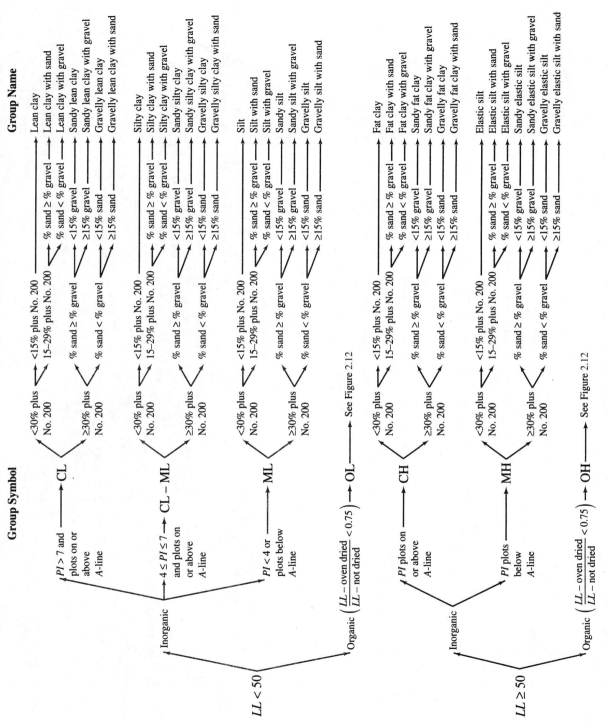

FIGURE 2.14 Flowchart group names for inorganic silty and clayey soils (After ASTM, 1998)

Group Symbol   **Group Name**

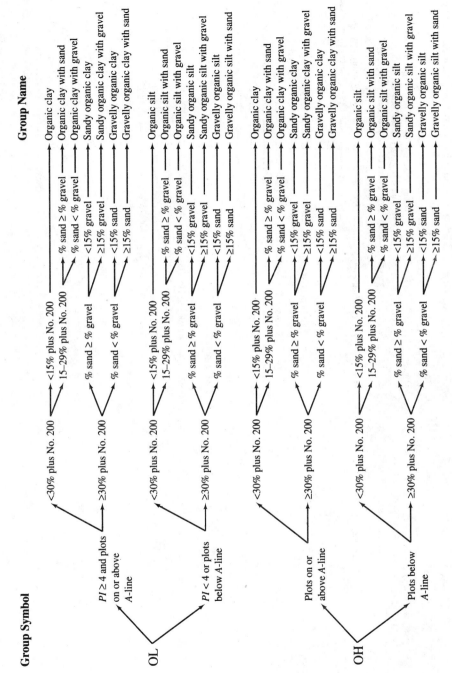

FIGURE 2.15 Flowchart group names for organic silty and clayey soils (After ASTM, 1998)

**EXAMPLE
2.5**

A given soil has the following values:

- Gravel fraction (retained on No. 4 sieve) = 30%
- Sand fraction (passing No. 4 sieve but retained on No. 200 sieve) = 40%
- Silt and clay (passing No. 200 sieve) = 30%
- Liquid limit = 33
- Plasticity index = 12

Classify the soil by the Unified Classification System, giving the group symbol and the group name.

Solution   We are given $F = 30$ (i.e., $< 50\%$); hence, it is a coarse-grained soil. Also $F_1 = 40$, so

$$F_1 = 40 > \frac{100 - F}{2} = \frac{100 - 30}{2} = 35$$

and it is a sandy soil. From Table 2.6 and Figure 2.12, we see that the soil is **SC**. Since the soil has more than 15% gravel (Figure 2.13), its group name is **clayey sand with gravel**.   ■

**EXAMPLE
2.6**

A soil has these values:

- Gravel fraction (retained on No. 4 sieve) = 10%
- Sand fraction (passing No. 4 sieve but retained on No. 200 sieve) = 82%
- Silt and clay (passing No. 200 sieve) = 8%
- Liquid limit = 39
- Plasticity index = 8
- $C_u = 3.9$
- $C_z = 2.1$

Classify the soil by the Unified Classification System, giving the group symbol and the group name.

Solution   We are given $F = 8$, so it is a coarse-grained soil. Also, we have

$$F_1 = 82 > \frac{100 - F}{2} = 46$$

Hence, it is a sandy soil. Since F is between 5 and 12, dual symbols are necessary. From Table 2.6 and Figure 2.12, since C_u is less than 6, the soil is **SP-SM**. Now, from Figure 2.13, since the soil contains less than 15% gravel, its group name is **poorly graded sand with silt**.   ■

EXAMPLE 2.7

For a given soil:

- Percent passing No. 4 sieve = 100
- Percent passing No. 200 sieve = 86
- Liquid limit = 55
- Plasticity index = 28

Classify the soil using the Unified Classification System, giving the group symbol and the group name.

Solution   We are given that the percent passing the No. 200 sieve is $F = 86$ (i.e., > 50%), so it is a fine-grained soil. From Table 2.7 and Figure 2.12, the group symbol is **CH**. From Figure 2.14, the group name is **fat clay**.   ∎

Problems

2.1   The moist weight of 2.83×10^{-3} m³ of soil is 54.3 N. If the moisture content is 12% and the specific gravity of soil solids is 2.72, find the following:
  a. Moist unit weight (kN/m³)
  b. Dry unit weight (kN/m³)
  c. Void ratio
  d. Porosity
  e. Degree of saturation (%)
  f. Volume occupied by water (m³)

2.2   The dry density of a sand with a porosity of 0.387 is 1600 kg/m³. Find the specific gravity of soil solids and the void ratio of the soil.

2.3   The moist unit weight of a soil is 19.2 kN/m³. Given $G_s = 2.69$ and moisture content $w = 9.8\%$, determine these values:
  a. Dry unit weight (kN/m³)
  b. Void ratio
  c. Porosity
  d. Degree of saturation (%)

2.4   For a saturated soil, given $w = 40\%$ and $G_s = 2.71$, determine the saturated and dry unit weights in kN/m³.

2.5   The mass of a moist soil sample collected from the field is 465 g, and its oven dry mass is 405.76 g. The specific gravity of the soil solids was determined in the laboratory to be 2.68. If the void ratio of the soil in the natural state is 0.83, find the following:
  a. The moist density of the soil in the field (kg/m³)
  b. The dry density of the soil in the field (kg/m³)
  c. The mass of water, in kilograms, to be added per cubic meter of soil in the field for saturation

2.6   A soil has a unit weight of 19.9 kN/m³. Given $G_s = 2.67$ and $w = 12.6\%$, determine these values:
  a. Dry unit weight
  b. Void ratio

      c. Porosity

      d. The weight of water per cubic meter of soil needed for full saturation

2.7 The saturated unit weight of a soil is 20.1 kN/m³. Given $G_s = 2.74$, determine these values:

      a. γ_{dry}

      b. e

      c. n

      d. w (%)

2.8 For a soil, given $e = 0.86$, $w = 28\%$, and $G_s = 2.72$, determine the following:

      a. Moist unit weight

      b. Degree of saturation (%)

2.9 For a saturated soil, given $\gamma_d = 15.3$ kN/m³ and $w = 21\%$, determine these values:

      a. γ_{sat}

      b. e

      c. G_s

      d. γ_{moist} when the degree of saturation is 50%

2.10 Show that, for any soil, $\gamma_{sat} = \gamma_w(e/w) [(1 + w)/(1 + e)]$.

2.11 The maximum and minimum void ratios of a sand are 0.8 and 0.41, respectively. What is the void ratio of the soil corresponding to a relative density of 48%?

2.12 For a sand, the maximum and minimum possible void ratios were determined in the laboratory to be 0.94 and 0.33, respectively. Find the moist unit weight of sand compacted in the field at a relative density of 60% and moisture content of 10%. Given $G_s = 2.65$, also calculate the maximum and minimum possible dry unit weights that the sand can have.

2.13 A saturated soil with a volume of 19.65 cm³ has a mass of 36 g. When the soil was dried, its volume and mass were 13.5 cm³ and 25 g, respectively. Determine the shrinkage limit for the soil.

2.14 The sieve analysis of ten soils and the liquid and plastic limits of the fraction passing through the No. 40 sieve are given in the table. Classify the soils by the AASHTO classification system and give the group indexes.

Soil no.	Sieve analysis, percent finer			Liquid limit	Plastic limit
	No. 10	No. 40	No. 200		
1	98	80	50	38	29
2	100	92	80	56	23
3	100	88	65	37	22
4	85	55	45	28	20
5	92	75	62	43	28
6	97	60	30	25	16
7	100	55	8	—	NP
8	94	80	63	40	21
9	83	48	20	20	15
10	100	92	86	70	38

2.15   Classify soils 1–6 given in Problem 2.14 by the Unified Classification System. Give the group symbol and group name for each soil.

2.16   Classify the soils listed in the table using the AASHTO classification system. Give group indexes also.

Soil	Sieve analysis, percent finer			Liquid limit	Plasticity index
	No. 10	No. 40	No. 200		
A	48	28	6	—	NP
B	87	62	30	32	8
C	90	76	34	37	12
D	100	78	8	—	NP
E	92	74	32	44	9

2.17   Classify the following soils using the Unified Classification System. Give the group symbol and group name for each soil.

Sieve size	Percent finer				
	A	B	C	D	E
No. 4	94	98	100	100	100
No. 10	63	86	100	100	100
No. 20	21	50	98	100	100
No. 40	10	28	93	99	94
No. 60	7	18	88	95	82
No. 100	5	14	83	90	66
No. 200	3	10	77	86	45
0.01 mm	—	—	65	42	26
0.002 mm	—	—	60	47	21
Liquid limit	—	—	63	55	36
Plasticity index	NP	NP	25	28	22

2.18   Classify the soils given in Problem 2.17 by the AASHTO classification system. Give group indexes.

2.19   Classify the soils listed in the table by the Unified Classification System. Give the group symbol and group name for each soil.

| Soil | Sieve analysis, percent finer | | Liquid limit | Plasticity index |
	No. 4	No. 200		
A	92	48	30	8
B	60	40	26	4
C	99	76	60	32
D	90	60	41	12
E	80	35	24	2

References

American Association of State Highway and Transportation Officials (1982). *AASHTO Materials, Part I, Specifications,* Washington, D.C.

American Society for Testing and Materials (1998). *ASTM Book of Standards,* Sec. 4, Vol. 04.08, West Conshohocken, PA.

Casagrande, A. (1932). "Research of Atterberg Limits of Soils," *Public Roads,* Vol. 13, No. 8, 121–136.

Casagrande, A. (1948). "Classification and Identification of Soils," *Transactions,* ASCE, Vol. 113, 901–930.

Mitchell, J. K. (1976). *Fundamentals of Soil Behavior,* Wiley, New York.

Seed, H. B., Woodward, R. J., and Lundgren, R. (1964). "Fundamental Aspects of the Atterberg Limits," *Journal of the Soil Mechanics and Foundations Division,* ASCE, Vol. 90, No. SM6, 75–105.

Skempton, A. W. (1953). "The Colloidal Activity of Clays," *Proceedings,* 3rd International Conference on Soil Mechanics and Foundation Engineering, London, Vol. 1, 57–61.

Supplementary References for Further Study

Collins, K., and McGown, A. (1974). "The Form and Function of Microfabric Features in a Variety of Natural Soils," *Geotechnique,* Vol. 24, No. 2, 223–254.

Lambe, T. W. (1958). "The Structure of Compacted Clay," *Journal of the Soil Mechanics and Foundations Division,* ASCE, Vol. 85, No. SM2, 1654-1–1654-35.

Pusch, R. (1978). "General Report on Physico-Chemical Processes Which Affect Soil Structure and Vice Versa," *Proceedings,* International Symposium on Soil Structure, Gothenburg, Sweden, Appendix, 33.

Yong, R. N., and Sheeran, D. E. (1973). "Fabric Unit Interaction and Soil Behavior," *Proceedings,* International Symposium on Soil Structure, Gothenburg, Sweden, 176–183.

Youd, T. L. (1973). "Factors Controlling Maximum and Minimum Densities of Sand," *Special Technical Publication No. 523,* ASTM, 98–122.

3

Soil Compaction

In the construction of highway embankments, earth dams, and many other engineering structures, loose soils must be compacted to increase their unit weights. Compaction increases the strength characteristics of soils, thereby increasing the bearing capacity of foundations constructed over them. Compaction also decreases the amount of undesirable settlement of structures and increases the stability of slopes of embankments. Smooth-wheel rollers, sheepsfoot rollers, rubber-tired rollers, and vibratory rollers are generally used in the field for soil compaction. Vibratory rollers are used mostly for the densification of granular soils. This chapter discusses the principles of soil compaction in the laboratory and in the field.

3.1 Compaction—General Principles

Compaction, in general, is the densification of soil by removal of air, which requires mechanical energy. The degree of compaction of a soil is measured in terms of its dry unit weight. When water is added to the soil during compaction, it acts as a softening agent on the soil particles. The soil particles slip over each other and move into a densely packed position. The dry unit weight after compaction first increases as the moisture content increases (Figure 3.1). Note that at a moisture content $w = 0$, the moist unit weight (γ) is equal to the dry unit weight (γ_d), or

$$\gamma = \gamma_{d(w=0)} = \gamma_1$$

When the moisture content is gradually increased and the same compactive effort is used for compaction, the weight of the soil solids in a unit volume gradually increases. For example, at $w = w_1$, the moist unit weight is equal to

$$\gamma = \gamma_2$$

However, the dry unit weight at this moisture content is given by

$$\gamma_{d(w=w_1)} = \gamma_{d(w=0)} + \Delta\gamma_d$$

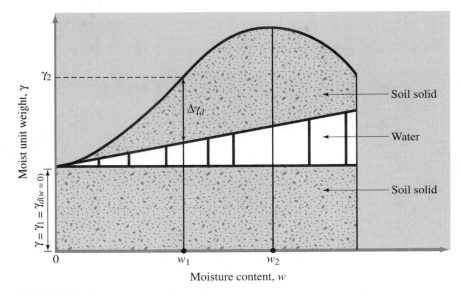

FIGURE 3.1 Principles of compaction

Beyond a certain moisture content $w = w_2$, (Figure 3.1), any increase in the moisture content tends to reduce the dry unit weight. This is because the water takes up the spaces that would have been occupied by the solid particles. The moisture content at which the maximum dry unit weight is attained is generally referred to as the *optimum moisture content.*

The laboratory test generally used to obtain the maximum dry unit weight of compaction and the optimum moisture content is called the *Proctor compaction test* (Proctor, 1933). The procedure for conducting this type of test is described in the following section.

3.2   *Standard Proctor Test*

In the Proctor test, the soil is compacted in a mold that has a volume of 943.3 cm³. The diameter of the mold is 101.6 mm. During the laboratory test, the mold is attached to a base plate at the bottom and to an extension at the top (Figure 3.2a). The soil is mixed with varying amounts of water and then compacted (Figure 3.3) in three equal layers by a hammer (Figure 3.2b) that delivers 25 blows to each layer. The hammer weighs 24.4 N and has a drop of 304.8 mm. For each test, the moist unit weight of compaction γ can be calculated as

$$\gamma = \frac{W}{V_{(m)}}$$ (3.1)

where W = weight of the compacted soil in the mold
$V_{(m)}$ = volume of the mold (= 943.3 cm³)

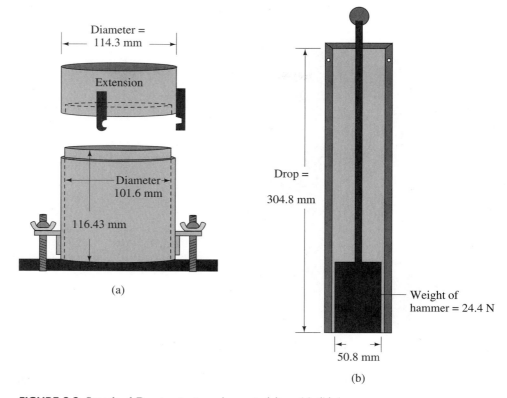

FIGURE 3.2 Standard Proctor test equipment: (a) mold; (b) hammer

For each test, the moisture content of the compacted soil is determined in the laboratory. With known moisture content, the dry unit weight γ_d can be calculated as

$$\gamma_d = \frac{\gamma}{1 + \dfrac{w\,(\%)}{100}} \tag{3.2}$$

where $w\,(\%)$ = percentage of moisture content.

The values of γ_d determined from Eq. (3.2) can be plotted against the corresponding moisture contents to obtain the maximum dry unit weight and the optimum moisture content for the soil. Figure 3.4 shows such a compaction for a silty clay soil.

The procedure for the standard Proctor test is given in ASTM Test Designation D-698 and AASHTO Test Designation T-99.

For a given moisture content, the theoretical maximum dry unit weight is obtained when there is no air in the void spaces—that is, when the degree of saturation equals 100%. Thus, the maximum dry unit weight at a given moisture

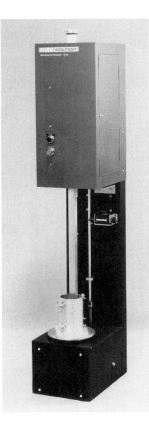

FIGURE 3.3 Standard Proctor test using a mechanical compactor (Courtesy of ELE International/Soiltest Products Division, Lake Bluff, Illinois)

content with zero air voids can be given by

$$\gamma_{zav} = \frac{G_s \gamma_w}{1 + e}$$

where γ_{zav} = zero-air-void unit weight
    γ_w = unit weight of water
    e = void ratio
    G_s = specific gravity of soil solids

For 100% saturation, $e = wG_s$, so

$$\gamma_{zav} = \frac{G_s \gamma_w}{1 + wG_s} = \frac{\gamma_w}{w + \dfrac{1}{G_s}} \tag{3.3}$$

where w = moisture content.

To obtain the variation of γ_{zav} with moisture content, use the following procedure:

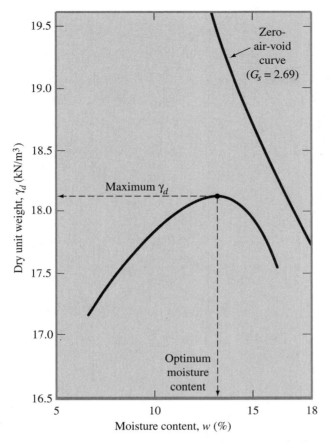

FIGURE 3.4 Standard Proctor compaction test results for a silty clay

1. Determine the specific gravity of soil solids.
2. Know the unit weight of water (γ_w).
3. Assume several values of w, such as 5%, 10%, 15%, and so on.
4. Use Eq. (3.3) to calculate γ_{zav} for various values of w.

Figure 3.4 also shows the variation of γ_{zav} with moisture content and its relative location with respect to the compaction curve. Under no circumstances should any part of the compaction curve lie to the right of the zero-air-void curve.

3.3 Factors Affecting Compaction

The preceding section showed that moisture content has a great influence on the degree of compaction achieved by a given soil. Besides moisture content, other important factors that affect compaction are soil type and compaction effort (energy per unit volume). The importance of each of these two factors is described in more detail in this section.

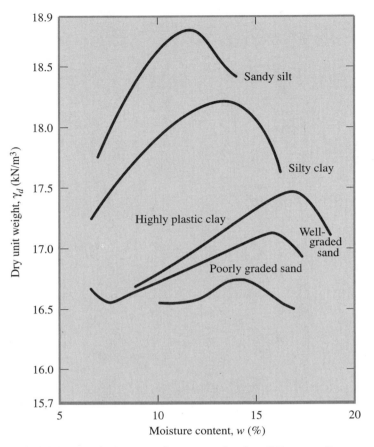

FIGURE 3.5 Typical compaction curves for five different soils
(ASTM D-698)

Effect of Soil Type

The soil type—that is, grain-size distribution, shape of the soil grains, specific gravity
of soil solids, and amount and type of clay minerals present—has a great influence
on the maximum dry unit weight and optimum moisture content. Figure 3.5 shows
typical compaction curves for five different soils. The laboratory tests were conducted in accordance with ASTM Test Designation D-698.

Note also that the bell-shaped compaction curve shown in Figure 3.4 is typical
of most clayey soils. Figure 3.5 shows that, for sands, the dry unit weight has a
general tendency first to decrease as moisture content increases and then to increase
to a maximum value with further increase of moisture. The initial decrease of dry
unit weight with increase of moisture content can be attributed to the capillary
tension effect. At lower moisture contents, the capillary tension in the pore water

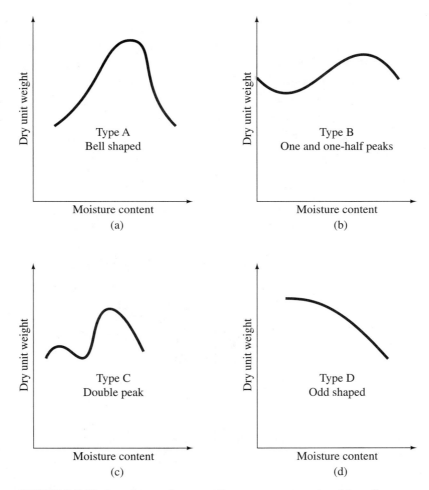

FIGURE 3.6 Various types of compaction curves encountered in soils

inhibits the tendency of the soil particles to move around and be densely compacted.

Lee and Suedkamp (1972) studied compaction curves for 35 different soil samples. They observed four different types of compaction curves. These curves are shown in Figure 3.6. Type A compaction curves are the ones that have a single peak. This type of curve is generally found in soils that have a liquid limit between 30 and 70. Curve type B is a one and one-half peak curve, and curve type C is a double peak curve. Compaction curves of types B and C can be found in soils that have a liquid limit less than about 30. Compaction curves of type D are ones that do not have a definite peak. They are termed odd-shaped. Soils with a liquid limit greater than about 70 may exhibit compaction curves of type C or D. Soils that produce C- and D-type curves are not very common.

Effect of Compaction Effort

The compaction energy per unit volume, E, used for the standard Proctor test described in Section 3.2 can be given as

$$E = \frac{\left(\begin{array}{c}\text{number}\\\text{of blows}\\\text{per layer}\end{array}\right) \times \left(\begin{array}{c}\text{number}\\\text{of}\\\text{layers}\end{array}\right) \times \left(\begin{array}{c}\text{weight}\\\text{of}\\\text{hammer}\end{array}\right) \times \left(\begin{array}{c}\text{height of}\\\text{drop of}\\\text{hammer}\end{array}\right)}{\text{volume of mold}} \qquad (3.4)$$

or

$$E = \frac{(25)(3)(24.4)(0.3048 \text{ m})}{943.3 \times 10^{-6} \text{ m}^3} = 591.3 \times 10^3 \text{ N-m/m}^3 = 591.3 \text{ kN-m/m}^3$$

If the compaction effort per unit volume of soil is changed, the moisture–unit weight curve will also change. This can be demonstrated with the aid of Figure 3.7,

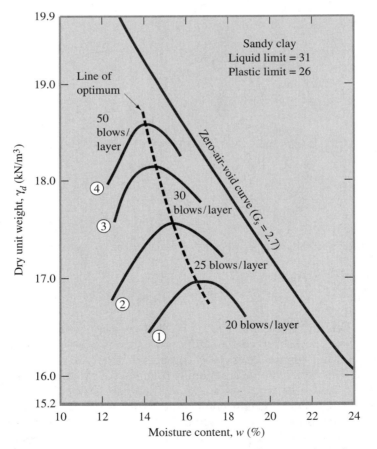

FIGURE 3.7 Effect of compaction energy on the compaction of a sandy clay

Table 3.1 Compaction energy for tests shown in Figure 3.7

Curve number in Figure 3.7	Number of blows/layer	Compaction energy (kN-m/m³)
1	20	473.0
2	25	591.3
3	30	709.6
4	50	1182.6

which shows four compaction curves for a sandy clay. The standard Proctor mold and hammer were used to obtain the compaction curves. The number of layers of soil used for compaction was kept at three for all cases. However, the number of hammer blows per each layer varied from 20 to 50. The compaction energy used per unit volume of soil for each curve can be calculated easily by using Eq. (3.4). These values are listed in Table 3.1.

From Table 3.1 and Figure 3.7, we can reach two conclusions:

1. As the compaction effort is increased, the maximum dry unit weight of compaction is also increased.
2. As the compaction effort is increased, the optimum moisture content is decreased to some extent.

The preceding statements are true for all soils. Note, however, that the degree of compaction is not directly proportional to the compaction effort.

3.4 Modified Proctor Test

With the development of heavy rollers and their use in field compaction, the standard Proctor test was modified to better represent field conditions. This is sometimes referred to as the *modified Proctor test* (ASTM Test Designation D-1557 and AASHTO Test Designation T-180). For conducting the modified Proctor test, the same mold is used, with a volume of 943.3 cm³, as in the case of the standard Proctor test. However, the soil is compacted in five layers by a hammer that weighs 44.5 N. The drop of the hammer is 457.2 mm. The number of hammer blows for each layer is kept at 25 as in the case of the standard Proctor test. Figure 3.8 shows a hammer used for the modified Proctor test. The compaction energy for unit volume of soil in the modified test can be calculated as

$$E = \frac{(25 \text{ blows/layer})(5 \text{ layers})(44.5 \times 10^{-3} \text{ kN})(0.4572 \text{ m})}{943.3 \times 10^{-6} \text{ m}^3} = 2696 \text{ kN-m/m}^3$$

Because it increases the compactive effort, the modified Proctor test results in an increase of the maximum dry unit weight of the soil. The increase of the maximum dry unit weight is accompanied by a decrease of the optimum moisture content.

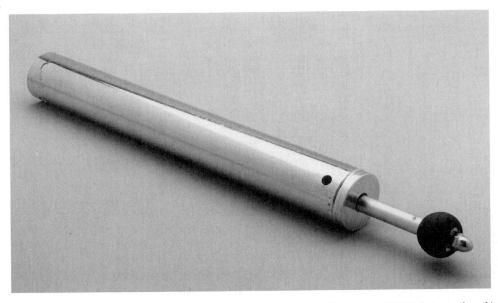

FIGURE 3.8 Hammer used for the modified Proctor test. (Courtesy of ELE International/ Soiltest Products Division, Lake Bluff, Illinois)

In the preceding discussions, the specifications given for Proctor tests adopted by ASTM and AASHTO regarding the volume of the mold (943.3 cm³) and the number of blows (25 blows/layer) are generally the ones adopted for fine-grained soils that pass the U.S. No. 4 sieve. However, under each test designation, three different suggested methods reflect the size of the mold, the number of blows per layer, and the maximum particle size in a soil aggregate used for testing. A summary of the test methods is given in Tables 3.2 and 3.3.

EXAMPLE 3.1

The laboratory test data for a standard Proctor test are given in the table. Find the maximum dry unit weight and the optimum moisture content.

Volume of Proctor mold (cm³)	Mass of wet soil in the mold (kg)	Moisture content (%)
943.3	1.76	12
943.3	1.86	14
943.3	1.92	16
943.3	1.95	18
943.3	1.93	20
943.3	1.90	22

Table 3.2 Specifications for standard Proctor test (Based on ASTM Test Designation 698-91)

Item	Method A	Method B	Method C
Diameter of mold	101.6 mm	101.6 mm	152.4 mm
Volume of mold	943.3 cm³	943.3 cm³	2124 cm³
Weight of hammer	24.4 N	24.4 N	24.4 N
Height of hammer drop	304.8 mm	304.8 mm	304.8 mm
Number of hammer blows per layer of soil	25	25	56
Number of layers of compaction	3	3	3
Energy of compaction	591.3 kN-m/m³	591.3 kN-m/m³	591.3 kN-m/m³
Soil to be used	Portion passing No. 4 (4.57 mm) sieve. May be used if 20% *or less* by weight of material is retained on No. 4 sieve.	Portion passing 9.5-mm sieve. May be used if soil retained on No. 4 sieve *is more* than 20%, and 20% *or less* by weight is retained on 9.5-mm sieve.	Portion passing 19-mm sieve. May be used if *more than* 20% by weight of material is retained on 9.5-mm sieve, and *less than* 30% by weight is retained on 19-mm sieve.

Table 3.3 Specifications for modified Proctor test (Based on ASTM Test Designation 1557-91)

Item	Method A	Method B	Method C
Diameter of mold	101.6 mm	101.6 mm	152.4 mm
Volume of mold	943.3 cm³	943.3 cm³	2124 cm³
Weight of hammer	44.5 N	44.5 N	44.5 N
Height of hammer drop	457.2 mm	457.2 mm	457.2 mm
Number of hammer blows per layer of soil	25	25	56
Number of layers of compaction	5	5	5
Energy of compaction	2696 kN-m/m³	2696 kN-m/m³	2696 kN-m/m³
Soil to be used	Portion passing No. 4 (4.57 mm) sieve. May be used if 20% *or less* by weight of material is retained on No. 4 sieve.	Portion passing 9.5-mm sieve. May be used if soil retained on No. 4 sieve *is more* than 20%, and 20% *or less* by weight is retained on 9.5-mm sieve.	Portion passing 19-mm sieve. May be used if *more than* 20% by weight of material is retained on 9.5-mm sieve, and *less than* 30% by weight is retained on 19-mm sieve.

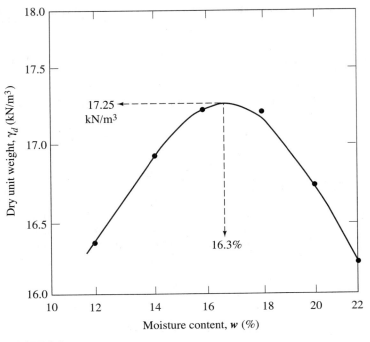

FIGURE 3.9

Solution   We can prepare the following table:

Volume, V (cm³)	Weight of wet soil, W* (N)	Moist unit weight, $\gamma^{\dagger}$ (kN/m³)	Moisture content, w (%)	Dry unit weight, $\gamma_d{}^{\ddagger}$ (kN/m³)
943.3	17.27	18.3	12	16.34
943.3	18.25	19.3	14	16.93
943.3	18.84	20.0	16	17.24
943.3	19.13	20.3	18	17.20
943.3	18.93	20.1	20	16.75
943.3	18.64	19.8	22	16.23

*W = mass (in kg) $\times$ 9.81

$^{\dagger}\gamma = \dfrac{W}{V}$

$^{\ddagger}\gamma_d = \dfrac{\gamma}{1 + \dfrac{w\%}{100}}$

The plot of γ_d against w is shown in Figure 3.9. From the graph, we observe:

Maximum dry unit weight = **17.25 kN/m³**

Optimum moisture content = **16.3%**   ■

3.5   *Structure of Compacted Cohesive Soil*

Lambe (1958) studied the effect of compaction on the structure of clay soils. The results of his study are illustrated in Figure 3.10. If clay is compacted with a moisture content on the dry side of the optimum, as represented by point *A*, it possesses a flocculent structure. This is because, at low moisture content, the diffuse double layers of ions surrounding the clay particles cannot be fully developed; hence, the interparticle repulsion is reduced. Less repulsion results in a more random particle orientation and a lower dry unit weight. When the moisture content of compaction is increased, as shown by point *B*, the diffuse double layers around the particles expand, thus increasing the repulsion between the clay particles and giving a lower degree of flocculation and a higher dry unit weight. A continued increase of moisture content from *B* to *C* expands the double layers more, which results in a continued increase of repulsion between the particles. Greater repulsion gives a still greater degree of particle orientation and a more or less dispersed structure. However, the

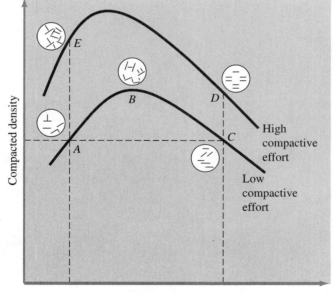

FIGURE 3.10 Effect of compaction on the structure of clay soils (Redrawn after Lambe, 1958)

dry unit weight decreases because the added water dilutes the concentration of soil solids per unit volume.

At a given moisture content, higher compactive effort tends to give a more parallel orientation to the clay particles, thereby giving a more dispersed structure. The particles are closer and the soil has a higher unit weight of compaction. This can be seen by comparing point *A* with point *E* in Figure 3.10.

The observations on the microstructure of compacted cohesive soil as discussed here have practical implications: Compacting dry of optimum produces a flocculated structure that typically gives higher strength, stiffness, brittleness, and hydraulic conductivity, whereas compacting wet of optimum produces a dispersed structure that typically gives lower strength, stiffness, brittleness, and hydraulic conductivity but more ductility. Consequently, compacting dry of optimum is usually more appropriate for the construction of foundations, whereas compacting wet of optimum is more appropriate for the construction of landfill liners and other seepage barriers.

3.6   *Field Compaction*

Most compaction in the field is done with rollers. There are four common types of rollers:

FIGURE 3.11 Smooth-wheel roller (Courtesy of David A. Carroll, Austin, Texas)

1. Smooth-wheel roller (or smooth-drum roller)
2. Pneumatic rubber-tired roller
3. Sheepsfoot roller
4. Vibratory roller

Smooth-wheel rollers (Figure 3.11) are suitable for proofrolling subgrades and for the finishing operation of fills with sandy and clayey soils. They provide 100% coverage under the wheels with ground contact pressures as high as 310–380 kN/m². They are not suitable for producing high unit weights of compaction when used on thicker layers.

Pneumatic rubber-tired rollers (Figure 3.12) are better in many respects than smooth-wheel rollers. The former are heavily loaded wagons with several rows of tires. These tires are closely spaced—four to six in a row. The contact pressure under the tires can range from 600 to 700 kN/m², and they produce 70% to 80% coverage. Pneumatic rollers can be used for sandy and clayey soil compaction. Compaction is achieved by a combination of pressure and kneading action.

FIGURE 3.12 Pneumatic rubber-tired roller (Courtesy of David A. Carroll, Austin, Texas)

FIGURE 3.13 Sheepsfoot roller (Courtesy of David A. Carroll, Austin, Texas)

Sheepsfoot rollers (Figure 3.13) are drums with a large number of projections. The area of each of these projections may range from 25 to 85 cm^2. Sheepsfoot rollers are most effective in compacting clayey soils. The contact pressure under the projections can range from 1380 to 6900 kN/m^2. During compaction in the field, the initial passes compact the lower portion of a lift. The top and middle portions of a lift are compacted at a later stage.

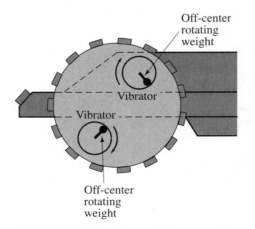

FIGURE 3.14 Principles of vibratory rollers

Vibratory rollers are very efficient in compacting granular soils. Vibrators can be attached to smooth-wheel, pneumatic rubber-tired, or sheepsfoot rollers to provide vibratory effects to the soil. Figure 3.14 demonstrates the principles of vibratory rollers. The vibration is produced by rotating off-center weights.

Hand-held vibrating plates can be used for effective compaction of granular soils over a limited area. Vibrating plates are also gang-mounted on machines, which can be used in less restricted areas.

In addition to soil type and moisture content, other factors must be considered to achieve the desired unit weight of compaction in the field. These factors include the thickness of lift, the intensity of pressure applied by the compacting equipment, and the area over which the pressure is applied. The pressure applied at the surface

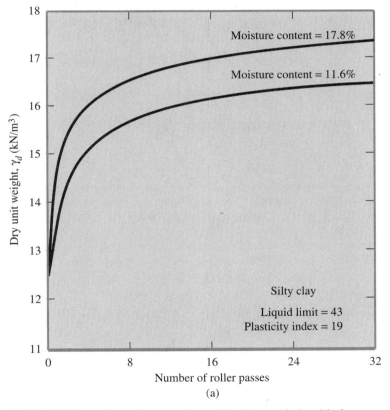

FIGURE 3.15 (a) Growth curves for a silty clay—relationship between dry unit weight and number of passes of 84.5-kN three-wheel roller when compacted in 229-mm loose layers at different moisture contents (Redrawn after Johnson and Sallberg, 1960); (b) Vibratory compaction of a sand—variation of dry unit weight with number of roller passes; thickness of lift = 2.44 m (Redrawn after D'Appolonia, Whitman, and D'Appolonia, 1969)

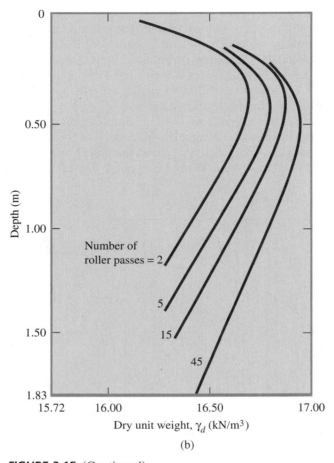

FIGURE 3.15 (Continued)

decreases with depth, resulting in a decrease in the degree of compaction of soil. During compaction, the dry unit weight of soil is also affected by the number of roller passes. Figure 3.15a shows the growth curves for a silty clay soil. The dry unit weight of a soil at a given moisture content will increase up to a certain point with the number of passes of the roller. Beyond this point, it will remain approximately constant. In most cases, about 10 to 15 roller passes yield the maximum dry unit weight economically attainable.

Figure 3.15b shows the variation in the unit weight of compaction with depth for a poorly graded dune sand for which compaction was achieved by a vibratory drum roller. Vibration was produced by mounting an eccentric weight on a single rotating shaft within the drum cylinder. The weight of the roller used for this compaction was 55.6 kN, and the drum diameter was 1.19 m. The lifts were kept at 2.44 m. Note that, at any given depth, the dry unit weight of compaction increases

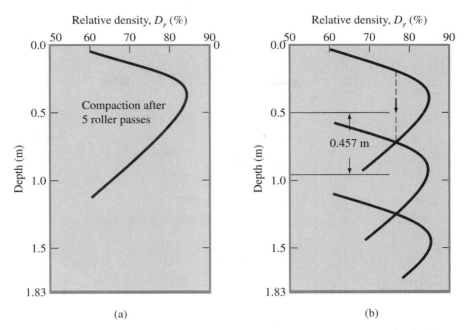

FIGURE 3.16 Estimation of compaction lift thickness for minimum required relative density of 75% with five roller passes (After D'Appolonia, Whitman, and D'Appolonia, 1969)

with the number of roller passes. However, the rate of increase of unit weight gradually decreases after about 15 passes. Another fact to note from Figure 3.15b is the variation of dry unit weight with depth for any given number of roller passes. The dry unit weight and hence the relative density, D_r, reach maximum values at a depth of about 0.5 m and gradually decrease at lesser depths. This decrease is because of the lack of confining pressure toward the surface. Once the relationship between depth and relative density (or dry unit weight) for a given soil with a given number of roller passes is determined, it is easy to estimate the approximate thickness of each lift. This procedure is shown in Figure 3.16 (D'Appolonia, Whitman, and D'Appolonia, 1969).

3.7 Specifications for Field Compaction

In most specifications for earth work, one stipulation is that the contractor must achieve a compacted field dry unit weight of 90% to 95% of the maximum dry unit weight determined in the laboratory by either the standard or modified Proctor test. This specification is, in fact, for relative compaction R, which can be expressed as

$$R\,(\%) = \frac{\gamma_{d(\text{field})}}{\gamma_{d(\text{max}-\text{lab})}} \times 100 \tag{3.5}$$

In the compaction of granular soils, specifications are sometimes written in terms of the required relative density D_r or compaction. Relative density should not be confused with relative compaction. From Chapter 2, we can write

$$D_r = \left[\frac{\gamma_{d(\text{field})} - \gamma_{d(\text{min})}}{\gamma_{d(\text{max})} - \gamma_{d(\text{min})}}\right]\left[\frac{\gamma_{d(\text{max})}}{\gamma_{d(\text{field})}}\right] \tag{3.6}$$

Comparing Eqs. (3.5) and (3.6), we can see that

$$R = \frac{R_0}{1 - D_r(1 - R_0)} \tag{3.7}$$

where

$$R_0 = \frac{\gamma_{d(\text{min})}}{\gamma_{d(\text{max})}} \tag{3.8}$$

Based on the observation of 47 soil samples, Lee and Singh (1971) gave a correlation between R and D_r for granular soils:

$$R = 80 + 0.2D_r \tag{3.9}$$

The specification for field compaction based on relative compaction or on relative density is an end-product specification. The contractor is expected to achieve a minimum dry unit weight regardless of the field procedure adopted. The most economical compaction condition can be explained with the aid of Figure 3.17. The compaction curves A, B, and C are for the same soil with varying compactive effort. Let curve A represent the conditions of maximum compactive effort that can be obtained from the existing equipment. Let it be required to achieve a minimum dry unit weight of $\gamma_{d(\text{field})} = R\gamma_{d(\text{max})}$. To achieve this, the moisture content w needs to be between w_1 and w_2. However, as can be seen from compaction curve C, the required $\gamma_{d(\text{field})}$ can be achieved with a lower compactive effort at a moisture content $w = w_3$. However, in practice, a compacted field unit weight of $\gamma_{d(\text{field})} = R\gamma_{d(\text{max})}$ cannot be achieved by the minimum compactive effort because it allows no margin for error considering the variability of field conditions. Hence, equipment with slightly more than the minimum compactive effort should be used. The compaction curve B represents this condition. Now it can be seen from Figure 3.17 that the most economical moisture content is between w_3 and w_4. Note that $w = w_4$ is the optimum moisture content for curve A, which is for the maximum compactive effort.

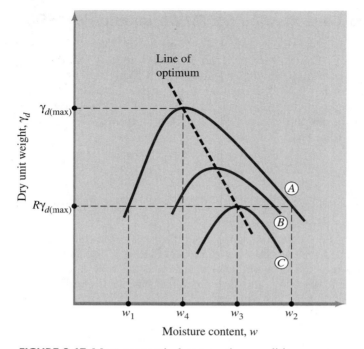

FIGURE 3.17 Most economical compaction condition

The concept described in the preceding paragraph, along with Figure 3.17, is historically attributed to Seed (1964), who was a prominent figure in modern geotechnical engineering. The idea is elaborated on in more detail in Holtz and Kovacs (1981).

3.8 *Determination of Field Unit Weight After Compaction*

When the compaction work is progressing in the field, it is useful to know whether or not the unit weight specified is achieved. Three standard procedures are used for determining the field unit weight of compaction:

1. Sand cone method
2. Rubber balloon method
3. Nuclear method

Following is a brief description of each of these methods.

Sand Cone Method (ASTM Designation D-1556)

The sand cone device consists of a glass or plastic jar with a metal cone attached at its top (Figure 3.18). The jar is filled with very uniform dry Ottawa sand. The weight of the jar, the cone, and the sand filling the jar is determined (W_1). In the field, a small hole is excavated in the area where the soil has been compacted. If the weight of the moist soil excavated from the hole (W_2) is determined and the moisture content of the excavated soil is known, the dry weight of the soil (W_3) can be found as

$$W_3 = \frac{W_2}{1 + \dfrac{w\,(\%)}{100}} \tag{3.10}$$

where w = moisture content.

After excavation of the hole, the cone with the sand-filled jar attached to it is inverted and placed over the hole (Figure 3.19). Sand is allowed to flow out of the jar into the hole and the cone. Once the hole and cone are filled, the weight of the jar, the cone, and the remaining sand in the jar is determined (W_4), so

$$W_5 = W_1 - W_4 \tag{3.11}$$

where W_5 = weight of sand to fill the hole and cone.

FIGURE 3.18 Plastic jar and the metal cone for the sand cone device (*Note:* The jar is filled with Ottawa sand.)

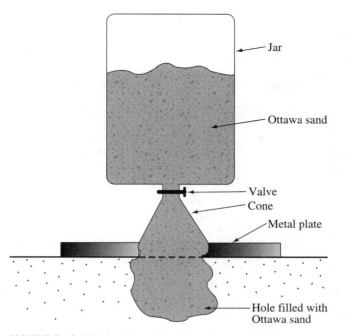

FIGURE 3.19 Field unit weight by sand cone method

The volume of the hole excavated can now be determined as

$$V = \frac{W_5 - W_c}{\gamma_{d(\text{sand})}}$$ (3.12)

where W_c = weight of sand to fill the cone only
$\gamma_{d(\text{sand})}$ = dry unit weight of Ottawa sand used

The values of W_c and $\gamma_{d(\text{sand})}$ are determined from the calibration done in the laboratory. The dry unit weight of compaction made in the field can now be determined as

$$\gamma_d = \frac{\text{dry weight of the soil excavated from the hole}}{\text{volume of the hole}} = \frac{W_3}{V}$$ (3.13)

Rubber Balloon Method (ASTM Designation D-2167)

The procedure for the rubber balloon method is similar to that for the sand cone method; a test hole is made and the moist weight of the soil removed from the hole and its moisture content are determined. However, the volume of the hole is determined by introducing a rubber balloon filled with water from a calibrated vessel into the hole, from which the volume can be read directly. The dry unit

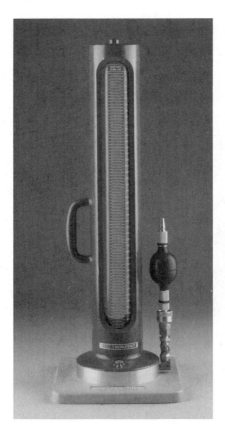

FIGURE 3.20 Calibrated vessel for the rubber balloon method for determination of field unit weight (Courtesy of ELE International/Soiltest Products Division, Lake Bluff, Illinois)

weight of the compacted soil can be determined by using Eq. (3.13). Figure 3.20 shows a calibrated vessel used in this method.

Nuclear Method

Nuclear density meters are now used often to determine the compacted dry unit weight of soil. The density meters operate either in drilled holes or from the ground surface. The instrument measures the weight of wet soil per unit volume and also the weight of water present in a unit volume of soil. The dry unit weight of compacted soil can be determined by subtracting the weight of water from the moist unit weight of soil. Figure 3.21 shows a photograph of a nuclear density meter.

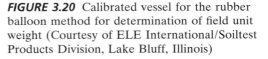

EXAMPLE 3.2

Following are the results of a field unit weight determination test using the sand cone method:

- Calibrated dry unit weight of Ottawa sand = 16.35 kN/m^3
- Mass of Ottawa sand to fill the cone = 0.117 kg
- Mass of jar + cone + sand (before use) = 6.005 kg

FIGURE 3.21 Nuclear density meter (Courtesy of David A. Carroll, Austin, Texas)

- Mass of jar + cone + sand (after use) = 2.818 kg
- Mass of moist soil from hole = 3.318 kg
- Moisture content of moist soil = 11.6%

Determine the dry unit weight of compaction in the field.

Solution   The mass of the sand needed to fill the hole and cone is

$$6.005 - 2.818 = 3.187 \text{ kg}$$

The mass of the sand used to fill the hole is

$$3.187 - 0.117 = 3.07 \text{ kg}$$

So the volume of the hole is

$$V = \frac{3.07 \times 9.81 \times 10^{-3}}{\text{dry unit weight of Ottawa sand}} = \frac{3.07 \times 9.81 \times 10^{-3}}{16.35} = 1.842 \times 10^{-3} \text{ m}^3$$

From Eq. (3.10), the dry weight of soil from the field is

$$W_3 = \frac{W_2}{1 + \dfrac{w \, (\%)}{100}} = \frac{3.318 \times 9.81 \times 10^{-3}}{1 + \dfrac{11.6}{100}} = 29.17 \times 10^{-3} \text{ kN}$$

Hence, the dry unit weight of compaction is

$$\gamma_d = \frac{W_3}{V} = \frac{29.17 \times 10^{-3}}{1.842 \times 10^{-3}} = \textbf{15.83 kN/m}^3$$

■

Problems

3.1 Calculate the zero-air-void unit weight for a soil (in kN/m³) at $w = 5\%$, 8%, 10%, 12%, and 15%, given $G_s = 2.68$.

3.2 For a slightly organic soil, $G_s = 2.54$. For this soil, calculate and plot the variation of γ_{zav} (in kN/m³) against w (in percent) with w varying from 5% to 20%.

3.3 **a.** Derive an equation for obtaining the theoretical dry unit weight for different degrees of saturation, S (i.e., γ_d as a function of G_s, γ_w, S, and w), for a soil.

b. For a given soil, if $G_s = 2.6$, calculate the theoretical variation of γ_d with w for 90% saturation.

3.4 For a compacted soil, given $G_s = 2.72$, $w = 18\%$, and $\gamma_d = 0.9\gamma_{zav}$, determine the dry unit weight of the compacted soil.

3.5 The results of a standard Proctor test are given in the table. Determine the maximum dry unit weight of compaction and the optimum moisture content. Also determine the moisture content required to achieve 95% of $\gamma_{d(max)}$.

Volume of Proctor mold (cm³)	Weight of wet soil in the mold (kg)	Moisture content (%)
943.3	1.65	10
943.3	1.75	12
943.3	1.83	14
943.3	1.81	16
943.3	1.76	18
943.3	1.70	20

3.6 Repeat Problem 3.5 with the following values:

Weight of wet soil in standard Proctor mold (kg)	Moisture content (%)
1.48	8.4
1.89	10.2
2.12	12.3
1.83	14.6
1.53	16.8

Volume of mold = 943.3 cm³.

3.7   A field unit weight determination test for the soil described in Problem 3.5 gave the following data: moisture content = 15% and moist unit weight = 16.8 kN/m³.

a. Determine the relative compaction.

b. If G_s is 2.68, what was the degree of saturation in the field?

3.8   The maximum and minimum dry unit weights of a sand were determined in the laboratory to be 16.3 kN/m³ and 14.6 kN/m³, respectively. What would be the relative compaction in the field if the relative density is 78%?

3.9   The maximum and minimum dry unit weights of a sand were determined in the laboratory to be 16.5 kN/m³ and 14.5 kN/m³, respectively. In the field, if the relative density of compaction of the same sand is 70%, what are its relative compaction and dry unit weight?

3.10   The relative compaction of a sand in the field is 94%. The maximum and minimum dry unit weights of the sand are 16.2 kN/m³ and 14.9 kN/m³, respectively. For the field condition, determine:

a. Dry unit weight

b. Relative density of compaction

c. Moist unit weight at a moisture content of 8%

3.11   Laboratory compaction test results on a clayey silt are listed in the table.

Moisture content (%)	Dry unit weight (kN/m³)
6	14.80
8	17.45
9	18.52
11	18.9
12	18.5
14	16.9

Following are the results of a field unit weight determination test on the same soil with the sand cone method:

- Calibrated dry density of Ottawa sand = 1570 kg/m³
- Calibrated mass of Ottawa sand to fill the cone = 0.545 kg
- Mass of jar + cone + sand (before use) = 7.59 kg
- Mass of jar + cone + sand (after use) = 4.78 kg
- Mass of moist soil from hole = 3.007 kg
- Moisture content of moist soil = 10.2%

Determine:

a. Dry unit weight of compaction in the field

b. Relative compaction in the field

References

American Association of State Highway and Transportation Officials (1982). *AASHTO Materials, Part II,* Washington, D.C.

American Society for Testing and Materials (1998). *ASTM Standards,* Vol. 04.08, West Conshohocken, PA.

D'Appolonia, D. J., Whitman, R. V., and D'Appolonia, E. D. (1969). "Sand Compaction with Vibratory Rollers," *Journal of the Soil Mechanics and Foundations Division,* ASCE, Vol. 95, No. SM1, 263–284.

Holtz, R. D., and Kovacs, W. D. (1981). *An Introduction to Geotechnical Engineering,* Prentice-Hall, Englewood Cliffs, NJ.

Johnson, A. W., and Sallberg, J. R. (1960). "Factors That Influence Field Compaction of Soil," Highway Research Board, *Bulletin No. 272.*

Lambe, T. W. (1958). "The Structure of Compacted Clay," *Journal of the Soil Mechanics and Foundations Division,* ASCE, Vol. 84, No. SM2, 1654-1–1654-34.

Lee, K. W., and Singh, A. (1971). "Relative Density and Relative Compaction," *Journal of the Soil Mechanics and Foundations Division,* ASCE, Vol. 97, No. SM7, 1049–1052.

Lee, P. Y., and Suedkamp, R. J. (1972). "Characteristics of Irregularly Shaped Compaction Curves of Soils," *Highway Research Record No. 381,* National Academy of Sciences, Washington, D.C., 1–9.

Proctor, R. R. (1933). "Design and Construction of Rolled Earth Dams," *Engineering News Record,* Vol. 3, 245–248, 286–289, 348–351, 372–376.

Seed, H. B. (1964). Lecture Notes, CE 271, Seepage and Earth Dam Design, University of California, Berkeley.

Supplementary References for Further Study

Brown, E. (1977). "Vibroflotation Compaction of Cohesionless Soils," *Journal of the Geotechnical Engineering Division,* ASCE, Vol. 103, No. GT12, 1437–1451.

Franklin, A. F., Orozco, L. F., and Semrau, R. (1973). "Compaction of Slightly Organic Soils," *Journal of the Soil Mechanics and Foundations Division,* ASCE, Vol. 99, No. SM7, 541–557.

Lancaster, J., Waco, R., Towle, J., and Chaney, R. (1996). "The Effect of Organic Content on Soil Compaction," *Proceedings,* 3rd International Symposium on Environmental Geotechnology, San Diego, 152–161.

Leonards, G. A, Cutter, W. A, and Holtz, R. D. (1980). "Dynamic Compaction of Granular Soils," *Journal of the Geotechnical Engineering Division,* ASCE, Vol. 106, No. GT1, 35–44.

Mitchell, J. K. (1970). "In-Place Treatment of Foundation Soils," *Journal of the Soil Mechanics and Foundations Division,* ASCE, Vol. 96, No. SM1, 73–110.

Mitchell, J. K., Hooper, D. R., and Campanella, R. G. (1965). "Permeability of Compacted Clay," *Journal of the Soil Mechanics and Foundations Division,* ASCE, Vol. 91, No. SM4, 41–65.

Moo-Young, H. K., and Zimmie, T. F. (1996). "Geotechnical Properties of Paper Mill Sludges for Use in Landfill Covers," *Journal of Geotechnical Engineering,* ASCE, Vol. 122, No. 9, 768–775.

Seed, H. B., and Chan, C. K. (1959). "Structure and Stength Characteristics of Compacted Clays," *Journal of the Soil Mechanics and Foundations Division,* ASCE, Vol. 85, No. SM5, 87–128.

4

Movement of Water Through Soil— Hydraulic Conductivity and Seepage

Soils have interconnected voids through which water can flow from points of high energy to points of low energy. The study of the flow of water through porous soil media is important in soil mechanics. It is necessary for estimating the quantity of underground seepage under various hydraulic conditions, for investigating problems involving the pumping of water for underground construction, and for making stability analyses of earth dams and earth-retaining structures that are subject to seepage forces.

HYDRAULIC CONDUCTIVITY

4.1 Bernoulli's Equation

From fluid mechanics we know that, according to Bernoulli's equation, the total head at a point in water under motion can be given by the sum of the pressure, velocity, and elevation heads, or

$$h = \underset{\substack{\uparrow \\ \text{Pressure} \\ \text{head}}}{\frac{u}{\gamma_w}} + \underset{\substack{\uparrow \\ \text{Velocity} \\ \text{head}}}{\frac{v^2}{2g}} + \underset{\substack{\uparrow \\ \text{Elevation} \\ \text{head}}}{Z} \tag{4.1}$$

where h = total head
u = pressure

v = velocity

g = acceleration due to gravity

γ_w = unit weight of water

Note that the elevation head, Z, is the vertical distance of a given point above or below a datum plane. The pressure head is the water pressure, u, at that point divided by the unit weight of water, γ_w.

If Bernoulli's equation is applied to the flow of water through a porous soil medium, the term containing the velocity head can be neglected because the seepage velocity is small. Then the total head at any point can be adequately represented by

$$h = \frac{u}{\gamma_w} + Z \tag{4.2}$$

Figure 4.1 shows the relationship among the pressure, elevation, and total heads for the flow of water through soil. Open standpipes called *piezometers* are installed at points A and B. The levels to which water rises in the piezometer tubes situated at points A and B are known as the *piezometric levels* of points A and B, respectively. The pressure head at a point is the height of the vertical column of water in the piezometer installed at that point.

The loss of head between two points, A and B, can be given by

$$\Delta h = h_A - h_B = \left(\frac{u_A}{\gamma_w} + Z_A\right) - \left(\frac{u_B}{\gamma_w} + Z_B\right) \tag{4.3}$$

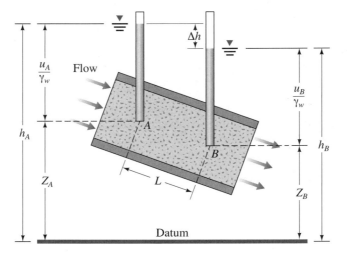

FIGURE 4.1 Pressure, elevation, and total heads for flow of water through soil

The head loss, Δh, can be expressed in a nondimensional form as

$$i = \frac{\Delta h}{L}$$ (4.4)

where i = hydraulic gradient

L = distance between points A and B—that is, the length of flow over which the loss of head occurred

In general, the variation of the velocity, v, with the hydraulic gradient, i, is as shown in Figure 4.2. This figure is divided into three zones:

1. Laminar flow zone (Zone I)
2. Transition zone (Zone II)
3. Turbulent flow zone (Zone III)

When the hydraulic gradient is gradually increased, the flow remains laminar in Zones I and II, and the velocity, v, bears a linear relationship to the hydraulic gradient. At a higher hydraulic gradient, the flow becomes turbulent (Zone III). When the hydraulic gradient is decreased, laminar flow conditions exist only in Zone I.

In most soils, the flow of water through the void spaces can be considered laminar; thus,

$v \propto i$ (4.5)

In fractured rock, stones, gravels, and very coarse sands, turbulent flow conditions may exist, and Eq. (4.5) may not be valid.

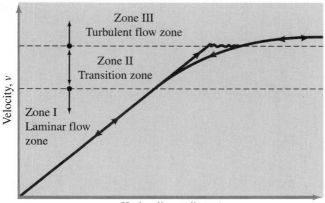

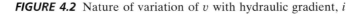

FIGURE 4.2 Nature of variation of v with hydraulic gradient, i

4.2   *Darcy's Law*

In 1856, Henri Philibert Gaspard Darcy published a simple empirical equation for the discharge velocity of water through saturated soils. This equation was based primarily on Darcy's observations about the flow of water through clean sands and is given as

$$v = ki \tag{4.6}$$

where v = *discharge velocity*, which is the quantity of water flowing in unit time through a unit gross cross-sectional area of soil at right angles to the direction of flow

k = hydraulic conductivity (otherwise known as the coefficient of permeability)

Hydraulic conductivity is expressed in cm/sec or m/sec, and discharge is in m³. It needs to be pointed out that the length is expressed in mm or m, so, in that sense, hydraulic conductivity should be expressed in mm/sec rather than cm/sec. However, geotechnical engineers continue to use cm/sec as the unit for hydraulic conductivity.

Note that Eq. (4.6) is similar to Eq. (4.5); both are valid for laminar flow conditions and applicable for a wide range of soils. In Eq. (4.6), v is the discharge velocity of water based on the gross cross-sectional area of the soil. However, the actual velocity of water (that is, the seepage velocity) through the void spaces is greater than v. A relationship between the discharge velocity and the seepage velocity can be derived by referring to Figure 4.3, which shows a soil of length L with a gross cross-sectional area A. If the quantity of water flowing through the soil in unit time is q, then

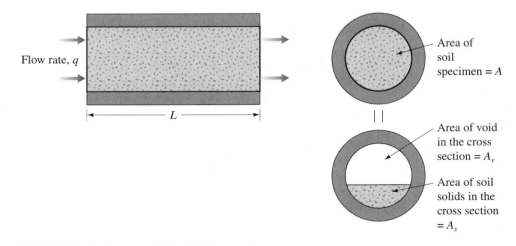

FIGURE 4.3 Derivation of Eq. (4.10)

$$q = vA = A_v v_s \tag{4.7}$$

where v_s = *seepage velocity*
  A_v = area of void in the cross section of the specimen

However,

$$A = A_v + A_s \tag{4.8}$$

where A_s = area of soil solids in the cross section of the specimen. Combining Eqs. (4.7) and (4.8) gives

$$q = v(A_v + A_s) = A_v v_s$$

or

$$v_s = \frac{v(A_v + A_s)}{A_v} = \frac{v(A_v + A_s)L}{A_v L} = \frac{v(V_v + V_s)}{V_v} \tag{4.9}$$

where V_v = volume of voids in the specimen
  V_s = volume of soil solids in the specimen

Equation (4.9) can be rewritten as

$$v_s = v \left[\frac{1 + \left(\dfrac{V_v}{V_s} \right)}{\dfrac{V_v}{V_s}} \right] = v \left(\frac{1 + e}{e} \right) = \frac{v}{n} \tag{4.10}$$

where e = void ratio
  n = porosity

Keep in mind that the terms *actual velocity* and *seepage velocity* are defined in an average sense. The actual and seepage velocities will vary with location within the pore volume of the soil.

4.3 *Hydraulic Conductivity*

The hydraulic conductivity of soils depends on several factors: fluid viscosity, pore-size distribution, grain-size distribution, void ratio, roughness of mineral particles, and degree of soil saturation. In clayey soils, structure plays an important role in hydraulic conductivity. Other major factors that affect the hydraulic conductivity of clays are the ionic concentration and the thickness of layers of water held to the clay particles.

Table 4.1 Typical values of hydraulic conductivity for saturated soils

Soil type	k (cm/sec)
Clean gravel	100–1
Coarse sand	1.0–0.01
Fine sand	0.01–0.001
Silty clay	0.001–0.00001
Clay	<0.000001

The value of hydraulic conductivity, k, varies widely for different soils. Some typical values for saturated soils are given in Table 4.1. The hydraulic conductivity of unsaturated soils is lower and increases rapidly with the degree of saturation.

The hydraulic conductivity of a soil is also related to the properties of the fluid flowing through it by the following equation:

$$k = \frac{\gamma_w}{\eta} \overline{K} \qquad\qquad (4.11)$$

where γ_w = unit weight of water
  η = viscosity of water
  $\overline{K}$ = absolute permeability

The *absolute permeability*, $\overline{K}$, is expressed in units of length squared (that is, cm^2).

4.4   *Laboratory Determination of Hydraulic Conductivity*

Two standard laboratory tests are used to determine the hydraulic conductivity of soil: the constant head test and the falling head test. The constant head test is used primarily for coarse-grained soils. For fine-grained soils, however, the flow rates through the soil are too small and, therefore, falling head tests are preferred. A brief description of each follows.

Constant Head Test

A typical arrangement of the constant head permeability test is shown in Figure 4.4. In this type of laboratory setup, the water supply at the inlet is adjusted in such a way that the difference of head between the inlet and the outlet remains constant during the test period. After a constant flow rate is established, water is collected in a graduated flask for a known duration.

The total volume of water, Q, collected may be expressed as

$$Q = Avt = A(ki)t \qquad\qquad (4.12)$$

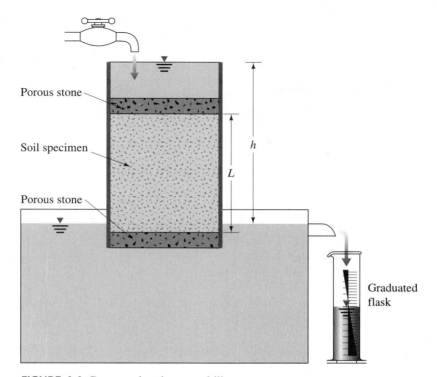

Porous stone

Soil specimen

Porous stone

Graduated flask

FIGURE 4.4 Constant head permeability test

where A = area of cross section of the soil specimen
t = duration of water collection

Also, because

$$i = \frac{h}{L} \tag{4.13}$$

where L = length of the specimen, Eq. (4.13) can be substituted into Eq. (4.12) to yield

$$Q = A\left(k\frac{h}{L}\right)t \tag{4.14}$$

or

$$k = \frac{QL}{Aht} \tag{4.15}$$

Falling Head Test

A typical arrangement of the falling head permeability test is shown in Figure 4.5. Water from a standpipe flows through the soil. The initial head difference, h_1, at time $t = 0$ is recorded, and water is allowed to flow through the soil specimen such that the final head difference at time $t = t_2$ is h_2.

The rate of flow of the water, q, through the specimen at any time t can be given by

$$q = k\frac{h}{L}A = -a\frac{dh}{dt} \tag{4.16}$$

where a = cross-sectional area of the standpipe
A = cross-sectional area of the soil specimen

Rearranging Eq. (4.16) gives

$$dt = \frac{aL}{Ak}\left(-\frac{dh}{h}\right) \tag{4.17}$$

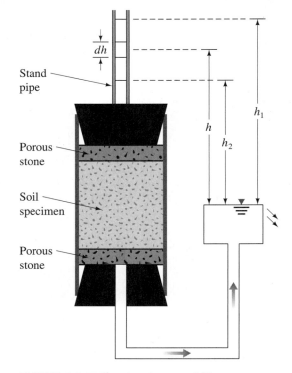

Stand
pipe

Porous
stone

Soil
specimen

Porous
stone

FIGURE 4.5 Falling head permeability test

Integration of the left side of Eq. (4.17) with limits of time from 0 to t and the right side with limits of head difference from h_1 to h_2 gives

$$t = \frac{aL}{Ak} \log_e \frac{h_1}{h_2}$$

or

$$k = 2.303 \frac{aL}{At} \log_{10} \frac{h_1}{h_2} \qquad (4.18)$$

EXAMPLE 4.1

A permeable soil layer is underlain by an impervious layer, as shown in Figure 4.6a. With $k = 4.8 \times 10^{-3}$ cm/sec for the permeable layer, calculate the rate of seepage through it in m³/hr/m width if $H = 3$ m and $\alpha = 5°$.

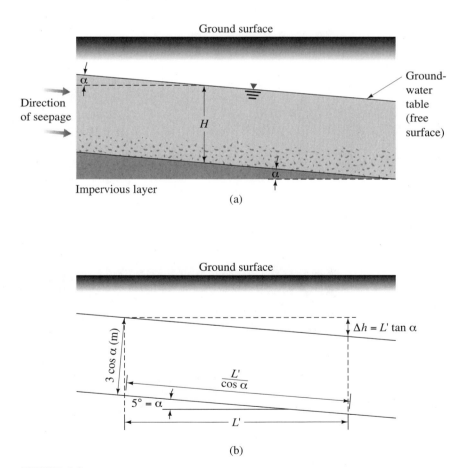

FIGURE 4.6

Solution   From Figure 4.6 and Eqs. (4.13) and (4.14), we have

$$i = \frac{\text{head loss}}{\text{length}} = \frac{L' \tan \alpha}{\left(\dfrac{L'}{\cos \alpha}\right)} = \sin \alpha$$

$$q = kiA = (k)(\sin \alpha)(3 \cos \alpha)(1)$$

$$k = 4.8 \times 10^{-3}\,\text{cm/sec} = 4.8 \times 10^{-5}\,\text{m/sec}$$

$$q = (4.8 \times 10^{-5})(\sin 5°)(3 \cos 5°)\,(3600) = \textbf{0.045 m}^3\textbf{/hr/m}$$

↑
To change to
m/hr ∎

**EXAMPLE
4.2**

Find the flow rate in m³/hr/m length (at right angles to the cross section shown) through the permeable soil layer shown in Figure 4.7 given $H = 3$ m, $H_1 = 1.1$ m, $h = 1.4$ m, $L = 40$ m, $\alpha = 14°$, and $k = 0.5 \times 10^{-3}$ m/sec.

Solution

$$\text{Hydraulic gradient, } i = \frac{h}{\dfrac{L}{\cos \alpha}}$$

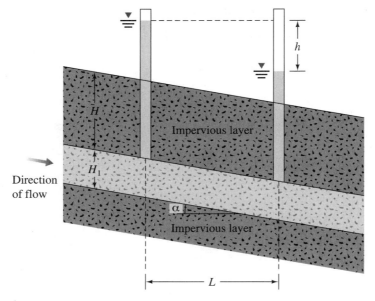

FIGURE 4.7

From Eqs. (4.13) and (4.14), we have

$$q = kiA = k \left(\frac{h \cos \alpha}{L} \right) (H_1 \cos \alpha \times 1)$$

$$= (0.5 \times 10^{-3} \times 60 \times 60 \text{ m/hr}) \left(\frac{1.4 \cos 14°}{40} \right) (1.1 \cos 14° \times 1)$$

$$= \textbf{0.065 m}^3\textbf{/hr/m}$$ ∎

EXAMPLE 4.3

A constant head laboratory permeability test on a fine sand gives the following values (refer to Figure 4.4):

- Length of specimen = 254 mm
- Diameter of specimen = 63.5 mm
- Head difference = 457 mm
- Water collected in 2 min = 0.51 cm³

Determine these values:

a. Hydraulic conductivity, k, of the soil (cm/sec)
b. Discharge velocity (cm/sec)
c. Seepage velocity (cm/sec)

The void ratio of the soil specimen is 0.46.

Solution

a. From Eq. (4.15),

$$k = \frac{QL}{Aht} = \frac{(0.51 \times 10^3)(254)}{\left(\frac{\pi}{4} 63.5^2 \right) (457)(2)} = 4.48 \times 10^{-2} \text{ mm/min} = \textbf{7.46} \times \textbf{10}^{-5} \textbf{ cm/sec}$$

b. From Eq. (4.6),

$$v = ki = (7.46 \times 10^{-5}) \left(\frac{457}{254} \right) = \textbf{13.42} \times \textbf{10}^{-5} \textbf{ cm/sec}$$

c. From Eq. (4.10),

$$v_s = v \left(\frac{1+e}{e} \right) = (13.42 \times 10^{-5}) \left(\frac{1+0.46}{0.46} \right) = \textbf{42.59} \times \textbf{10}^{-5} \textbf{ cm/sec}$$ ∎

EXAMPLE   For a constant head permeability test, the following values are given:
4.4

- $L = 300$ mm
- $A =$ specimen area $= 32$ cm^2
- $k = 0.0244$ cm/sec

The head difference was slowly changed in steps to 800, 700, 600, 500, and 400 mm. Calculate and plot the rate of flow, q, through the specimen, in cm^3/sec, against the head difference.

Solution   From Eq. (4.14), and given that $L = 300$ mm, we have

$$q = kiA = (0.0244)\left(\frac{h}{L}\right)(32) = 0.7808\left(\frac{h}{300}\right)$$

Now, we can prepare the following table:

h (mm)	q (cm³/sec)
800	2.08
700	1.82
600	1.56
500	1.30
400	1.04

The plot of q versus h is shown in Figure 4.8. ∎

EXAMPLE   For a variable head permeability test, the following values are given: length of
4.5   specimen $= 380$ mm, area of specimen $= 19.4$ cm^2, and $k = 2.92 \times 10^{-3}$ cm/sec. What should be the area of the standpipe for the head to drop from 640 to 320 mm in 8 min?

Solution   From Eq. (4.18), we have

$$k = 2.303\frac{aL}{At}\log_{10}\frac{h_1}{h_2}$$

$$2.92 \times 10^{-3} = 2.303\left(\frac{a \times 38\text{ cm}}{19.4 \times 480\text{ sec}}\right)\log_{10}\left(\frac{64\text{ cm}}{32\text{ cm}}\right)$$

$$a = \textbf{1.03 cm}^2$$ ∎

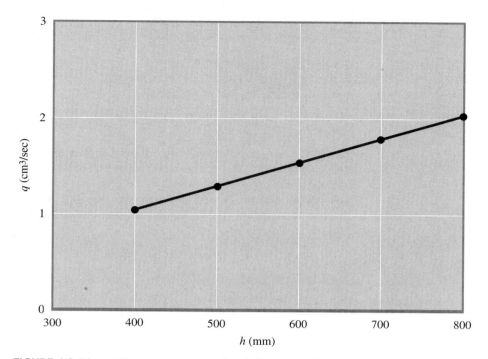

FIGURE 4.8 Plot of flow rate, q, versus head difference, h

**EXAMPLE
4.6**

The hydraulic conductivity of a clayey soil is 3×10^{-7} cm/sec. The viscosity of water at 25°C is 0.0911×10^{-4} g·sec/cm². Calculate the absolute permeability, $\overline{K}$, of the soil.

Solution From Eq. (4.11), we have

$$k = \frac{\gamma_w}{\eta}\overline{K} = 3 \times 10^{-7} \text{ cm/sec}$$

so

$$3 \times 10^{-7} = \left(\frac{1 \text{ g/cm}^3}{0.0911 \times 10^{-4}}\right)\overline{K}$$

$$\overline{K} = \mathbf{0.2733 \times 10^{-11} \ cm^2}$$

■

4.5 *Empirical Relations for Hydraulic Conductivity*

Several empirical equations for estimating hydraulic conductivity have been proposed over the years. Some of these are discussed briefly in this section.

For fairly uniform sand (that is, a small uniformity coefficient), Hazen (1930) proposed an empirical relationship for hydraulic conductivity in the form

$$k \text{ (cm/sec)} = cD_{10}^2 \qquad (4.19)$$

where c = a constant that varies from 1.0 to 1.5
          D_{10} = the effective size (mm)

Equation (4.19) is based primarily on Hazen's observations of loose, clean, filter sands. A small quantity of silts and clays, when present in a sandy soil, may change the hydraulic conductivity substantially.

Casagrande, in an unpublished report, proposed a simple relationship for hydraulic conductivity for fine to medium clean sand in the following form:

$$k = 1.4e^2 k_{0.85} \qquad (4.20)$$

where k = hydraulic conductivity at a void ratio e
          $k_{0.85}$ = the corresponding value at a void ratio of 0.85

Another form of equation that gives fairly good results in estimating the hydraulic conductivity of sandy soils is based on the Kozeny-Carman equation. The derivation of this equation is not presented here; you may refer to any advanced soil mechanics book (for example, Das, 1997). An application of the Kozeny-Carman equation yields

$$k \propto \frac{e^3}{1+e} \qquad (4.21)$$

where k = hydraulic conductivity at a void ratio of e. This equation can be rewritten as

$$k = C_1 \frac{e^3}{1+e} \qquad (4.22)$$

where C_1 = a constant.

It was stated at the end of Section 4.1 that turbulent flow conditions may exist in very coarse sands and gravels, and that Darcy's law may not be valid for these materials. However, under a low hydraulic gradient, laminar flow conditions usually exist. Kenney, Lau, and Ofoegbu (1984) conducted laboratory tests on granular soils in which the particle sizes in various specimens ranged from 0.074 to 25.4 mm. The uniformity coefficients, C_u, of these specimens ranged from 1.04 to 12. All permeability tests were conducted at a relative density of 80% or more. These tests showed that for laminar flow conditions,

$$\overline{K} \text{ (mm}^2) = (0.05 \text{ to } 1)D_5^2 \qquad (4.23)$$

where D_5 = diameter (mm) through which 5% of soil passes.

According to their experimental observations, Samarasinghe, Huang, and Drnevich (1982) suggested that the hydraulic conductivity of normally consolidated clays (see Chapter 6 for definition) can be given by the following equation:

$$k = C_3 \left(\frac{e^n}{1+e} \right) \tag{4.24}$$

where C_3 and n are constants to be determined experimentally. This equation can be rewritten as

$$\log[k(1+e)] = \log C_3 + n \log e \tag{4.25}$$

EXAMPLE 4.7

For a normally consolidated clay soil, the following values are given:

Void ratio	k (cm/sec)
1.1	0.302×10^{-7}
0.9	$0.12 \ \times 10^{-7}$

Estimate the hydraulic conductivity of the clay at a void ratio of 0.75. Use Eq. (4.24).

Solution From Eq. (4.24), we have

$$k = C_3 \left(\frac{e^n}{1+e} \right)$$

$$\frac{k_1}{k_2} = \frac{\left(\dfrac{e_1^n}{1+e_1} \right)}{\left(\dfrac{e_2^n}{1+e_2} \right)} \qquad \textit{(Note: } k_1 \text{ and } k_2 \text{ are hydraulic conductivities at void ratios } e_1 \text{ and } e_2 \text{, respectively.)}$$

$$\frac{0.302 \times 10^{-7}}{0.12 \times 10^{-7}} = \frac{\dfrac{(1.1)^n}{1+1.1}}{\dfrac{(0.9)^n}{1+0.9}}$$

$$2.517 = \left(\frac{1.9}{2.1} \right) \left(\frac{1.1}{0.9} \right)^n$$

$$2.782 = (1.222)^n$$

$$n = \frac{\log(2.782)}{\log(1.222)} = \frac{0.444}{0.087} = 5.1$$

so

$$k = C_3 \left(\frac{e^{5.1}}{1+e} \right)$$

To find C_3, we perform the following calculation:

$$0.302 \times 10^{-7} = C_3 \left(\frac{(1.1)^{5.1}}{1 + 1.1} \right) = \left(\frac{1.626}{2.1} \right) C_3$$

$$C_3 = \frac{(0.302 \times 10^{-7})(2.1)}{1.626} = 0.39 \times 10^{-7}$$

Hence,

$$k = (0.39 \times 10^{-7} \text{ cm/sec}) \left(\frac{e^n}{1 + e} \right)$$

At a void ratio of 0.75, we have

$$k = (0.39 \times 10^{-7}) \left(\frac{0.75^{5.1}}{1 + 0.75} \right) = \mathbf{0.514 \times 10^{-8} \text{ cm/sec}} \qquad \blacksquare$$

4.6   *Permeability Test in the Field by Pumping from Wells*

In the field, the average hydraulic conductivity of a soil deposit in the direction of flow can be determined by performing pumping tests from wells. Figure 4.9 shows a case where the top permeable layer, whose hydraulic conductivity has to be

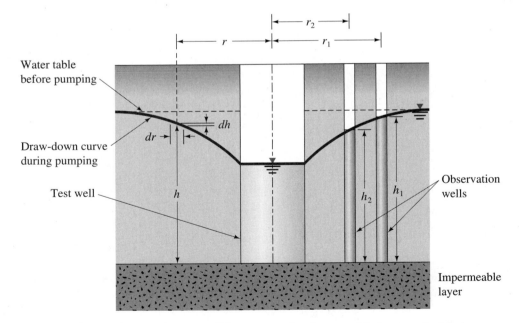

FIGURE 4.9 Pumping test from a well in an unconfined permeable layer underlain by an impermeable stratum

determined, is unconfined and underlain by an impermeable layer. During the test, water is pumped out at a constant rate from a test well that has a perforated casing. Several observation wells at various radial distances are made around the test well. Continuous observations of the water level in the test well and in the observation wells are made after the start of pumping, until a steady state is reached. The steady state is established when the water level in the test and observation wells becomes constant. The expression for the rate of flow of groundwater, q, into the well, which is equal to the rate of discharge from pumping, can be written as

$$q = k \left(\frac{dh}{dr} \right) 2\pi rh \tag{4.26}$$

or

$$\int_{r_2}^{r_1} \frac{dr}{r} = \left(\frac{2\pi k}{q} \right) \int_{h_2}^{h_1} h\, dh$$

Thus,

$$k = \frac{2.303q \log_{10} \left(\dfrac{r_1}{r_2} \right)}{\pi(h_1^2 - h_2^2)} \tag{4.27}$$

From field measurements, if q, r_1, r_2, h_1, and h_2 are known, then the hydraulic conductivity can be calculated from the simple relationship presented in Eq. (4.27).

The average hydraulic conductivity for a confined aquifer can also be determined by conducting a pumping test from a well with a perforated casing that penetrates the full depth of the aquifer and by observing the piezometric level in a number of observation wells at various radial distances (Figure 4.10). Pumping is continued at a uniform rate q until a steady state is reached.

Because water can enter the test well only from the aquifer of thickness H, the steady state of discharge is

$$q = k \left(\frac{dh}{dr} \right) 2\pi rH \tag{4.28}$$

or

$$\int_{r_2}^{r_1} \frac{dr}{r} = \int_{h_2}^{h_1} \frac{2\pi kH}{q}\, dh$$

This gives the hydraulic conductivity in the direction of flow as

$$k = \frac{q \log_{10} \left(\dfrac{r_1}{r_2} \right)}{2.727\, H(h_1 - h_2)} \tag{4.29}$$

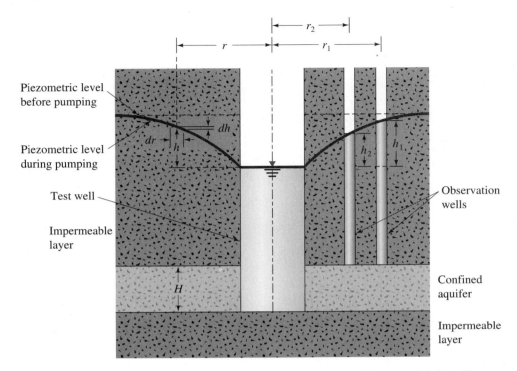

FIGURE 4.10 Pumping test from a well penetrating the full depth in a confined aquifer

---

EXAMPLE 4.8

Consider the case of pumping from a well in an unconfined permeable layer underlain by an impermeable stratum (see Figure 4.9). These values are given:

- $q = 0.74$ m³/min
- $h_1 = 5.5$ m at $r_1 = 60$ m
- $h_2 = 4.8$ m at $r_2 = 30$ m

Calculate the hydraulic conductivity (in cm/sec) of the permeable layer.

Solution   From Eq. (4.27), we have

$$k = \frac{2.303q \log_{10}\left(\dfrac{r_1}{r_2}\right)}{\pi(h_1^2 - h_2^2)} = \frac{(2.303)(0.74) \log_{10}\left(\dfrac{60}{30}\right)}{\pi(5.5^2 - 4.8^2)}$$

$$= 2.26 \times 10^{-2} \text{ m/min} = \mathbf{3.77 \times 10^{-2} \text{ cm/sec}}$$

∎

SEEPAGE

In the preceding sections of this chapter, we considered some simple cases for which direct application of Darcy's law was required to calculate the flow of water through soil. In many instances, the flow of water through soil is not in one direction only, and it is not uniform over the entire area perpendicular to the flow. In such cases, the groundwater flow is generally calculated by the use of graphs referred to as *flow nets*. The concept of the flow net is based on *Laplace's equation of continuity*, which governs the steady flow condition for a given point in the soil mass. The following sections explain the derivation of Laplace's equation of continuity and its application to drawing flow nets.

4.7 *Laplace's Equation of Continuity*

To derive the Laplace differential equation of continuity, we consider a single row of sheet piles that have been driven into a permeable soil layer, as shown in Figure 4.11a. The row of sheet piles is assumed to be impervious. The steady-state flow of water from the upstream to the downstream side through the permeable layer is a two-dimensional flow. For flow at a point A, we consider an elemental soil block. The block has dimensions dx, dy, and dz (length dy is perpendicular to the plane of the paper); it is shown in an enlarged scale in Figure 4.11b. Let v_x and v_z be the components of the discharge velocity in the horizontal and vertical directions, respectively. The rate of flow of water into the elemental block in the horizontal direction is equal to $v_x\, dz\, dy$, and in the vertical direction it is $v_z\, dx\, dy$. The rates of outflow from the block in the horizontal and vertical directions are

$$\left(v_x + \frac{\partial v_x}{\partial x}\, dx\right) dz\, dy$$

and

$$\left(v_z + \frac{\partial v_z}{\partial z}\, dz\right) dx\, dy$$

respectively. Assuming that water is incompressible and that no volume change in the soil mass occurs, we know that the total rate of inflow should equal the total rate of outflow. Thus,

$$\left[\left(v_x + \frac{\partial v_x}{\partial x}\, dx\right) dz\, dy + \left(v_z + \frac{\partial v_z}{\partial z}\, dz\right) dx\, dy\right] - [v_x\, dz\, dy + v_z\, dx\, dy] = 0$$

or

$$\frac{\partial v_x}{\partial x} + \frac{\partial v_z}{\partial z} = 0 \tag{4.30}$$

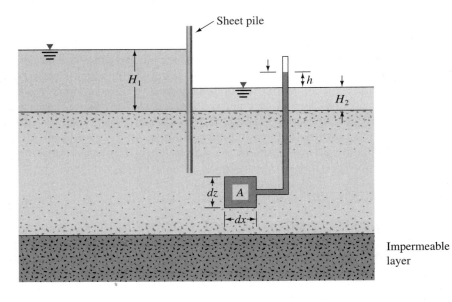

(a)

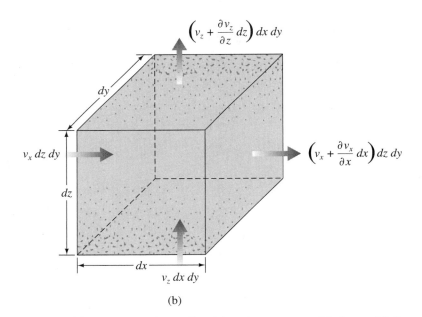

(b)

FIGURE 4.11 (a) Single-row sheet piles driven into a permeable layer; (b) flow at *A*

With Darcy's law, the discharge velocities can be expressed as

$$v_x = k_x i_x = k_x \left(-\frac{\partial h}{\partial x} \right) \tag{4.31}$$

and

$$v_z = k_z i_z = k_z \left(-\frac{\partial h}{\partial z} \right) \tag{4.32}$$

where k_x and k_z are the hydraulic conductivities in the vertical and horizontal directions, respectively.

From Eqs. (4.30), (4.31), and (4.32), we can write

$$k_x \frac{\partial^2 h}{\partial x^2} + k_z \frac{\partial^2 h}{\partial z^2} = 0 \tag{4.33}$$

If the soil is isotropic with respect to the hydraulic conductivity—that is, $k_x = k_z$—the preceding continuity equation for two-dimensional flow simplifies to

$$\frac{\partial^2 h}{\partial x^2} + \frac{\partial^2 h}{\partial z^2} = 0 \tag{4.34}$$

4.8 *Flow Nets*

The continuity equation [Eq. (4.34)] in an isotropic medium represents two orthogonal families of curves: the flow lines and the equipotential lines. A *flow line* is a line along which a water particle will travel from the upstream to the downstream side in the permeable soil medium. An *equipotential line* is a line along which the potential head at all points is equal. Thus, if piezometers are placed at different points along an equipotential line, the water level will rise to the same elevation in all of them. Figure 4.12a demonstrates the definition of flow and equipotential lines for flow in the permeable soil layer around the row of sheet piles shown in Figure 4.11 (for $k_x = k_z = k$).

A combination of a number of flow lines and equipotential lines is called a *flow net*. Flow nets are constructed to calculate groundwater flow in the media. To complete the graphic construction of a flow net, one must draw the flow and equipotential lines in such a way that the equipotential lines intersect the flow lines at right angles and the flow elements formed are approximate squares.

Figure 4.12b shows an example of a completed flow net. Another example of a flow net in an isotropic permeable layer is shown in Figure 4.13. In these figures, N_f is the number of flow channels in the flow net, and N_d is the number of potential drops (defined later in this chapter).

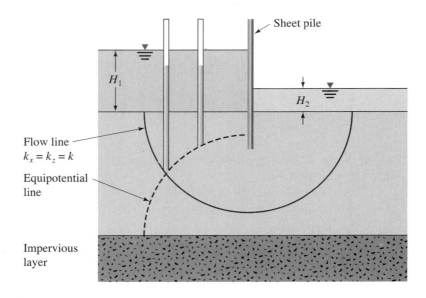

(a)

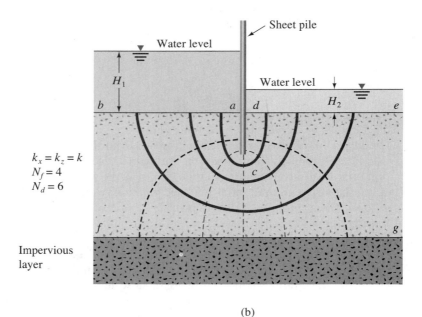

(b)

FIGURE 4.12 (a) Definition of flow lines and equipotential lines; (b) completed flow net

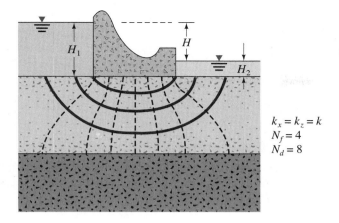

FIGURE 4.13 Flow net under a dam

Drawing a flow net takes several trials. While constructing the flow net, keep the boundary conditions in mind. For the flow net shown in Figure 4.12b, the following four boundary conditions apply:

1. The upstream and downstream surfaces of the permeable layer (lines *ab* and *de*) are equipotential lines.
2. Because *ab* and *de* are equipotential lines, all the flow lines intersect them at right angles.
3. The boundary of the impervious layer—that is, line *fg*—is a flow line, and so is the surface of the impervious sheet pile, line *acd*.
4. The equipotential lines intersect *acd* and *fg* at right angles.

Seepage Calculation from a Flow Net

In any flow net, the strip between any two adjacent flow lines is called a *flow channel*. Figure 4.14 shows a flow channel with the equipotential lines forming

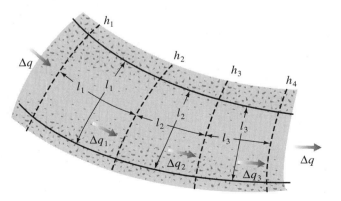

FIGURE 4.14 Seepage through a flow channel with square elements

square elements. Let $h_1, h_2, h_3, h_4, \ldots, h_n$ be the piezometric levels corresponding to the equipotential lines. The rate of seepage through the flow channel per unit length (perpendicular to the vertical section through the permeable layer) can be calculated as follows: Because there is no flow across the flow lines,

$$\Delta q_1 = \Delta q_2 = \Delta q_3 = \cdots = \Delta q \tag{4.35}$$

From Darcy's law, the flow rate is equal to kiA. Thus, Eq. (4.35) can be written as

$$\Delta q = k \left(\frac{h_1 - h_2}{l_1} \right) l_1 = k \left(\frac{h_2 - h_3}{l_2} \right) l_2 = k \left(\frac{h_3 - h_4}{l_3} \right) l_3 = \cdots \tag{4.36}$$

Equation (4.36) shows that if the flow elements are drawn as approximate squares, then the drop in the piezometric level between any two adjacent equipotential lines is the same. This is called the *potential drop*. Thus,

$$h_1 - h_2 = h_2 - h_3 = h_3 - h_4 = \cdots = \frac{H}{N_d} \tag{4.37}$$

and

$$\Delta q = k \frac{H}{N_d} \tag{4.38}$$

where H = head difference between the upstream and downstream sides
$\quad\quad N_d$ = number of potential drops

In Figure 4.12b, for any flow channel, $H = H_1 - H_2$ and $N_d = 6$.

If the number of flow channels in a flow net is equal to N_f, the total rate of flow through all the channels per unit length can be given by

$$q = k \frac{H N_f}{N_d} \tag{4.39}$$

Although drawing square elements for a flow net is convenient, it is not always necessary. Alternatively, one can draw a rectangular mesh for a flow channel, as shown in Figure 4.15, provided that the width-to-length ratios for all the rectangular

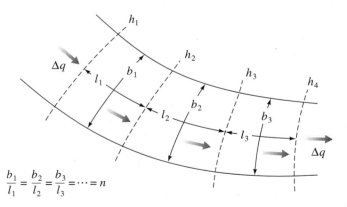

FIGURE 4.15 Seepage through a flow channel with rectangular elements

elements in the flow net are the same. In this case, Eq. (4.36) for rate of flow through the channel can be modified to

$$\Delta q = k \left(\frac{h_1 - h_2}{l_1}\right) b_1 = k \left(\frac{h_2 - h_3}{l_2}\right) b_2 = k \left(\frac{h_3 - h_4}{l_3}\right) b_3 = \cdots \tag{4.40}$$

If $b_1/l_1 = b_2/l_2 = b_3/l_3 = \cdots = n$ (i.e., the elements are not square), Eqs. (4.38) and (4.39) can be modified:

$$\Delta q = kH \left(\frac{n}{N_d}\right) \tag{4.41}$$

or

$$q = kH \left(\frac{N_f}{N_d}\right) n \tag{4.42}$$

Figure 4.16 shows a flow net for seepage around a single row of sheet piles. Note that flow channels 1 and 2 have square elements. Hence, the rate of flow through these two channels can be obtained from Eq. (4.38):

$$\Delta q_1 + \Delta q_2 = k \frac{H}{N_d} + k \frac{H}{N_d} = 2k \frac{H}{N_d}$$

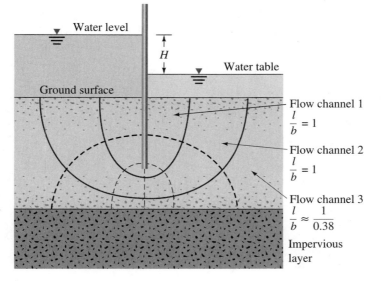

FIGURE 4.16 Flow net for seepage around a single row of sheet piles

However, flow channel 3 has rectangular elements. These elements have a width-to-length ratio of about 0.38; hence, from Eq. (4.41), we have

$$\Delta q_3 = kH \left(\frac{0.38}{N_d} \right)$$

So, the total rate of seepage can be given as

$$q = \Delta q_1 + \Delta q_2 + \Delta q_3 = 2.38 \frac{kH}{N_d}$$

EXAMPLE 4.9

A flow net for flow around a single row of sheet piles in a permeable soil layer is shown in Figure 4.17. We are given that $k_x = k_z = k = 5 \times 10^{-3}$ cm/sec.

    a. How high (above the ground surface) will the water rise if piezometers are placed at points a, b, c, and d?

    b. What is the rate of seepage through flow channel II per unit length (perpendicular to the section shown)?

Solution

    a. From Figure 4.17, we see that $N_f = 3$ and $N_d = 6$. The head difference between the upstream and downstream sides is 3.33 m, so the head loss for each drop is $3.33/6 = 0.555$ m. Point a is located on equipotential line

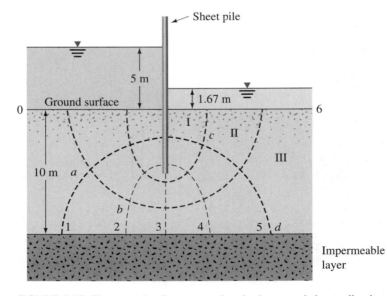

FIGURE 4.17 Flow net for flow around a single row of sheet piles in a permeable soil layer

1, which means that the potential drop at a is 1×0.555 m. The water in the piezometer at a will rise to an elevation of $(5 - 0.555) =$ **4.445 m above the ground surface.** Similarly, we can calculate the other piezometric levels:

$$b = (5 - 2 \times 0.555) = \textbf{3.89 m above the ground surface}$$

$$c = (5 - 5 \times 0.555) = \textbf{2.225 m above the ground surface}$$

$$d = (5 - 5 \times 0.555) = \textbf{2.225 m above the ground surface}$$

b. From Eq. (4.38), we have

$$\Delta q = k \frac{H}{N_d}$$

$$k = 5 \times 10^{-3} \text{ cm/sec} = 5 \times 10^{-5} \text{ m/sec}$$

$$\Delta q = (5 \times 10^{-5})(0.555) = \textbf{2.775} \times \textbf{10}^{-5} \textbf{ m}^3\textbf{/sec/m}$$ ■

4.9 Capillary Rise in Soils

The continuous void spaces in soil can act as bundles of capillary tubes with variable cross sections; therefore, because of the surface tension effect, movement of water in soil may take place by capillary rise. Figure 4.18 shows the fundamental concept

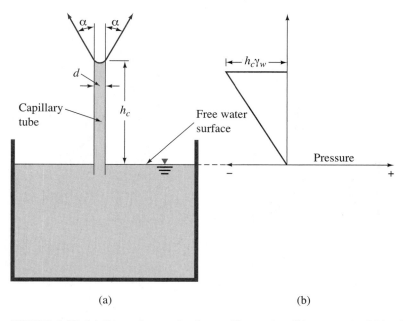

(a)   (b)

FIGURE 4.18 (a) Rise of water in the capillary tube; (b) pressure within the height of rise in the capillary tube (atmospheric pressure taken as datum)

of the height of rise in a capillary tube. The height of rise of water in the capillary tube can be found by summing the forces in the vertical direction, or

$$\left(\frac{\pi}{4}d^2\right)h_c\gamma_w = \pi\, dT\cos\alpha \tag{4.43}$$

$$h_c = \frac{4T\cos\alpha}{d\gamma_w}$$

where T = surface tension
α = angle of contact
d = diameter of capillary tube
γ_w = unit weight of water

From Eq. (4.43), we can see that, with T, α, and γ_w remaining constant,

$$h_c \propto \frac{1}{d} \tag{4.44}$$

The pressure at any point in the capillary tube above the free water surface is negative with respect to the atmospheric pressure, and the magnitude may be given by $h\gamma_w$ (where h = height above the free water surface).

Although the concept of capillary rise as demonstrated for an ideal capillary tube can be applied to soils, it must be realized that the capillary tubes formed in soils because of the continuity of voids have variable cross sections. The results of the nonuniformity on capillary rise can be seen when a dry column of sandy soil is placed in contact with water (Figure 4.19). After a given amount of time has

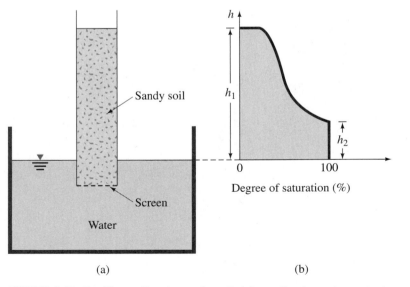

(a)   (b)

FIGURE 4.19 Capillary effect in sandy soil: (a) a soil column in contact with water; (b) variation of degree of saturation in the soil column

Table 4.2 Approximate range of capillary rise in soils

Soil type	Range of capillary rise (m)
Coarse sand	0.1–0.15
Fine sand	0.3–1.2
Silt	0.75–7.5
Clay	7.5–20

elapsed, the variation of the degree of saturation with the height of the soil column caused by capillary rise will be roughly as shown in Figure 4.19b. The degree of saturation is about 100% up to a height of h_2, and this corresponds to the largest voids. Beyond the height h_2, water can occupy only the smaller voids; hence, the degree of saturation will be less than 100%. The maximum height of capillary rise corresponds to the smallest voids. Hazen (1930) gave a formula for the approximate determination of the height of capillary rise in the form

$$h_1 \text{ (mm)} = \frac{C}{eD_{10}} \tag{4.45}$$

where D_{10} = effective size (mm)

e = void ratio

C = a constant that varies from 10 to 50 mm²

Equation (4.45) has an approach similar to that of Eq. (4.44). With the decrease of D_{10}, the pore size in soil will decrease, causing higher capillary rise. Table 4.2 shows the approximate range of capillary rise in various types of soils.

Capillary rise is important in the formation of some types of soils such as *caliche*, which can be found in the desert Southwest of the United States. Caliche is a mixture of sand, silt, and gravel bonded together by calcareous deposits. The calcareous deposits are brought to the surface by a net upward migration of water by capillary action. The water evaporates in the high local temperature. Because of sparse rainfall, the carbonates are not washed out of the top soil layer.

Problems

4.1 Refer to Figure 4.20. Find the flow rate in m³/sec/m length (at right angles to the cross section shown) through the permeable soil layer given $H = 4$ m, $H_1 = 2$ m, $h = 3.1$ m, $L = 30$ m, $\alpha = 14°$, and $k = 0.05$ cm/sec.

4.2 Repeat Problem 4.1 with the following values: $H = 2.2$ m, $H_1 = 1.5$ m, $h = 2.7$ m, $L = 5$ m, $\alpha = 20°$, and $k = 1.12 \times 10^{-5}$ m/sec. The flow rate should be given in m³/hr/m width (at right angles to the cross section shown).

4.3 Refer to the constant head arrangement shown in Figure 4.4. For a test, these values are given:

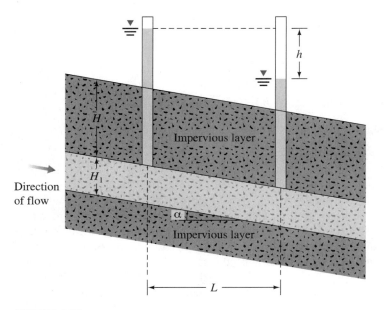

FIGURE 4.20

- L = 460 mm
- A = area of the specimen = 22.6 cm^2
- Constant head difference = h = 700 mm
- Water collected in 3 min = 354 cm^3

Calculate the hydraulic conductivity in cm/sec.

4.4 Refer to Figure 4.4. For a constant head permeability test in a sand, the following values are given:

- L = 350 mm
- A = 125 cm^2
- h = 420 mm
- Water collected in 3 min = 580 cm^3
- Void ratio of sand = 0.61

Determine:
a. Hydraulic conductivity, k (cm/sec)
b. Seepage velocity

4.5 In a constant head permeability test in the laboratory, the following values are given: L = 250 mm and A = 105 cm^2. If the value of k = 0.014 cm/sec and a flow rate of 120 cm^3/min must be maintained through the soil, what is the head difference, h, across the specimen? Also, determine the discharge velocity under the test conditions.

4.6 For a variable head permeabiltity test, these values are given:

  • Length of the soil specimen = 381 mm
  • Area of the soil specimen = 19.4 cm²
  • Area of the standpipe = 0.97 cm²
  • Head difference at time t = 0 is 635 mm
  • Head difference at time t = 8 min is 305 mm

  a. Determine the hydraulic conductivity of the soil in cm/sec.
  b. What was the head difference at time t = 4 min?

4.7 For a variable head permeability test, these values are given:

  • Length of the soil specimen = 200 mm
  • Area of the soil specimen = 1000 mm²
  • Area of the standpipe = 40 mm²
  • Head difference at time t = 0 is 500 mm
  • Head difference at time t = 3 min is 300 mm

  a. Determine the hydraulic conductivity of the soil in cm/sec.
  b. What was the head difference at time t = 100 sec?

4.8 The hydraulic conductivity, k, of a soil is 0.832×10^{-5} cm/sec at a temperature of 20°C. Determine its absolute permeability at 20°C, given that at 20°C, $\gamma_w = 9.789$ kN/m³ and $\eta = 1.005 \times 10^{-3}$ N·sec/m² (Newton-second per square meter).

4.9 The hydraulic conductivity of a sand at a void ratio of 0.62 is 0.03 cm/sec. Estimate its hydraulic conductivity at a void ratio of 0.48. Use Eqs. (4.20) and (4.21).

4.10 For a sand, we have porosity (n) = 0.31 and k = 0.066 cm/sec. Determine k when n = 0.4. Use Eqs. (4.20) and (4.21).

4.11 The maximum dry unit weight determined in the laboratory for a quartz sand is 16.0 kN/m³. In the field, if the relative compaction is 90%, determine the hydraulic conductivity of the sand in the field compaction condition (given that k for the sand at the maximum dry unit weight condition is 0.03 cm/sec and G_s = 2.7). Use Eq. (4.21).

4.12 For a sandy soil, we have e_{max} = 0.66, e_{min} = 0.36, and k at a relative density of 90% = 0.008 cm/sec. Determine k at a relative density of 50%. Use Eqs. (4.20) and (4.21).

4.13 A normally consolidated clay has the values given in the table:

Void ratio, e	k (cm/sec)
0.8	1.2×10^{-6}
1.4	3.6×10^{-6}

Estimate the hydraulic conductivity of the clay at a void ratio (e) of 0.62. Use Eq. (4.24).

4.14 A normally consolidated clay has the following values:

Void ratio, e	k (cm/sec)
1.2	0.2×10^{-6}
1.9	0.91×10^{-6}

Estimate the magnitude of k of the clay at a void ratio (e) of 0.9. Use Eq. (4.24).

4.15 Refer to Figure 4.21 and use these values:

- $H_1 = 7$ m, $\quad D = 3.5$ m
- $H_2 = 1.75$ m, $\quad D_1 = 7$ m

Draw a flow net. Calculate the seepage loss per meter length of the sheet pile (at a right angle to the cross section shown).

4.16 Draw a flow net for the single row of sheet piles driven into a permeability layer as shown in Figure 4.21, given the following:

- $H_1 = 5$ m, $\quad D = 4$ m
- $H_2 = 0.7$ m, $\quad D_1 = 10$ m

Calculate the seepage loss per meter length of the sheet pile (at right angles to the cross section shown).

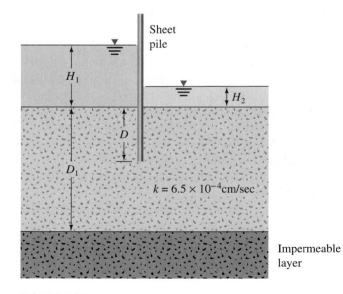

FIGURE 4.21

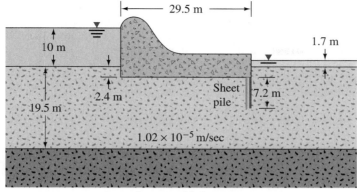

FIGURE 4.22

4.17 Draw a flow net for the weir shown in Figure 4.22. Calculate the rate of seepage under the weir.

References

Darcy, H. (1856). *Les Fontaines Publiques de la Ville de Dijon*, Dalmont, Paris.

Das, B. M. (1997). *Advanced Soil Mechanics*, 2nd edition, Taylor and Francis, Washington, D.C.

Hazen, A. (1930). "Water Supply," in *American Civil Engineers Handbook*, Wiley, New York.

Kenney, T. C., Lau, D., and Ofoegbu, G. I. (1984). "Permeability of Compacted Granular Materials," *Canadian Geotechnical Journal*, Vol. 21, No. 4, 726–729.

Samarasinghe, A. M., Huang, Y. H., and Drnevich, V. P. (1982). "Permeability and Consolidation of Normally Consolidated Soils," *Journal of the Geotechnical Engineering Division*, ASCE, Vol. 108, No. GT6, 835–850.

Supplementary References for Further Study

Ahmad, S., Lacroix, Y., and Steinback, J. (1975). "Pumping Tests in an Unconfined Aquifer," *Proceedings*, Conference on *in situ* Measurement of Soil Properties, ASCE, Vol. 1, 1–21.

Al-Tabbaa, A., and Wood, D. M. (1987). "Some Measurements of the Permeability of Kaolin," *Geotechnique*, Vol. 37, 499–503.

Amer, A. M., and Awad, A. A. (1974). "Permeability of Cohesionless Soils," *Journal of the Geotechnical Engineering Division*, ASCE, Vol. 100, No. GT12, 1309–1316.

Basak, P. (1972). "Soil Structure and Its Effects on Hydraulic Conductivity," *Soil Science*, Vol. 114, No. 6, 417–422.

Benson, C. H., and Daniel, D. E. (1990). "Influence of Clods on Hydraulic Conductivity of Compacted Clay," *Journal of Geotechnical Engineering*, ASCE, Vol. 116, No. 8, 1231–1248.

Chan, H. T., and Kenney, T. C. (1973). "Laboratory Investigation of Permeability Ratio of New Liskeard Varved Soil," *Canadian Geotechnical Journal*, Vol. 10, No. 3, 453–472.

Chapuis, R. P., Gill, D. E., and Baass, K. (1989). "Laboratory Permeability Tests on Sand: Influence of Compaction Method on Anisotropy," *Canadian Geotechnical Journal*, Vol. 26, 614–622.

Daniel, D. E., and Benson, C. H. (1990). "Water Content–Density Criteria for Compacted Soil Liners," *Journal of Geotechnical Engineering*, ASCE, Vol. 116, No. 12, 1811–1830.

Hansbo, S. (1960). "Consolidation of Clay with Special Reference to Influence of Vertical Sand Drains," Swedish Geotechnical Institute, *Proc. No. 18*, 41–61.

Tavenas, F., Jean, P., Leblond, F. T. P., and Leroueil, S. (1983). "The Permeability of Natural Soft Clays. Part II: Permeability Characteristics," *Canadian Geotechnical Journal*, Vol. 20, No. 4, 645–660.

5

Stresses in a Soil Mass

As described in Chapter 2, soils are multiphase systems. In a given volume of soil, the solid particles are distributed randomly with void spaces in between. The void spaces are continuous and are occupied by water, air, or both. To analyze problems such as compressibility of soils, bearing capacity of foundations, stability of embankments, and lateral pressure on earth-retaining structures, engineers need to know the nature of the distribution of stress along a given cross section of the soil profile—that is, what fraction of the normal stress at a given depth in a soil mass is carried by water in the void spaces and what fraction is carried by the soil skeleton at the points of contact of the soil particles. This issue is referred to as the *effective stress concept,* and it is discussed in the first part of this chapter.

When a foundation is constructed, changes take place in the soil under the foundation. The net stress usually increases. This net stress increase in the soil depends on the load per unit area to which the foundation is subjected, the depth below the foundation at which the stress estimation is made, and other factors. It is necessary to estimate the net increase of vertical stress in soil that occurs as a result of the construction of a foundation so that settlement can be calculated. The second part of this chapter discusses the principles for estimating the *vertical stress increase* in soil caused by various types of loading, based on the theory of elasticity. Although natural soil deposits are not fully elastic, isotropic, or homogeneous materials, calculations for estimating increases in vertical stress yield fairly good results for practical work.

EFFECTIVE STRESS CONCEPT

5.1 *Stresses in Saturated Soil Without Seepage*

Figure 5.1a shows a column of saturated soil mass with no seepage of water in any direction. The total stress at the elevation of point A, σ, can be obtained from the saturated unit weight of the soil and the unit weight of water above it. Thus,

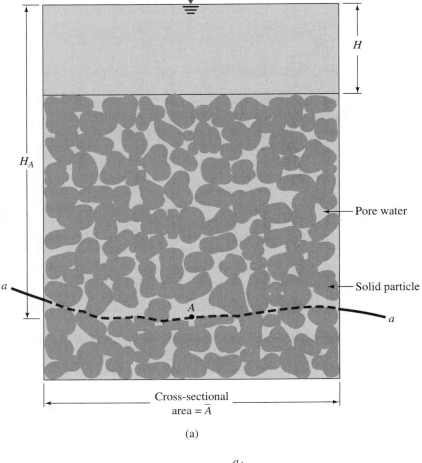

(a)

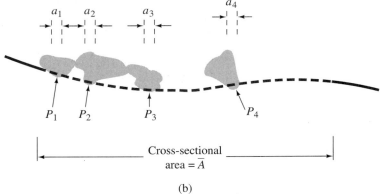

(b)

FIGURE 5.1 (a) Effective stress consideration for a saturated soil column without seepage; (b) forces acting at the points of contact of soil particles at the level of point A

$$\sigma = H\gamma_w + (H_A - H)\gamma_{sat} \tag{5.1}$$

where γ_w = unit weight of water

γ_{sat} = saturated unit weight of the soil

H = height of water table from the top of the soil column

H_A = distance between point A and the water table

The total stress, σ, given by Eq. (5.1) can be divided into two parts:

1. A portion is carried by water in the continuous void spaces. This portion acts with equal intensity in all directions.
2. The rest of the total stress is carried by the soil solids at their points of contact. The sum of the vertical components of the forces developed at the points of contact of the solid particles per unit cross-sectional area of the soil mass is called the *effective stress*.

The concept of effective stress can be illustrated by drawing a wavy line, *a-a*, through the point A that passes through only the points of contacts of the solid particles. Let $P_1, P_2, P_3, \ldots, P_n$ be the forces that act at the points of contact of the soil particles (Figure 5.1b). The sum of the vertical components of all such forces over the unit cross-sectional area is equal to the effective stress, σ', or

$$\sigma' = \frac{P_{1(v)} + P_{2(v)} + P_{3(v)} + \cdots + P_{n(v)}}{\overline{A}} \tag{5.2}$$

where $P_{1(v)}, P_{2(v)}, P_{3(v)}, \ldots, P_{n(v)}$ are the vertical components of $P_1, P_2, P_3, \ldots, P_n$, respectively, and $\overline{A}$ is the cross-sectional area of the soil mass under consideration.

Again, if a_s is the cross-sectional area occupied by solid-to-solid contacts (that is, $a_s = a_1 + a_2 + a_3 + \cdots + a_n$), then the space occupied by water equals $(\overline{A} - a_s)$. So we can write

$$\sigma = \sigma' + \frac{u(\overline{A} - a_s)}{\overline{A}} = \sigma' + u(1 - a_s') \tag{5.3}$$

where $u = H_A\gamma_w$ = pore water pressure (that is, the hydrostatic pressure at A)

$a_s' = a_s/\overline{A}$ = fraction of unit cross-sectional area of the soil mass occupied by solid-to-solid contacts

The value of a_s' is very small and can be neglected for the pressure ranges generally encountered in practical problems. Thus, Eq. (5.3) can be approximated by

$$\sigma = \sigma' + u \tag{5.4}$$

where u is also referred to as *neutral stress*. Substituting Eq. (5.1) for σ in Eq. (5.4) gives

$$\sigma' = [H\gamma_w + (H_A - H)\gamma_{sat}] - H_A\gamma_w$$

$$= (H_A - H)(\gamma_{sat} - \gamma_w)$$

$$= (\text{height of the soil column}) \times \gamma' \tag{5.5}$$

where $\gamma' = \gamma_{sat} - \gamma_w$ is the submerged unit weight of soil. Thus, it is clear that the effective stress at any point A is independent of the depth of water, H, above the submerged soil.

The principle of effective stress [Eq. (5.4)] was first developed by Terzaghi (1925, 1936). Skempton (1960) extended the work of Terzaghi and proposed the relationship between total and effective stress in the form of Eq. (5.3).

EXAMPLE 5.1

A soil profile is shown in Figure 5.2. Calculate the total stress, pore water pressure, and effective stress at points A, B, C, and D.

Solution   At A: Total stress: $\sigma'_A = 0$
Pore water pressure: $u_A = 0$
Effective stress: $\sigma'_A = 0$

At B: $\sigma_B = 3\gamma_{dry(sand)} = 3 \times 16.5 = \textbf{49.5 kN/m}^2$

$u_B = \textbf{0 kN/m}^2$

$\sigma'_B = 49.5 - 0 = \textbf{49.5 kN/m}^2$

At C: $\sigma_C = 6\gamma_{dry(sand)} = 6 \times 16.5 = \textbf{99 kN/m}^2$

$u_C = \textbf{0 kN/m}^2$

$\sigma'_C = 99 - 0 = \textbf{99 kN/m}^2$

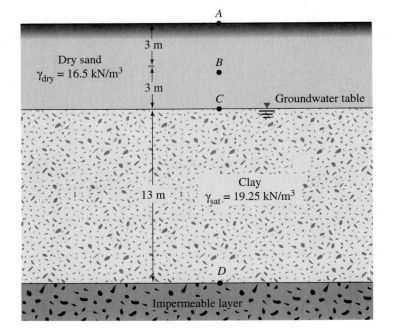

FIGURE 5.2

At D: $\sigma_D = 6\gamma_{dry(sand)} + 13\gamma_{sat(clay)}$
$= 6 \times 16.5 + 13 \times 19.25$
$= 99 + 250.25 = \mathbf{349.25\ kN/m^2}$

$u_D = 13\gamma_w = 13 \times 9.81 = \mathbf{127.53\ kN/m^2}$

$\sigma'_D = 349.25 - 127.53 = \mathbf{221.72\ kN/m^2}$ ■

5.2 *Stresses in Saturated Soil with Seepage*

If water is seeping, the effective stress at any point in a soil mass will be different from the static case. It will increase or decrease, depending on the direction of seepage.

Upward Seepage

Figure 5.3a shows a layer of granular soil in a tank where upward seepage is caused by adding water through the valve at the bottom of the tank. The rate of water supply is kept constant. The loss of head caused by upward seepage between the levels of points A and B is h. Keeping in mind that the total stress at any point in the soil mass is determined solely by the weight of soil and the water above it, we find the effective stress calculations at points A and B:

At A
- Total stress: $\sigma_A = H_1\gamma_w$
- Pore water pressure: $u_A = H_1\gamma_w$
- Effective stress: $\sigma'_A = \sigma_A - u_A = 0$

At B
- Total stress: $\sigma_B = H_1\gamma_w + H_2\gamma_{sat}$
- Pore water pressure: $u_B = (H_1 + H_2 + h)\gamma_w$
- Effective stress: $\sigma'_B = \sigma_B - u_B$
$= H_2(\gamma_{sat} - \gamma_w) - h\gamma_w$
$= H_2\gamma' - h\gamma_w$

Similarly, we can calculate the effective stress at a point C located at a depth z below the top of the soil surface:

At C
- Total stress: $\sigma_C = H_1\gamma_w + z\gamma_{sat}$
- Pore water pressure: $u_C = \left(H_1 + z + \dfrac{h}{H_2}z\right)\gamma_w$
- Effective stress: $\sigma'_C = \sigma_C - u_C$
$= z(\gamma_{sat} - \gamma_w) - \dfrac{h}{H_2}z\gamma_w$
$= z\gamma' - \dfrac{h}{H_2}z\gamma_w$

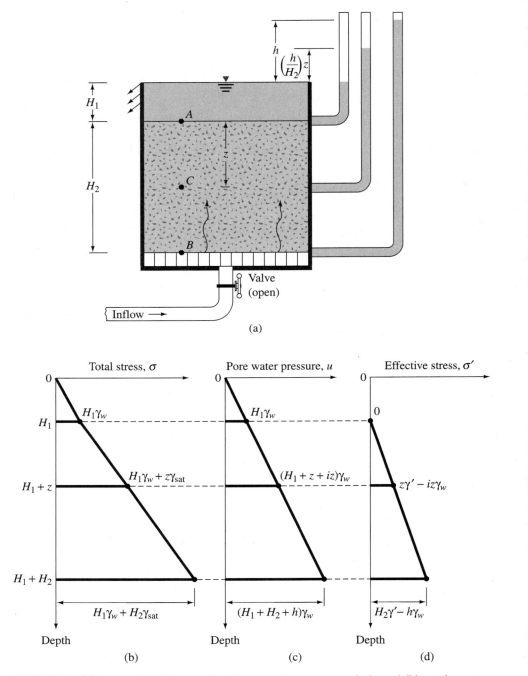

FIGURE 5.3 (a) Layer of soil in a tank with upward seepage; variation of (b) total stress; (c) pore water pressure; (d) effective stress with depth in a soil layer with upward seepage

Note that h/H_2 is the hydraulic gradient i caused by the flow, and so

$$\sigma'_C = z\gamma' - iz\gamma_w \tag{5.6}$$

The variations of total stress, pore water pressure, and effective stress with depth are plotted in Figures 5.3b, c, and d, respectively. If the rate of seepage and thereby the hydraulic gradient are gradually increased, a limiting condition will be reached, at which point

$$\sigma'_C = z\gamma' - i_{cr}z\gamma_w = 0 \tag{5.7}$$

where i_{cr} = critical hydraulic gradient (for zero effective stress). In such a situation, the stability of the soil will be lost. This is generally referred to as *boiling*, or *quick condition*.

From Eq. (5.7), we have

$$i_{cr} = \frac{\gamma'}{\gamma_w} \tag{5.8}$$

For most soils, the value of i_{cr} varies from 0.9 to 1.1, with an average of 1.

Downward Seepage

The condition of downward seepage is shown in Figure 5.4a. The level of water in the soil tank is held constant by adjusting the supply from the top and the outflow at the bottom.

The hydraulic gradient caused by the downward seepage is $i = h/H_2$. The total stress, pore water pressure, and effective stress at any point C are, respectively,

$$\sigma_C = H_1\gamma_w + z\gamma_{sat}$$

$$u_C = (H_1 + z - iz)\gamma_w$$

$$\sigma'_C = (H_1\gamma_w + z\gamma_{sat}) - (H_1 + z - iz)\gamma_w$$

$$= z\gamma' + iz\gamma_w \tag{5.9}$$

The variations of total stress, pore water pressure, and effective stress with depth are also shown graphically in Figures 5.4b, c, and d.

EXAMPLE 5.2

An exploratory drill hole was made in a saturated stiff clay (Figure 5.5). It was observed that the sand layer underlying the clay was under artesian pressure. Water in the drill hole rose to a height of H_1 above the top of the sand layer. If an open excavation is to be made in the clay, how deep can the excavation proceed before the bottom heaves? We are given $H = 8$ m, $H_1 = 4$ m, and $w = 32\%$.

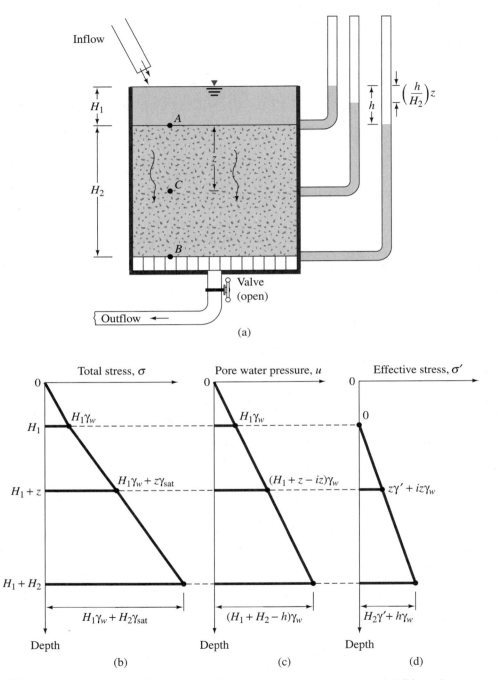

FIGURE 5.4 (a) Layer of soil in a tank with downward seepage; variation of (b) total stress; (c) pore water pressure; (d) effective stress with depth in a soil layer with downward seepage

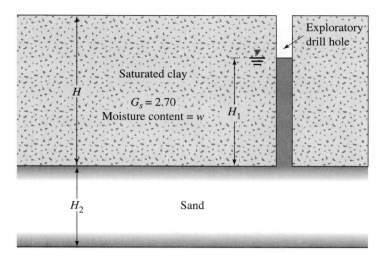

FIGURE 5.5

Solution Consider a point at the sand–clay interface. For heaving, $\sigma' = 0$, so

$$(H - H_{exc})\gamma_{sat(clay)} - H_1\gamma_w = 0$$

$$\gamma_{sat(clay)} = \frac{G_s\gamma_w + wG_s\gamma_w}{1 + e} = \frac{[2.70 + (0.32)(2.70)](9.81)}{1 + (0.32)(2.70)}$$

$$= 18.76 \text{ kN/m}^3$$

Thus,

$$(8 - H_{exc})(18.76) - (3)(9.81) = 0$$

$$H_{exc} = 8 - \frac{(3)(9.81)}{18.76} = \textbf{6.43 m}$$

■

5.3 *Effective Stress in Partially Saturated Soil*

In partially saturated soil, water in the void spaces is not continuous, and it is a three-phase system—that is, solid, pore water, and pore air (Figure 5.6). Hence, the total stress at any point in a soil profile is made up of intergranular, pore air, and pore water pressures. From laboratory test results, Bishop et al. (1960) gave the following equation for effective stress, σ', in partially saturated soils:

$$\sigma' = \sigma - u_a + \chi(u_a - u_w) \tag{5.10}$$

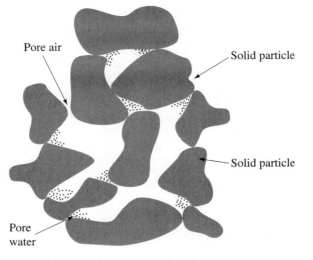

Pore air

Solid particle

Solid particle

Pore water

FIGURE 5.6 Partially saturated soil

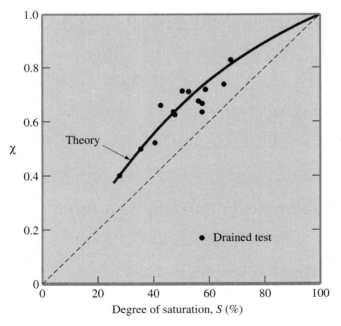

FIGURE 5.7 Relationship between the parameter χ and the degree of saturation for Bearhead silt (After Bishop et al., 1960)

where σ = total stress
u_a = pore air pressure
u_w = pore water pressure

In Eq. (5.10), χ represents the fraction of a unit cross-sectional area of the soil occupied by water. For dry soil $\chi = 0$, and for saturated soil $\chi = 1$.

Bishop et al. pointed out that the intermediate values of χ depend primarily on the degree of saturation, S. However, these values are also influenced by factors such as soil structure. The nature of variation of χ with the degree of saturation, S, for a silt is shown in Figure 5.7.

VERTICAL STRESS INCREASE DUE TO VARIOUS TYPES OF LOADING

5.4   *Stress Caused by a Point Load*

Boussinesq (1883) solved the problem of stresses produced at any point in a homogeneous, elastic, and isotropic medium as the result of a point load applied on the surface of an infinitely large half-space. According to Figure 5.8, Boussinesq's solution for normal stresses at a point A caused by the point load P is

$$\Delta\sigma_x = \frac{P}{2\pi}\left\{\frac{3x^2z}{L^5} - (1 - 2\mu)\left[\frac{x^2 - y^2}{Lr^2(L + z)} + \frac{y^2z}{L^3r^2}\right]\right\} \tag{5.11a}$$

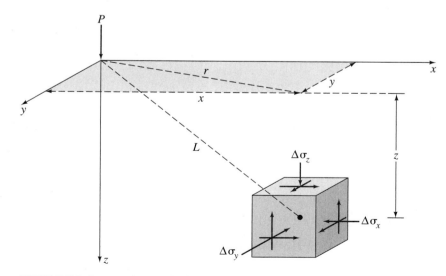

FIGURE 5.8 Stresses in an elastic medium caused by a point load

$$\Delta\sigma_y = \frac{P}{2\pi}\left\{\frac{3y^2z}{L^5} - (1 - 2\mu)\left[\frac{y^2 - x^2}{Lr^2(L + z)} + \frac{x^2z}{L^3r^2}\right]\right\}$$ (5.11b)

and

$$\Delta\sigma_z = \frac{3P}{2\pi}\frac{z^3}{L^5} = \frac{3P}{2\pi}\frac{z^3}{(r^2 + z^2)^{5/2}}$$ (5.12)

where $r = \sqrt{x^2 + y^2}$
$L = \sqrt{x^2 + y^2 + z^2} = \sqrt{r^2 + z^2}$
μ = Poisson's ratio

Note that Eqs. (5.11a) and (5.11b), which are the expressions for horizontal normal stresses, are dependent on Poisson's ratio of the medium. However, the relationship for the vertical normal stress, $\Delta\sigma_z$, as given by Eq. (5.12), is independent of Poisson's ratio. The relationship for $\Delta\sigma_z$ can be rewritten in the following form:

$$\Delta\sigma_z = \frac{P}{z^2}\left\{\frac{3}{2\pi}\frac{1}{[(r/z)^2 + 1]^{5/2}}\right\} = \frac{P}{z^2}I_1$$ (5.13)

where $I_1 = \dfrac{3}{2\pi}\dfrac{1}{[(r/z)^2 + 1]^{5/2}}.$ (5.14)

The variation of I_1 for various values of r/z is given in Table 5.1.

Table 5.1 Variation of I_1 [Eq. (5.14)]

r/z	I_1	r/z	I_1
0	0.4775	0.9	0.1083
0.1	0.4657	1.0	0.0844
0.2	0.4329	1.5	0.0251
0.3	0.3849	1.75	0.0144
0.4	0.3295	2.0	0.0085
0.5	0.2733	2.5	0.0034
0.6	0.2214	3.0	0.0015
0.7	0.1762	4.0	0.0004
0.8	0.1386	5.0	0.00014

Table 5.2 Representative values of Poisson's ratio

Type of soil	Poisson's ratio, μ
Loose sand	0.2–0.4
Medium sand	0.25–0.4
Dense sand	0.3–0.45
Silty sand	0.2–0.4
Soft clay	0.15–0.25
Medium clay	0.2–0.5

Typical values of Poisson's ratio for various soils are listed in Table 5.2.

EXAMPLE 5.3

Consider a point load $P = 4.5$ kN (Figure 5.8). Plot the variation of the vertical stress increase $\Delta\sigma_z$ with depth caused by the point load below the ground surface, with $x = 1$ m and $y = 1.5$ m.

Solution We have

$$r = \sqrt{x^2 + y^2} = \sqrt{1^2 + 1.5^2} = 1.8 \text{ m}$$

We can now prepare the following table:

r (m)	z (m)	$\dfrac{r}{z}$	I_1*	$\Delta\sigma_z$[†] (kN/m²)
1.8	0.5	3.6	0.0007	0.013
	1	1.8	0.013	0.059
	2	0.9	0.108	0.122
	3	0.6	0.221	0.111
	4	0.45	0.301	0.085
	5	0.36	0.352	0.063

*Eq. (5.14)
[†]Eq. (5.13)

Figure 5.9 shows the variation of $\Delta\sigma_z$ with depth, z. ∎

5.5 *Vertical Stress Caused by a Line Load*

Figure 5.10 shows a flexible line load of infinite length that has an intensity q per unit length on the surface of a semi-infinite soil mass. The vertical stress increase,

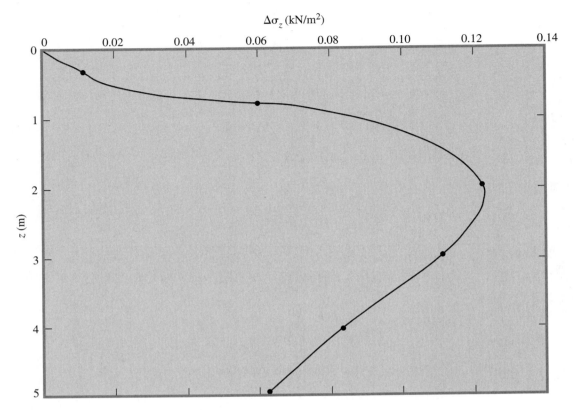

FIGURE 5.9

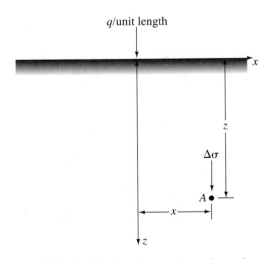

FIGURE 5.10 Line load over the surface of a semi-infinite soil mass

Table 5.3 Variation of $\Delta\sigma/(q/z)$ with x/z [Eq. (5.16)]

x/z	$\dfrac{\Delta\sigma}{q/z}$
0	0.637
0.1	0.624
0.2	0.589
0.3	0.536
0.4	0.473
0.5	0.407
0.6	0.344
0.7	0.287
0.8	0.237
0.9	0.194
1.0	0.159
1.5	0.060
2.0	0.025
3.0	0.006

$\Delta\sigma$, inside the soil mass can be determined by using the principles of the theory of elasticity, or

$$\Delta\sigma = \frac{2qz^3}{\pi(x^2 + z^2)^2} \tag{5.15}$$

The preceding equation can be rewritten as

$$\Delta\sigma = \frac{2q}{\pi z[(x/z)^2 + 1]^2}$$

or

$$\frac{\Delta\sigma}{(q/z)} = \frac{2}{\pi[(x/z)^2 + 1]^2} \tag{5.16}$$

Note that Eq. (5.16) is in a nondimensional form. Using this equation, we can calculate the variation of $\Delta\sigma/(q/z)$ with x/z. The variation is given in Table 5.3. The value of $\Delta\sigma$ calculated by using Eq. (5.16) is the additional stress on soil caused by the line load. The value of $\Delta\sigma$ does not include the overburden pressure of the soil above the point A.

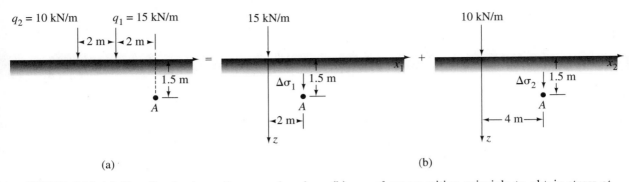

FIGURE 5.11 (a) Two line loads on the ground surface; (b) use of superposition principle to obtain stress at point *A*

**EXAMPLE
5.4**

Figure 5.11a shows two line loads on the ground surface. Determine the increase in the stress at point *A*.

Solution   Refer to Figure 5.11b. The total stress at point *A* is

$$\Delta\sigma = \Delta\sigma_1 + \Delta\sigma_2$$

or

$$\Delta\sigma = \frac{2q_1 z^3}{\pi(x_1^2 + z^2)^2} + \frac{2q_2 z^3}{\pi(x_2^2 + z^2)^2}$$

$$= \frac{(2)(15)(1.5)^3}{\pi[(2)^2 + (1.5)^2]^2} + \frac{(2)(10)(1.5)^3}{\pi[(4)^2 + (1.5)^2]^2}$$

$$= 0.825 + 0.065 = \textbf{0.89 kN/m} \qquad \blacksquare$$

 5.6   ## Vertical Stress Caused by a Strip Load (Finite Width and Infinite Length)

The fundamental equation for the vertical stress increase at a point in a soil mass as the result of a line load (see Section 5.5) can be used to determine the vertical stress at a point caused by a flexible strip load of width *B* (Figure 5.12). Let the load per unit area of the strip shown in Figure 5.12 be equal to *q*. If we consider an elemental strip of width *dr*, the load per unit length of this strip will be equal to *q dr*. This elemental strip can be treated as a line load. Equation (5.15) gives the vertical stress increase, *dσ*, at point *A* inside the soil mass caused by this

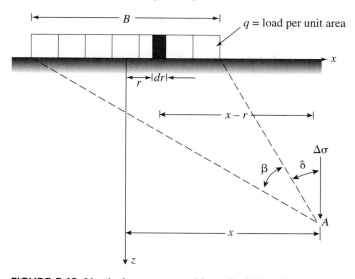

FIGURE 5.12 Vertical stress caused by a flexible strip load
(*Note*: Angles measured in counterclockwise direction are
taken as positive.)

elemental strip load. To calculate the vertical stress increase, we need to substitute
$q \, dr$ for q and $(x - r)$ for x. So,

$$d\sigma = \frac{2(q \, dr)z^3}{\pi[(x - r)^2 + z^2]^2} \tag{5.17}$$

The total increase in the vertical stress ($\Delta\sigma$) at point A caused by the entire strip
load of width B can be determined by the integration of Eq. (5.17) with limits of
r from $-B/2$ to $+B/2$, or

$$\Delta\sigma = \int d\sigma = \int_{-B/2}^{+B/2} \left(\frac{2q}{\pi}\right) \left\{\frac{z^3}{[(x - r)^2 + z^2]^2}\right\} dr$$

$$= \frac{q}{\pi} \left\{ \tan^{-1} \left[\frac{z}{x - (B/2)}\right] - \tan^{-1} \left[\frac{z}{x + (B/2)}\right] \right.$$

$$\left. - \frac{Bz[x^2 - z^2 - (B^2/4)]}{[x^2 + z^2 - (B^2/4)]^2 + B^2 z^2} \right\} \tag{5.18}$$

Equation (5.18) can be simplified to the form

$$\Delta\sigma = \frac{q}{\pi} [\beta + \sin \beta \cos(\beta + 2\delta)] \tag{5.19}$$

The angles β and δ are defined in Figure 5.12.

Table 5.4 Variation of $\Delta\sigma/q$ with $2z/B$ and $2x/B$

	2x/B				
2z/B	0	0.5	1.0	1.5	2.0
0	1.000	1.000	0.500	—	—
0.5	0.959	0.903	0.497	0.089	0.019
1.0	0.818	0.735	0.480	0.249	0.078
1.5	0.668	0.607	0.448	0.270	0.146
2.0	0.550	0.510	0.409	0.288	0.185
2.5	0.462	0.437	0.370	0.285	0.205
3.0	0.396	0.379	0.334	0.273	0.211
3.5	0.345	0.334	0.302	0.258	0.216
4.0	0.306	0.298	0.275	0.242	0.205
4.5	0.274	0.268	0.251	0.226	0.197
5.0	0.248	0.244	0.231	0.212	0.188

Table 5.4 shows the variation of $\Delta\sigma/q$ with $2z/B$ for $2x/B$ equal to 0, 0.5, 1.0, 1.5, and 2.0. This table can be conveniently used to calculate the vertical stress at a point caused by a flexible strip load. The net increase given by Eq. (5.19) can also be used to calculate stresses at various grid points under the load. Stress *isobars* can then be drawn. Stress isobars are contours of equal stress increase. Some vertical pressure isobars are plotted in Figure 5.13.

EXAMPLE 5.5

With reference to Figure 5.12, we are given $q = 200$ kN/m², $B = 6$ m, and $z = 3$ m. Determine the vertical stress increase at $x = \pm9$ m, ±6 m, ±3 m, and 0 m. Plot a graph of $\Delta\sigma$ against x.

Solution   We create the following table:

x (m)	2x/B	2z/B	$\Delta\sigma/q$*	$\Delta\sigma^\dagger$ kN/m²
±9	±3	1	0.0171	3.42
±6	±2	1	0.078	15.6
±3	±1	1	0.480	96.0
0	0	1	0.8183	163.66

*From Table 5.4
†$q = 200$ kN/m²

The plot of $\Delta\sigma$ versus x is given in Figure 5.14.   ∎

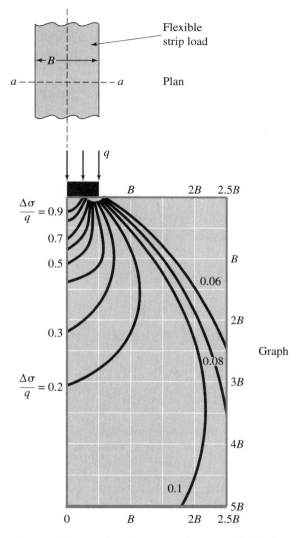

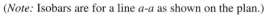

(*Note:* Isobars are for a line *a-a* as shown on the plan.)

FIGURE 5.13 Vertical pressure isobars under a flexible strip load

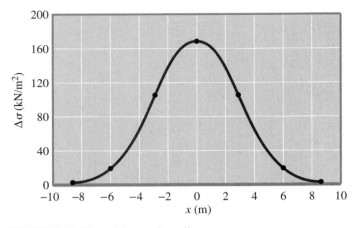

FIGURE 5.14 Plot of $\Delta\sigma$ against distance x

5.7   *Vertical Stress Below the Center of a Uniformly Loaded Circular Area*

Using Boussinesq's solution for vertical stress $\Delta\sigma$ caused by a point load [Eq. (5.12)], we can also develop an expression for the vertical stress below the center of a uniformly loaded flexible circular area.

From Figure 5.15, let the intensity of pressure on the circular area of radius R be equal to q. The total load on the elemental area (shaded in the figure) = $qr\, dr\, d\alpha$. The vertical stress, $d\sigma$, at point A caused by the load on the elemental area (which may be assumed to be a concentrated load) can be obtained from Eq. (5.12):

$$d\sigma = \frac{3(qr\, dr\, d\alpha)}{2\pi}\, \frac{z^3}{(r^2 + z^2)^{5/2}} \tag{5.20}$$

The increase in the stress at point A caused by the entire loaded area can be found by integrating Eq. (5.20), or

$$\Delta\sigma = \int d\sigma = \int_{\alpha=0}^{\alpha=2\pi} \int_{r=0}^{r=R} \frac{3q}{2\pi}\, \frac{z^3 r}{(r^2 + z^2)^{5/2}}\, dr\, d\alpha$$

So

$$\Delta\sigma = q\left\{1 - \frac{1}{[(R/z)^2 + 1]^{3/2}}\right\} \tag{5.21}$$

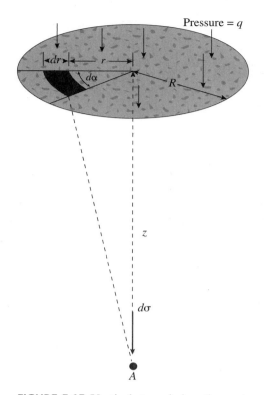

FIGURE 5.15 Vertical stress below the center of a uniformly loaded flexible circular area

The variation of $\Delta\sigma/q$ with z/R obtained from Eq. (5.21) is given in Table 5.5. A plot of this variation is shown in Figure 5.16. The value of $\Delta\sigma$ decreases rapidly with depth, and, at $z = 5R$, it is about 6% of q, which is the intensity of pressure at the ground surface.

5.8 *Vertical Stress Caused by a Rectangularly Loaded Area*

Boussinesq's solution can also be used to calculate the vertical stress increase below a flexible rectangular loaded area, as shown in Figure 5.17. The loaded area is located at the ground surface and has length L and width B. The uniformly distributed load per unit area is equal to q. To determine the increase in the vertical stress $\Delta\sigma$ at

Table 5.5 Variation of $\Delta\sigma/q$ with z/R [Eq. (5.21)]

z/R	Δσ/q
0	1
0.02	0.9999
0.05	0.9998
0.10	0.9990
0.2	0.9925
0.4	0.9488
0.5	0.9106
0.8	0.7562
1.0	0.6465
1.5	0.4240
2.0	0.2845
2.5	0.1996
3.0	0.1436
4.0	0.0869
5.0	0.0571

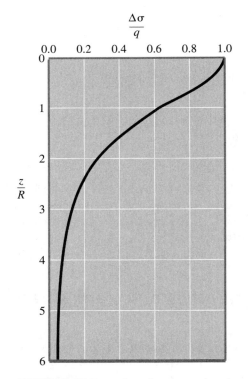

FIGURE 5.16 Intensity of stress under the center of a uniformly loaded circular flexible area

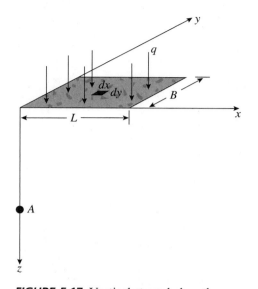

FIGURE 5.17 Vertical stress below the corner of a uniformly loaded flexible rectangular area

point A located at depth z below the corner of the rectangular area, we need to consider a small elemental area $dx\,dy$ of the rectangle (Figure 5.17). The load on this elemental area can be given by

$$dq = q\,dx\,dy \tag{5.22}$$

The increase in the stress $d\sigma$ at point A caused by the load dq can be determined by using Eq. (5.12). However, we need to replace P with $dq = q\,dx\,dy$ and r^2 with $x^2 + y^2$. Thus,

$$d\sigma = \frac{3q\,dx\,dy\,z^3}{2\pi(x^2 + y^2 + z^2)^{5/2}} \tag{5.23}$$

The increase in the stress $\Delta\sigma$ at point A caused by the entire loaded area can now be determined by integrating the preceding equation:

$$\Delta\sigma = \int d\sigma = \int_{y=0}^{B} \int_{x=0}^{L} \frac{3qz^3(dx\,dy)}{2\pi(x^2 + y^2 + z^2)^{5/2}} = qI_2 \tag{5.24}$$

where $I_2 = \dfrac{1}{4\pi}\left[\dfrac{2mn\sqrt{m^2 + n^2 + 1}}{m^2 + n^2 + m^2n^2 + 1}\left(\dfrac{m^2 + n^2 + 2}{m^2 + n^2 + 1}\right)\right.$

$$\left. + \tan^{-1}\left(\frac{2mn\sqrt{m^2 + n^2 + 1}}{m^2 + n^2 - m^2n^2 + 1}\right)\right] \tag{5.25}$$

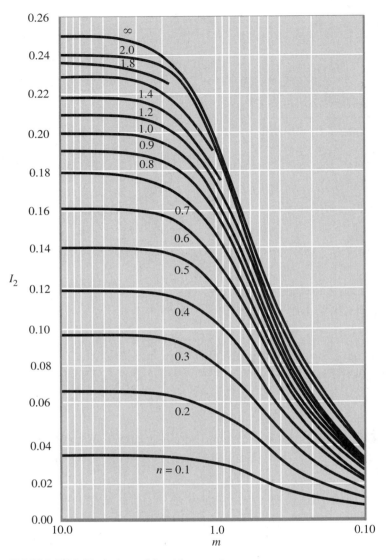

FIGURE 5.18 Variation of I_2 with m and n

$$m = \frac{B}{z} \qquad (5.26)$$

$$n = \frac{L}{z} \qquad (5.27)$$

The variation of I_2 with m and n is shown in Figure 5.18.

The increase in the stress at any point below a rectangularly loaded area can be found by using Eq. (5.24) and Figure 5.18. This concept can further be explained

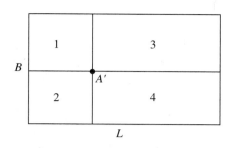

FIGURE 5.19 Increase of stress at any point below a rectangularly loaded flexible area

by referring to Figure 5.19. Let us determine the stress at a point below point A' at depth z. The loaded area can be divided into four rectangles as shown. The point A' is the corner common to all four rectangles. The increase in the stress at depth z below point A' due to each rectangular area can now be calculated by using Eq. (5.24). The total stress increase caused by the entire loaded area can be given by

$$\Delta\sigma = q[I_{2(1)} + I_{2(2)} + I_{2(3)} + I_{2(4)}] \tag{5.28}$$

where $I_{2(1)}$, $I_{2(2)}$, $I_{2(3)}$, and $I_{2(4)}$ = values of I_2 for rectangles 1, 2, 3, and 4, respectively.

As shown in Figure 5.13 (which is for a strip-loading case), Eq. (5.24) can be used to calculate the stress increase at various grid points. From those grid points, stress isobars can be plotted. Figure 5.20 shows such a plot for a uniformly loaded square area. Note that the stress isobars are valid for a vertical plane drawn through line a–a as shown at the top of Figure 5.20. Figure 5.21 is a nondimensional plot of $\Delta\sigma/q$ below the center of a rectangularly loaded area with $L/B = 1$, 1.5, 2, and ∞, which has been calculated by using Eq. (5.24).

EXAMPLE 5.6

The flexible area shown in Figure 5.22 is uniformly loaded. Given $q = 150 \text{ kN/m}^2$, determine the vertical stress increase at point A.

Solution   The flexible area shown in Figure 5.22 is divided into three parts in Figure 5.23. At point A,

$$\Delta\sigma = \Delta\sigma_1 + \Delta\sigma_2 + \Delta\sigma_3$$

From Eq. (5.21), we have

$$\Delta\sigma_1 = \left(\frac{1}{2}\right)q\left\{1 - \frac{1}{[(R/z)^2 + 1]^{3/2}}\right\}$$

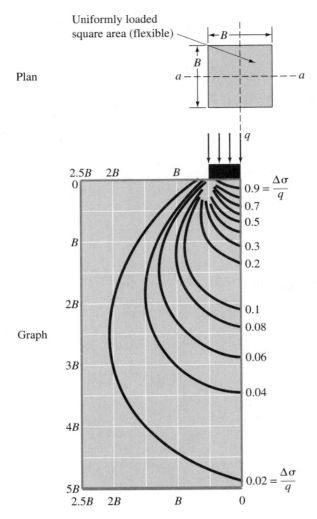

Uniformly loaded
square area (flexible)

Plan

Graph

(*Note:* Isobars are for a line *a-a* as shown on the plan.)

FIGURE 5.20 Vertical pressure isobars under a uniformly
loaded square area

We have $R = 1.5$ m, $z = 3$ m, and $q = 150$ kN/m^2, so

$$\Delta\sigma_1 = \frac{150}{2}\left\{1 - \frac{1}{[(1.5/3)^2 + 1]^{3/2}}\right\} = 21.3 \text{ kN/m}^2$$

It can be seen that $\Delta\sigma_2 = \Delta\sigma_3$. From Eqs. (5.26) and (5.27), we have

$$m = \frac{1.5}{3} = 0.5$$

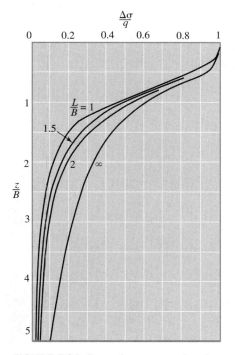

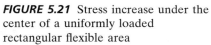

FIGURE 5.21 Stress increase under the center of a uniformly loaded rectangular flexible area

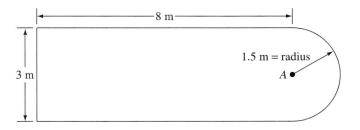

Plan

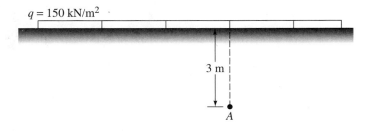

FIGURE 5.22

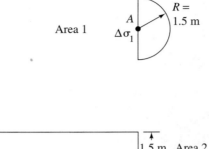

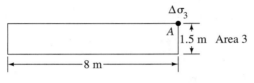

FIGURE 5.23

$$n = \frac{8}{3} = 2.67$$

From Figure 5.18, for $m = 0.5$ and $n = 2.67$, the magnitude of $I_2 = 0.138$. Thus, from Eq. (5.24), we have

$$\Delta\sigma_2 = \Delta\sigma_3 = qI_2 = (150)(0.138) = 20.7 \text{ kN/m}^2$$

so

$$\Delta\sigma = 21.3 + 20.7 + 20.7 = \textbf{62.7 kN/m}^2 \qquad \blacksquare$$

5.9   *Influence Chart for Vertical Pressure*

Equation (5.21) can be rearranged and written in the form

$$\frac{R}{z} = \sqrt{\left(1 - \frac{\Delta\sigma}{q}\right)^{-2/3} - 1} \qquad (5.29)$$

Note that R/z and $\Delta\sigma/q$ in the preceding equation are nondimensional quantities. The values of R/z that correspond to various pressure ratios are given in Table 5.6.

Using the values of R/z obtained from Eq. (5.29) for various pressure ratios, Newmark (1942) developed an influence chart that can be used to determine the vertical pressure at any point below a uniformly loaded flexible area of any shape.

We can follow the procedure presented by Newmark. Figure 5.24 shows an influence chart that was constructed by drawing concentric circles. The radii of the circles are equal to the R/z values corresponding to $\Delta\sigma/q = 0, 0.1, 0.2, \ldots, 1$. (*Note:* For $\Delta\sigma/q = 0$, $R/z = 0$, and for $\Delta\sigma/q = 1$, $R/z = \infty$, so nine circles are shown.) The unit length for plotting the circles is $\overline{AB}$. The circles are divided by several equally spaced radial lines. The influence value of the chart is given by $1/N$, where N is equal to the number of elements in the chart. In Figure 5.24, there are 200 elements; hence, the influence value is 0.005.

The procedure for finding the vertical pressure at any point below a loaded area is as follows:

1. Determine the depth z below the uniformly loaded area at which the stress increase is required.
2. Plot the plan of the loaded area with a scale of z equal to the unit length of the chart ($\overline{AB}$).
3. Place the plan (plotted in Step 2) on the influence chart in such a way that the point below which the stress is to be determined is located at the center of the chart.
4. Count the number of elements (M) of the chart enclosed by the plan of the loaded area.

Table 5.6 Values of R/z for various pressure ratios

$\Delta\sigma/q$	R/z	$\Delta\sigma/q$	R/z
0	0	0.55	0.8384
0.05	0.1865	0.60	0.9176
0.10	0.2698	0.65	1.0067
0.15	0.3383	0.70	1.1097
0.20	0.4005	0.75	1.2328
0.25	0.4598	0.80	1.3871
0.30	0.5181	0.85	1.5943
0.35	0.5768	0.90	1.9084
0.40	0.6370	0.95	2.5232
0.45	0.6997	1.00	∞
0.50	0.7664		

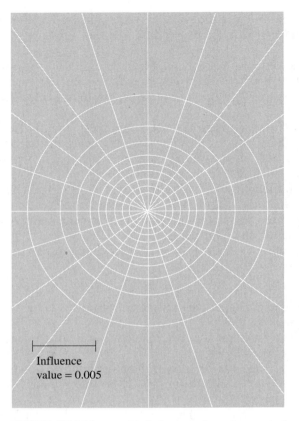

FIGURE 5.24 Newmark's influence chart for vertical pressure based on Boussinesq's theory

The increase in the pressure at the point under consideration is given by

$$\Delta\sigma = (IV)qM \qquad (5.30)$$

where IV = influence value

q = pressure on the loaded area

EXAMPLE 5.7   The cross section and plan of a column footing are shown in Figure 5.25. Find the increase in stress produced by the column footing at point A.

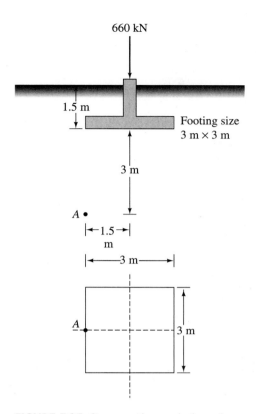

FIGURE 5.25 Cross section and plan of a column footing

Solution Point A is located at a depth 3 m below the bottom of the footing. The plan of the square footing has been replotted to a scale of $\overline{AB} = 3$ m and placed on the influence chart (Figure 5.26) in such a way that point A on the plan falls directly over the center of the chart. The number of elements inside the outline of the plan is about 48.5. Hence,

$$\Delta\sigma = (IV)qM = 0.005\left(\frac{660}{3 \times 3}\right)48.5 = \textbf{17.78 kN/m}^2$$ ■

Problems

5.1 A soil profile is shown in Figure 5.27. Calculate the values of σ, u, and σ' at points A, B, C, and D. Plot the variation of σ, u, and σ' with depth. We are given the values in the table.

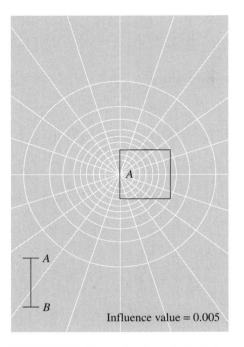

FIGURE 5.26 Determination of stress at
a point by use of Newmark's
influence chart

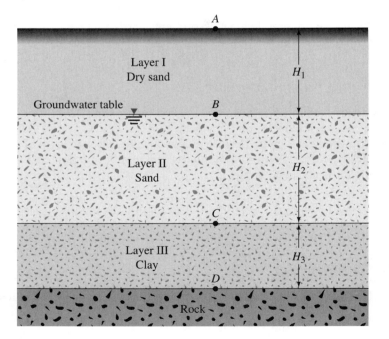

FIGURE 5.27

Layer no.	Thickness (m)	Unit weight (kN/m³)
I	$H_1 = 4$	$\gamma_d = 17.3$
II	$H_2 = 5$	$\gamma_{sat} = 18.9$
III	$H_3 = 6$	$\gamma_{sat} = 19.7$

5.2 Repeat Problem 5.1 with the following data:

Layer no.	Thickness (m)	Unit weight (kN/m³)
I	$H_1 = 4.5$	$\gamma_d = 15.0$
II	$H_2 = 10$	$\gamma_{sat} = 18.0$
III	$H_3 = 8.5$	$\gamma_{sat} = 19.0$

5.3 Repeat Problem 5.1 with these data:

Layer no.	Thickness (m)	Soil parameters
I	$H_1 = 3$	$e = 0.4, \ G_s = 2.62$
II	$H_2 = 4$	$e = 0.60, \ G_s = 2.68$
III	$H_3 = 2$	$e = 0.81, \ G_s = 2.73$

5.4 Plot the variation of total stress, pore water pressure, and effective stress with depth for the sand and clay layers shown in Figure 5.28 with $H_1 = 4$ m and $H_2 = 3$ m. Give numerical values.

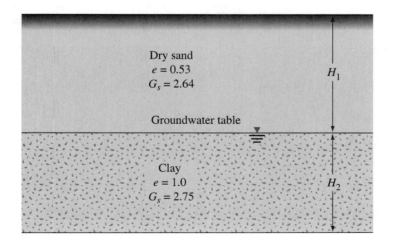

FIGURE 5.28

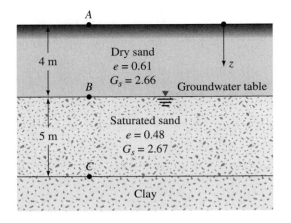

FIGURE 5.29

5.5   A soil profile is shown in Figure 5.29.
   a. Calculate the total stress, pore water pressure, and effective stress at points
         A, B, and C.
   b. How high should the groundwater table rise so that the effective stress at
         point C is 104 kN/m²?

5.6   A sand has $G_s = 2.66$. Calculate the hydraulic gradient that will cause boiling
         for $e = 0.35, 0.45, 0.55, 0.7,$ and 0.8. Plot a graph for i_{cr} versus e.

5.7   A 10-m-thick layer of stiff saturated clay is underlain by a layer of sand
         (Figure 5.30). The sand is under artesian pressure. Calculate the maximum
         depth of cut, H, that can be made in the clay.

5.8   A cut is made in a stiff saturated clay that is underlain by a layer of sand
         (Figure 5.31). What should be the height of the water, h, in the cut so that
         the stability of the saturated clay is not lost?

5.9   Refer to Figure 5.8. Given $P = 30$ kN, determine the vertical stress increase
         at a point with $x = 5$ m, $y = 4$ m, and $z = 6$ m. Use Boussinesq's solution.

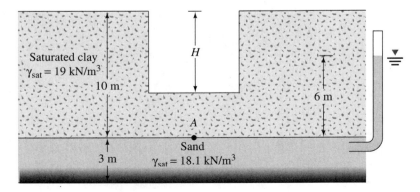

FIGURE 5.30

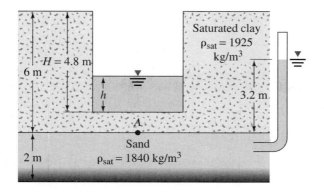

FIGURE 5.31

5.10 Refer to Figure 5.10. The magnitude of the line load q is 50 kN/m. Calculate and plot the variation of the vertical stress increase, $\Delta\sigma$, between the limits of $x = -8$ m and $x = +8$ m, given $z = 3$ m.

5.11 Refer to Figure 5.10. Assume $q = 65$ kN/m. Point A is located at a depth of 1.5 m below the ground surface. Because of the application of the point load, the vertical stress at point A increases by 24 kN/m². What is the horizontal distance between the line load and point A?

5.12 Refer to Figure 5.32. Determine the vertical stress increase, $\Delta\sigma$, at point A with the following values:

$$q_1 = 60 \text{ kN/m} \qquad x_1 = 1.5 \text{ m} \qquad z = 1.5 \text{ m}$$
$$q_2 = 0 \qquad\qquad x_2 = 0.5 \text{ m}$$

5.13 Repeat Problem 5.12 with the following values:

$$q_1 = 15 \text{ kN/m} \qquad x_1 = 5 \text{ m} \qquad z = 4 \text{ m}$$
$$q_2 = 9 \text{ kN/m} \qquad\, x_2 = 3 \text{ m}$$

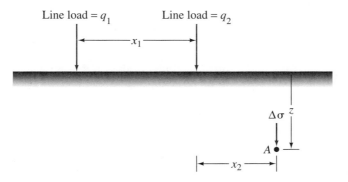

FIGURE 5.32 Stress at a point due to two line loads

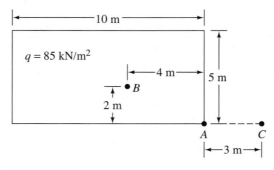

FIGURE 5.33

5.14 Refer to Figure 5.12. Given $B = 4$ m, $q = 20$ kN/m², $x = 1.5$ m, and $z = 2$ m, determine the vertical stress increase, $\Delta\sigma$, at point A.

5.15 Repeat Problem 5.14 for $q = 600$ kN/m², $B = 3$ m, $x = 1.5$ m, and $z = 3$ m.

5.16 Consider a circularly loaded flexible area on the ground surface. Given radius of circular area, $R = 2$ m, and uniformly distributed load, $q = 170$ kN/m², calculate the vertical stress increase, $\Delta\sigma$, at a point located 1.5 m below the ground surface (immediately below the center of the circular area).

5.17 Repeat Problem 5.16 with $R = 3$ m, $q = 250$ kN/m², and $z = 2.5$ m.

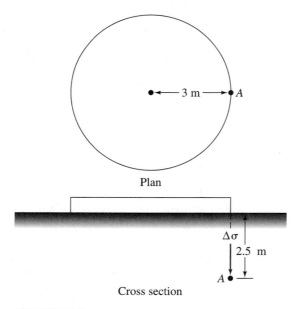

FIGURE 5.34

5.18 The plan of a flexible rectangular loaded area is shown in Figure 5.33. The uniformly distributed load on the flexible area, q, is 85 kN/m^2. Determine the increase in the vertical stress, $\Delta\sigma$, at a depth of $z = 5$ m below these points:
a. Point A
b. Point B
c. Point C

5.19 Repeat Problem 5.19. Use Newmark's influence chart for vertical pressure distribution.

5.20 Refer to Figure 5.34. The circular flexible area is uniformly loaded. Given $q = 250$ kN/m^2 and using Newmark's chart, determine the vertical stress increase, $\Delta\sigma$, at point A.

References

Bishop, A. W., Alpen, I., Blight, G. C., and Donald, I. B. (1960). "Factors Controlling the Strength of Partially Saturated Cohesive Soils," *Proceedings*, Research Conference on Shear Strength of Cohesive Soils, ASCE, 500–532.

Boussinesq, J. (1883). *Application des Potentials à L'Etude de L'Equilibre et du Mouvement des Solides Elastiques*, Gauthier–Villars, Paris.

Newmark, N. M. (1942). "Influence Charts for Computation of Stresses in Elastic Soil," University of Illinois Engineering Experiment Station, *Bulletin No. 338*.

Skempton, A. W. (1960). "Correspondence," *Geotechnique*, Vol. 10, No. 4, 186.

Terzaghi, K. (1925). *Erdbaumechanik auf Bodenphysikalischer Grundlage*, Deuticke, Vienna.

Terzaghi, K. (1936). "Relation between Soil Mechanics and Foundation Engineering: Presidential Address," *Proceedings*, First International Conference on Soil Mechanics and Foundation Engineering, Boston, Vol. 3, 13–18.

Supplementary References for Further Study

Ahlvin, R. G., and Ulery, H. H. (1962). "Tabulated Values of Determining the Complete Pattern of Stresses, Strains, and Deflection Beneath a Uniform Load on a Homogeneous Half Space," *Bulletin 342*, Highway Research Record, Washington, D.C., 1–13.

Giroud, J. P. (1970). "Stress Under Linearly Loaded Rectangular Area," *Journal of the Soil Mechanics and Foundations Division*, ASCE, Vol. 98, No. SM1, 263–268.

Griffiths, D. V. (1984). "A Chart for Estimating the Average Vertical Stress Increase in an Elastic Foundation Below a Uniformly Loaded Rectangular Area," *Canadian Geotechnical Journal*, Vol. 21, No. 4, 710–713.

Peattie, K. R. (1962). "Stresses and Strain Factors for Three-Layered Systems," *Bulletin 342*, Highway Research Record, Washington, D.C., 215–253.

Poulos, G. H., and Davis, E. H. (1974). *Elastic Solutions for Soil and Rock Mechanics*, Wiley, New York.

6

Consolidation

A stress increase caused by the construction of foundations or other loads compresses the soil layers. The compression is caused by (a) deformation of soil particles, (b) relocations of soil particles, and (c) expulsion of water or air from the void spaces. In general, the soil settlement caused by load may be divided into three broad categories:

1. *Immediate settlement,* which is caused by the elastic deformation of dry soil and of moist and saturated soils without any change in the moisture content. Immediate settlement calculations are generally based on equations derived from the theory of elasticity.
2. *Primary consolidation settlement,* which is the result of a volume change in saturated cohesive soils because of the expulsion of water that occupies the void spaces.
3. *Secondary consolidation settlement,* which is observed in saturated cohesive soils and is the result of the plastic adjustment of soil fabrics. It follows the primary consolidation settlement under a constant effective stress.

This chapter presents the fundamental principles for estimating the immediate and consolidation settlements of soil layers under superimposed loadings.

6.1 Fundamentals of Consolidation

When a saturated soil layer is subjected to a stress increase, the pore water pressure suddenly increases. In sandy soils that are highly permeable, the drainage caused by the increase in the pore water pressure is completed immediately. Pore water drainage is accompanied by a reduction in the volume of the soil mass, resulting in settlement. Because of the rapid drainage of the pore water in sandy soils, immediate settlement and consolidation take place simultaneously. This is not the case, however, for clay soils, which have low hydraulic conductivity. The consolidation settlement is time dependent.

Keeping this in mind, we can analyze the strain of a saturated clay layer subjected to a stress increase (Figure 6.1a). A layer of saturated clay of thickness H is confined between two layers of sand and is subjected to an instantaneous increase in *total stress* of $\Delta\sigma$. From Chapter 5, we know that

$$\Delta\sigma = \Delta\sigma' + \Delta u \qquad (6.1)$$

where $\Delta\sigma'$ = increase in the effective stress

$\quad\quad \Delta u$ = increase in the pore water pressure

Since clay has very low hydraulic conductivity and water is incompressible compared with the soil skeleton, at time $t = 0$, the entire incremental stress, $\Delta\sigma$, will be carried by water ($\Delta\sigma = \Delta u$) at all depths (Figure 6.1b). None will be carried by the soil skeleton (that is, incremental effective stress, $\Delta\sigma' = 0$).

After the application of incremental stress, $\Delta\sigma$, to the clay layer, the water in the void spaces will begin to be squeezed out and will drain in both directions into the sand layers. By this process, the excess pore water pressure at any depth on the clay layer will gradually decrease, and the stress carried by the soil solids (effective stress) will increase. Thus, at time $0 < t < \infty$,

$$\Delta\sigma = \Delta\sigma' + \Delta u \qquad (\Delta\sigma' > 0 \text{ and } \Delta u < \Delta\sigma)$$

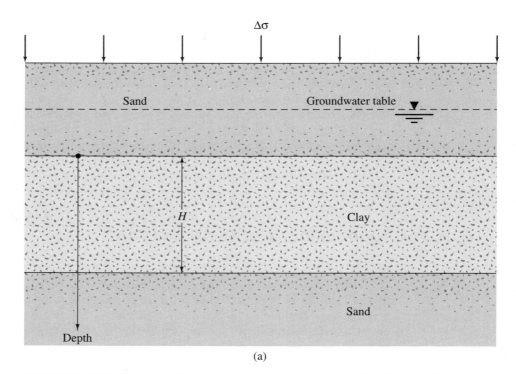

(a)

FIGURE 6.1 Variation of total stress, pore water pressure, and effective stress in a clay layer drained at top and bottom as the result of an added stress, $\Delta\sigma$

However, the magnitudes of $\Delta\sigma'$ and Δu at various depths will change (Figure 6.1c), depending on the minimum distance of the drainage path to either the top or bottom sand layer.

Theoretically, at time $t = \infty$, the entire excess pore water pressure would dissipate by drainage from all points of the clay layer, thus giving $\Delta u = 0$. Then the total stress increase, $\Delta\sigma$, would be carried by the soil structure (Figure 6.1d), so

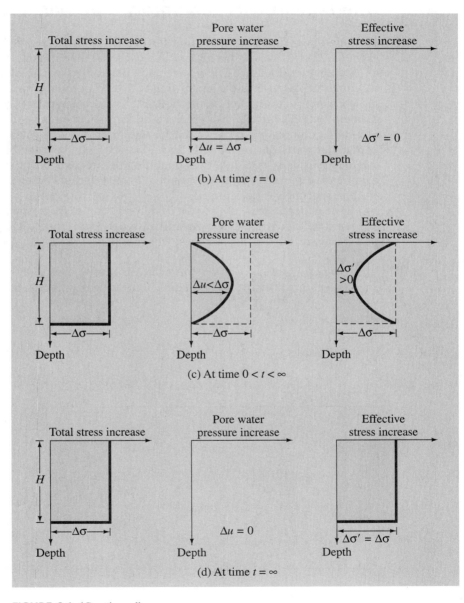

FIGURE 6.1 (Continued)

$$\Delta\sigma = \Delta\sigma'$$

This gradual process of drainage under the application of an additional load and the associated transfer of excess pore water pressure to effective stress causes the time-dependent settlement (consolidation) in the clay soil layer.

6.2   *One-Dimensional Laboratory Consolidation Test*

The one-dimensional consolidation testing procedure was first suggested by Ter-zaghi (1925). This test is performed in a consolidometer (sometimes referred to as an oedometer). Figure 6.2 is a schematic diagram of a consolidometer. The soil specimen is placed inside a metal ring with two porous stones, one at the top of the specimen and another at the bottom. The specimens are usually 63.5 mm in diameter and 25.4 mm thick. The load on the specimen is applied through a lever arm, and compression is measured by a micrometer dial gauge. The specimen is kept under water during the test. Each load is usually kept for 24 hours. After that, the load is usually doubled, thus doubling the pressure on the specimen, and the compression measurement is continued. At the end of the test, the dry weight of the test specimen is determined.

The general shape of the plot of deformation of the specimen versus time for a given load increment is shown in Figure 6.3. From the plot, it can be observed that there are three distinct stages, which may be described as follows:

Stage I: Initial compression, which is mostly caused by preloading.
Stage II: Primary consolidation, during which excess pore water pressure

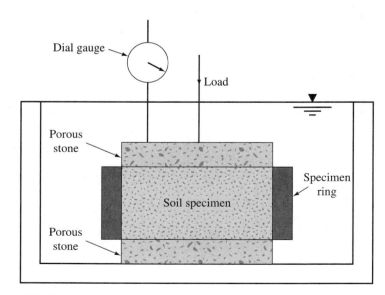

FIGURE 6.2 Consolidometer

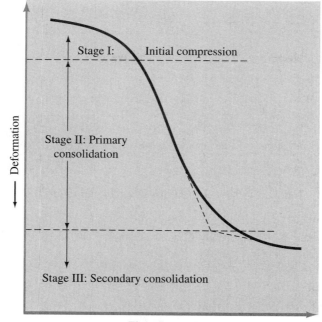

FIGURE 6.3 Time–deformation plot during consolidation for a given load increment

is gradually transferred into effective stress by the expulsion of pore water.

Stage III: Secondary consolidation, which occurs after complete dissipation of the excess pore water pressure, when some deformation of the specimen takes place because of the plastic readjustment of soil fabric.

6.3 *Void Ratio–Pressure Plots*

After the time–deformation plots for various loadings are obtained in the laboratory, it is necessary to study the change in the void ratio of the specimen with pressure. Following is a step-by-step procedure:

1. Calculate the height of solids, H_s, in the soil specimen (Figure 6.4):

$$H_s = \frac{W_s}{AG_s\gamma_w} \tag{6.2}$$

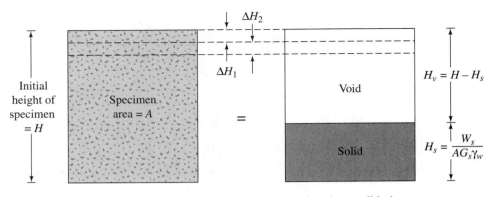

FIGURE 6.4 Change of height of specimen in one-dimensional consolidation test

where W_s = dry weight of the specimen
A = area of the specimen
G_s = specific gravity of soil solids
γ_w = unit weight of water

2. Calculate the initial height of voids, H_v:

$$H_v = H - H_s \tag{6.3}$$

where H = initial height of the specimen.

3. Calculate the initial void ratio, e_0, of the specimen:

$$e_0 = \frac{V_v}{V_s} = \frac{H_v}{H_s}\frac{A}{A} = \frac{H_v}{H_s} \tag{6.4}$$

4. For the first incremental loading σ_1 (total load/unit area of specimen), which causes deformation ΔH_1, calculate the change in the void ratio Δe_1:

$$\Delta e_1 = \frac{\Delta H_1}{H_s} \tag{6.5}$$

ΔH_1 is obtained from the initial and the final dial readings for the loading. At this time, the effective pressure on the specimen is $\sigma' = \sigma_1 = \sigma_1'$.

5. Calculate the new void ratio, e_1, after consolidation caused by the pressure increment σ_1:

$$e_1 = e_0 - \Delta e_1 \tag{6.6}$$

For the next loading, σ_2 (*note:* σ_2 equals the cumulative load per unit area of specimen), which causes additional deformation ΔH_2, the void ratio e_2 at the end of consolidation can be calculated as

$$e_2 = e_1 - \frac{\Delta H_2}{H_s} \tag{6.7}$$

Note that, at this time, the effective pressure on the specimen is $\sigma' = \sigma_2 = \sigma_2'$.

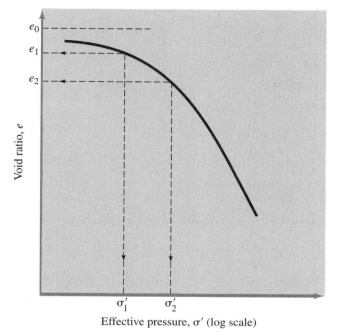

FIGURE 6.5 Typical plot of e versus log σ'

Proceeding in a similar manner, we can obtain the void ratios at the end of the consolidation for all load increments.

The effective pressures ($\sigma = \sigma'$) and the corresponding void ratios (e) at the end of consolidation are plotted on semilogarithmic graph paper. The typical shape of such a plot is shown in Figure 6.5.

EXAMPLE 6.1

Following are the results of a laboratory consolidation test on a soil specimen obtained from the field: dry mass of specimen = 116.74 g, height of specimen at beginning of test = 25.4 mm, G_s = 2.72, and diameter of specimen = 63.5 mm.

Pressure, σ' (kN/m²)	Final height of specimen at end of consolidation (mm)
0	25.4
50	25.19
100	25.00
200	24.29
400	23.22
800	22.06

Perform the necessary calculations and draw an e–log σ' curve.

Solution From Eq. (6.2), we have

$$H_s = \frac{W_s}{AG_s\gamma_w} = \frac{116.74 \text{ g}}{\left[\frac{\pi}{4}(6.35 \text{ cm})^2\right](2.72)(1 \text{ g/cm}^3)}$$

$$= 1.356 \text{ cm} = 13.56 \text{ mm}$$

Now we can prepare the following table:

Pressure, σ' (kN/m²)	Height at end of consolidation, H (mm)	$H_v = H - H_s$ (mm)	$e = H_v/H_s$
0	25.4	11.84	0.873
50	25.19	11.63	0.858
100	25.00	11.44	0.843
200	24.29	10.73	0.791
400	23.22	9.66	0.712
800	22.06	8.50	0.627

The e–log σ' plot is shown in Figure 6.6. ■

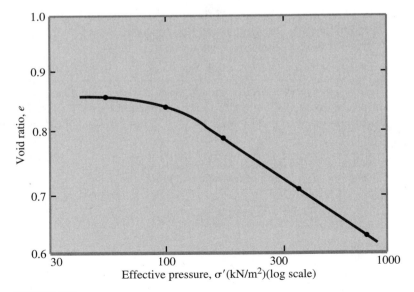

FIGURE 6.6

6.4 *Normally Consolidated and Overconsolidated Clays*

Figure 6.5 showed that the upper part of the e–log σ' plot is somewhat curved with a flat slope, followed by a linear relationship for the void ratio, with log σ' having a steeper slope. This can be explained in the following manner.

A soil in the field at some depth has been subjected to a certain maximum effective past pressure in its geologic history. This maximum effective past pressure may be equal to or greater than the existing overburden pressure at the time of sampling. The reduction of pressure in the field may be caused by natural geologic processes or human processes. During the soil sampling, the existing effective overburden pressure is also released, resulting in some expansion. When this specimen is subjected to a consolidation test, a small amount of compression (that is, a small change in the void ratio) will occur when the total pressure applied is less than the maximum effective overburden pressure in the field to which the soil has been subjected in the past. When the total applied pressure on the specimen is greater than the maximum effective past pressure, the change in the void ratio is much larger, and the e–log σ' relationship is practically linear with a steeper slope.

This relationship can be verified in the laboratory by loading the specimen to exceed the maximum effective overburden pressure, and then unloading and reloading again. The e–log σ' plot for such cases is shown in Figure 6.7, in which *cd* represents unloading and *dfg* represents the reloading process.

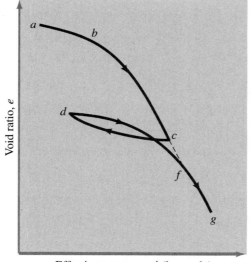

FIGURE 6.7 Plot of e versus log σ' showing loading, unloading, and reloading branches

This leads us to the two basic definitions of clay based on stress history:

1. *Normally consolidated:* The present effective overburden pressure is the maximum pressure to which the soil has been subjected in the past.
2. *Overconsolidated:* The present effective overburden pressure is less than that which the soil has experienced in the past. The maximum effective past pressure is called the *preconsolidation pressure.*

The past effective pressure cannot be determined explicitly because it is usually a function of geological processes and, consequently, it must be inferred from laboratory test results.

Casagrande (1936) suggested a simple graphic construction to determine the preconsolidation pressure, σ'_c, from the laboratory e–log σ' plot. The procedure follows (see Figure 6.8):

1. By visual observation, establish point a at which the e–log σ' plot has a minimum radius of curvature.
2. Draw a horizontal line ab.
3. Draw the line ac tangent at a.
4. Draw the line ad, which is the bisector of the angle bac.
5. Project the straight-line portion gh of the e–log σ' plot back to intersect ad at f. The abscissa of point f is the preconsolidation pressure, σ'_c.

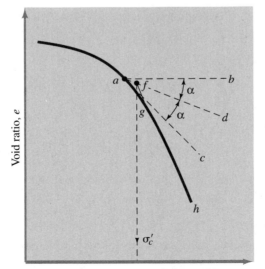

FIGURE 6.8 Graphic procedure for determining preconsolidation pressure

The overconsolidation ratio (OCR) for a soil can now be defined as

$$OCR = \frac{\sigma'_c}{\sigma'}$$

where σ'_c = preconsolidation pressure of a specimen
σ' = present effective vertical pressure

6.5 *Effect of Disturbance on Void Ratio–Pressure Relationship*

A soil specimen will be remolded when it is subjected to some degree of disturbance. This will affect the void ratio–pressure relationship for the soil. For a normally consolidated clayey soil of low to medium sensitivity (Figure 6.9) under an effective overburden pressure of σ'_o and with a void ratio of e_0, the change in the void ratio with an increase of pressure in the field will be roughly as shown by curve 1.

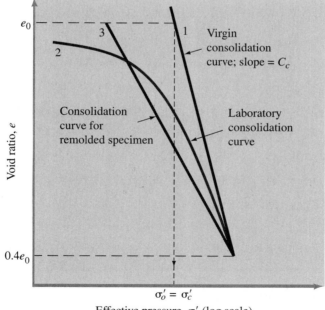

FIGURE 6.9 Consolidation characteristics of normally consolidated clay of low to medium sensitivity

This is the *virgin compression curve,* which is approximately a straight line on a semilogarithmic plot. However, the laboratory consolidation curve for a fairly undisturbed specimen of the same soil (curve 2) will be located to the left of curve 1. If the soil is completely remolded and a consolidation test is conducted on it, the general position of the *e*–log σ' plot will be represented by curve 3. Curves 1, 2, and 3 will intersect approximately at a void ratio of $e = 0.4e_0$ (Terzaghi and Peck, 1967).

For an overconsolidated clayey soil of low to medium sensitivity that has been subjected to a preconsolidation pressure of σ'_c (Figure 6.10) and for which the present effective overburden pressure and the void ratio are σ'_o and e_0, respectively, the field consolidation curve will take a path represented approximately by *cbd.* Note that *bd* is a part of the virgin compression curve. The laboratory consolidation test results on a specimen subjected to moderate disturbance will be represented by curve 2. Schmertmann (1953) concluded that the slope of line *cb,* which is the field recompression path, has approximately the same slope as the laboratory rebound curve *fg.*

Soils that exhibit high sensitivity have flocculent structures. In the field, they are generally somewhat overconsolidated. The consolidation characteristics of such soils are shown in Figure 6.11.

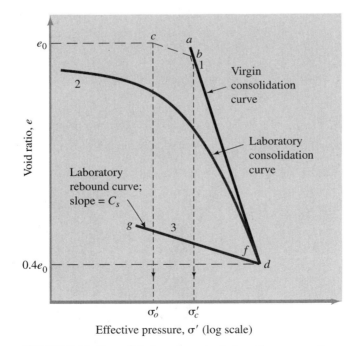

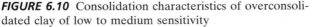

FIGURE 6.10 Consolidation characteristics of overconsolidated clay of low to medium sensitivity

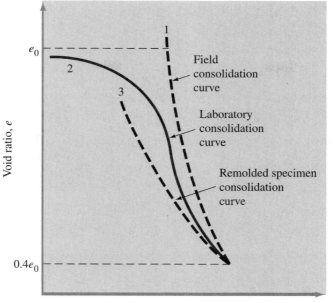

FIGURE 6.11 Consolidation characteristics of sensitive clays.

6.6 Calculation of Settlement from One-Dimensional Primary Consolidation

With the knowledge gained from the analysis of consolidation test results, we can now proceed to calculate the probable settlement caused by primary consolidation in the field, assuming one-dimensional consolidation.

Let us consider a saturated clay layer of thickness H and cross-sectional area A under an existing average effective overburden pressure σ_o'. Because of an increase of pressure, $\Delta\sigma$, let the primary settlement be S. At the end of consolidation, $\Delta\sigma = \Delta\sigma'$. Thus, the change in volume (Figure 6.12) can be given by

$$\Delta V = V_0 - V_1 = HA - (H - S)A = SA \tag{6.8}$$

where V_0 and V_1 are the initial and final volumes, respectively. However, the change in the total volume is equal to the change in the volume of voids, ΔV_v. Thus,

$$\Delta V = SA = V_{v0} - V_{v1} = \Delta V_v \tag{6.9}$$

where V_{v0} and V_{v1} are the initial and final void volumes, respectively. From the definition of the void ratio, we have

$$\Delta V_v = \Delta e V_s \tag{6.10}$$

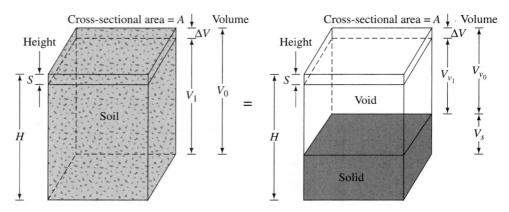

FIGURE 6.12 Settlement caused by one-dimensional consolidation

where Δe = change of void ratio. But

$$V_s = \frac{V_0}{1 + e_0} = \frac{AH}{1 + e_0}$$

(6.11)

where e_0 = initial void ratio at volume V_0. Thus, from Eqs. (6.8), (6.9), (6.10), and (6.11), we get

$$\Delta V = SA = \Delta e V_s = \frac{AH}{1 + e_0} \Delta e$$

or

$$S = H \frac{\Delta e}{1 + e_0}$$

(6.12)

For normally consolidated clays that exhibit a linear e–log σ' relationship (Figure 6.9) (*note:* $\Delta\sigma = \Delta\sigma'$ at the end of consolidation),

$$\Delta e = C_c[\log(\sigma'_o + \Delta\sigma') - \log \sigma'_o]$$

(6.13)

where C_c = slope of the e–log σ'_o plot and is defined as the compression index. Substituting Eq. (6.13) into Eq. (6.12) gives

$$S = \frac{C_c H}{1 + e_0} \log\left(\frac{\sigma'_o + \Delta\sigma'}{\sigma'_o}\right)$$

(6.14)

For a thicker clay layer, a more accurate measurement of settlement can be made if the layer is divided into a number of sublayers and calculations are made for each sublayer. Thus, the total settlement for the entire layer can be given as

$$S = \sum \left[\frac{C_c H_i}{1 + e_0} \log \left(\frac{\sigma'_{o(i)} + \Delta\sigma'_{(i)}}{\sigma'_{o(i)}} \right) \right]$$

where H_i = thickness of sublayer i

       $\sigma'_{o(i)}$ = initial average effective overburden pressure for sublayer i

       $\Delta\sigma'_{(i)}$ = increase of vertical pressure for sublayer i

In overconsolidated clays (Figure 6.10), for $\sigma'_o + \Delta\sigma' \le \sigma'_c$, field e–log σ' variation will be along the line *cb*, the slope of which will be approximately equal to the slope of the laboratory rebound curve. The slope of the rebound curve, C_s, is referred to as the *swell index,* so

$$\Delta e = C_s[\log(\sigma'_o + \Delta\sigma') - \log \sigma'_o] \tag{6.15}$$

From Eqs. (6.12) and (6.15), we have

$$S = \frac{C_s H}{1 + e_0} \log \left(\frac{\sigma'_o + \Delta\sigma'}{\sigma'_o} \right) \tag{6.16}$$

If $\sigma'_o + \Delta\sigma > \sigma'_c$, then

$$S = \frac{C_s H}{1 + e_0} \log \frac{\sigma'_c}{\sigma'_o} + \frac{C_c H}{1 + e_0} \log \left(\frac{\sigma'_o + \Delta\sigma'}{\sigma'_c} \right) \tag{6.17}$$

However, if the e–log σ' curve is given, it is possible simply to pick Δe off the plot for the appropriate range of pressures. This value may be substituted into Eq. (6.12) for calculating the settlement, S.

6.7 *Compression Index (C_c)*

We can determine the compression index for field settlement caused by consolidation by graphic construction (as shown in Figure 6.9) after obtaining laboratory test results for the void ratio and pressure.

Terzaghi and Peck (1967) suggested empirical expressions for the compression index. For undisturbed clays:

$$C_c = 0.009(LL - 10) \tag{6.18}$$

For remolded clays:

$$C_c = 0.007(LL - 10) \tag{6.19}$$

where LL = liquid limit (%). In the absence of laboratory consolidation data, Eq. (6.18) is often used for an approximate calculation of primary consolidation in the field. Several other correlations for the compression index are also available now.

Based on observations on several natural clays, Rendon-Herrero (1983) gave the relationship for the compression index in the form

$$C_c = 0.141 G_s^{1.2} \left(\frac{1 + e_0}{G_s} \right)^{2.38} \tag{6.20}$$

Nagaraj and Murty (1985) expressed the compression index as

$$C_c = 0.2343 \left[\frac{LL\,(\%)}{100} \right] G_s \tag{6.21}$$

6.8   *Swell Index (C_s)*

The swell index is appreciably smaller in magnitude than the compression index and can generally be determined from laboratory tests. In most cases,

$$C_s \approx \frac{1}{5} \quad \text{to} \quad \frac{1}{10} C_c \tag{6.22}$$

The swell index was expressed by Nagaraj and Murty (1985) as

$$C_s = 0.0463 \left[\frac{LL\,(\%)}{100} \right] G_s \tag{6.23}$$

Typical values of the liquid limit, plastic limit, virgin compression index, and swell index for some natural soils are given in Table 6.1.

EXAMPLE 6.2   Refer to the e–log σ' curve obtained in Example 6.1.

    a. Determine the preconsolidation pressure, σ_c'.
    b. Find the compression index, C_c.

Table 6.1 Compression and swell of natural soils

Soil	Liquid limit	Plastic limit	Compression index, C_c	Swell index, C_s
Boston blue clay	41	20	0.35	0.07
Chicago clay	60	20	0.4	0.07
Ft. Gordon clay, Georgia	51	26	0.12	—
New Orleans clay	80	25	0.3	0.05
Montana clay	60	28	0.21	0.05

Solution

a. The e–log σ' plot shown in Figure 6.6 is replotted in Figure 6.13. Using the procedure shown in Figure 6.8, we determine the preconsolidation pressure. From the plot, $\sigma'_c = $ **160 kN/m².**

b. From the e–log σ' plot, we find

$$\sigma'_1 = 400 \text{ kN/m}^2 \qquad e_1 = 0.712$$
$$\sigma'_2 = 800 \text{ kN/m}^2 \qquad e_2 = 0.627$$

So

$$C_c = \frac{e_1 - e_2}{\log(\sigma'_2/\sigma'_1)} = \frac{0.712 - 0.627}{\log(800/400)} = \textbf{0.282} \qquad \blacksquare$$

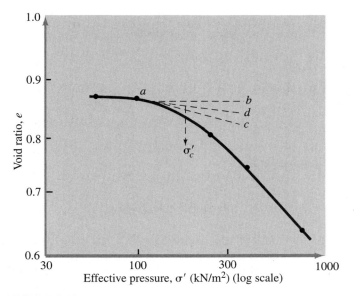

FIGURE 6.13

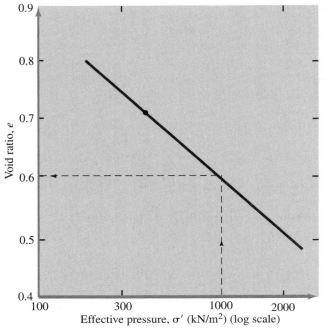

FIGURE 6.14

EXAMPLE 6.3

Refer to Examples 6.1 and 6.2. For the clay, what will the void ratio be for a pressure of 1000 kN/m²? (*Note:* $\sigma'_c = 160$ kN/m².)

Solution   From Example 6.1, we find the following values:

$$\sigma'_1 = 400 \text{ kN/m}^2 \qquad e_1 = 0.712$$
$$\sigma'_2 = 800 \text{ kN/m}^2 \qquad e_2 = 0.627$$

Also, from Example 6.2, $C_c = 0.282$. Referring to Figure 6.14, we have

$$C_c = \frac{e_1 - e_3}{\log \sigma'_3 - \log \sigma'_1}$$

$$e_3 = e_1 - C_c \log \left(\frac{\sigma'_3}{\sigma'_1}\right) = 0.712 - 0.282 \log \left(\frac{1000}{400}\right) = \mathbf{0.6} \qquad \blacksquare$$

EXAMPLE 6.4

A soil profile is shown in Figure 6.15. If a uniformly distributed load $\Delta\sigma$ is applied at the ground surface, what will be the settlement of the clay layer caused by primary consolidation? We are given that σ'_c for the clay is 125 kN/m² and $C_s = \frac{1}{6}C_c$.

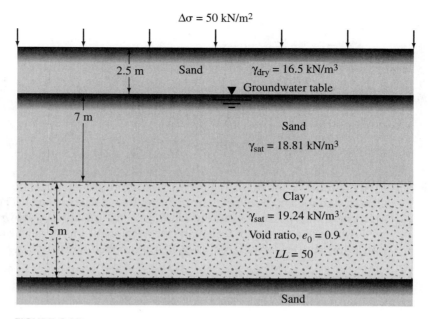

$\Delta\sigma = 50$ kN/m²

2.5 m   Sand $\gamma_{dry} = 16.5$ kN/m³

▼ Groundwater table

7 m

Sand

$\gamma_{sat} = 18.81$ kN/m³

Clay

$\gamma_{sat} = 19.24$ kN/m³

5 m   Void ratio, $e_0 = 0.9$

$LL = 50$

Sand

FIGURE 6.15

Solution The average effective stress at the middle of the clay layer is

$$\sigma_o' = 2.5\gamma_{dry(sand)} + (7 - 2.5)[\gamma_{sat(sand)} - \gamma_w] + \left(\frac{5}{2}\right)[\gamma_{sat(clay)} - \gamma_w]$$

or

$$\sigma_o' = (2.5)(16.5) + (4.5)(18.81 - 9.81) + (2.5)(19.24 - 9.81)$$

$$= 105.33 \text{ kN/m}^2$$

$$\sigma_c' = 125 \text{ kN/m}^2 > 105.33 \text{ kN/m}^2$$

$$\sigma_o' + \Delta\sigma' = 105.33 + 50 = 155.33 \text{ kN/m}^2 > \sigma_c'$$

(*Note:* $\Delta\sigma = \Delta\sigma'$ at the end of consolidation.) So we need to use Eq. (6.17):

$$S = \frac{C_s H}{1 + e_0} \log\left(\frac{\sigma_c'}{\sigma_o'}\right) + \frac{C_c H}{1 + e_0} \log\left(\frac{\sigma_o' + \Delta\sigma'}{\sigma_c'}\right)$$

We have $H = 5$ m and $e_0 = 0.9$. From Eq. (6.18),

$$C_c = 0.009(LL - 10) = 0.009(50 - 10) = 0.36$$

$$C_s = \frac{1}{6}C_c = \frac{0.36}{6} = 0.06$$

Thus,

$$S = \frac{5}{1 + 0.9}\left[0.06 \log\left(\frac{125}{105.33}\right) + 0.36 \log\left(\frac{105.33 + 50}{125}\right)\right]$$

$$= 0.1011 \text{ m} \approx \mathbf{101 \ mm}$$ ■

EXAMPLE 6.5

A soil profile is shown in Figure 6.16a. Laboratory consolidation tests were conducted on a specimen collected from the middle of the clay layer. The field consolidation curve interpolated from the laboratory test results (as shown in Figure 6.10) is shown in Figure 6.16b. Calculate the settlement in the field caused by primary consolidation for a surcharge of 48 kN/m² applied at the ground surface.

Solution

$$\sigma_o' = (5)(\gamma_{sat} - \gamma_w) = 5(18.0 - 9.81)$$

$$= 40.95 \text{ kN/m}^2$$

$$e_0 = 1.1$$

$$\Delta\sigma' = 48 \text{ kN/m}^2$$

$$\sigma_o' + \Delta\sigma' = 40.95 + 48 = 88.95 \text{ kN/m}^2$$

The void ratio corresponding to 88.95 kN/m² (Figure 6.16) is 1.045. Hence, $\Delta e = 1.1 - 1.045 = 0.055$. From Eq. (6.12) we have

$$\text{settlement}, S = H\frac{\Delta e}{1 + e_0}$$

so

$$S = 10\frac{0.055}{1 + 1.1} = 0.262 \text{ m} = \mathbf{262 \ mm}$$ ■

6.9   *Settlement from Secondary Consolidation*

Section 6.2 showed that at the end of primary consolidation (that is, after complete dissipation of excess pore water pressure) some settlement is observed because of the plastic adjustment of soil fabrics, which is usually termed *creep*. This stage of consolidation is called *secondary consolidation*. During secondary consolidation, the plot of deformation versus the log of time is practically linear (Figure 6.3). The variation of the void ratio *e* with time *t* for a given load increment will be similar to that shown in Figure 6.3. This variation is illustrated in Figure 6.17.

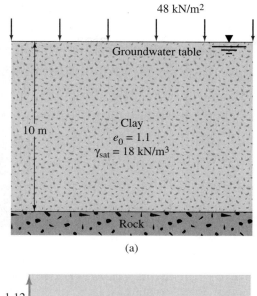

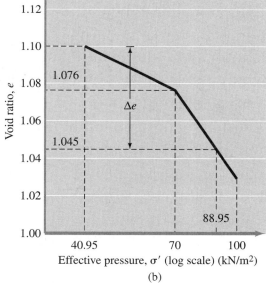

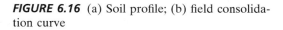

FIGURE 6.16 (a) Soil profile; (b) field consolidation curve

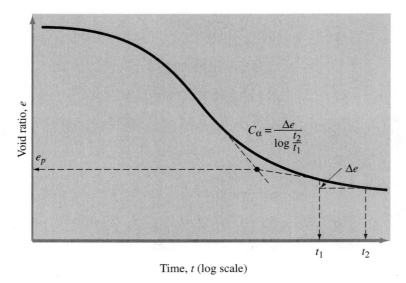

FIGURE 6.17 Variation of e with log t under a given load increment, and definition of secondary compression index

The secondary compression index can be defined from Figure 6.17 as

$$C_\alpha = \frac{\Delta e}{\log t_2 - \log t_1} = \frac{\Delta e}{\log (t_2/t_1)} \qquad (6.24)$$

where C_α = secondary compression index
$\quad\quad\Delta e$ = change of void ratio
$\quad t_1, t_2$ = time

The magnitude of the secondary consolidation can be calculated as

$$S_s = C'_\alpha H \log \left(\frac{t_1}{t_2}\right) \qquad (6.25)$$

where

$$C'_\alpha = \frac{C_\alpha}{1 + e_p} \qquad (6.26)$$

and

e_p = void ratio at the end of primary consolidation (Figure 6.17)

H = thickness of clay layer

Secondary consolidation settlement is more important than primary consolidation in organic and highly compressible inorganic soils. In overconsolidated inorganic clays, the secondary compression index is very small and has less practical significance.

Several factors might affect the magnitude of secondary consolidation, and some of them are not very clearly understood (Mesri, 1973). The ratio of secondary to primary compression for a given thickness of soil layer is dependent on the ratio of the stress increment ($\Delta\sigma'$) to the initial effective stress (σ'_o). For small $\Delta\sigma'/\sigma'_o$ ratios, the secondary-to-primary compression ratio is larger.

EXAMPLE 6.6

Refer to Example 6.4. Assume that the primary consolidation will be complete in 3.5 years. Estimate the secondary consolidation that will occur from 3.5 to 10 years after the load application. Given $C_\alpha = 0.022$, what is the total consolidation settlement after 10 years?

Solution   From Eq. (6.26), we have

$$C'_\alpha = \frac{C_\alpha}{1 + e_p}$$

The value of e_p can be calculated as

$$e_p = e_0 - \Delta e_{\text{primary}}$$

Combining Eqs. (6.12) and (6.17), we get

$$\Delta e = C_s \log \left(\frac{\sigma'_c}{\sigma'_o} \right) + C_s \log \left(\frac{\sigma'_o + \Delta\sigma'}{\sigma'_c} \right)$$

$$= 0.06 \log \left(\frac{125}{105.33} \right) + 0.36 \log \left(\frac{155.33}{125} \right) = 0.038$$

We are given $e_0 = 0.9$, so

$$e_p = 0.9 - 0.038 = 0.862$$

Hence,

$$C'_\alpha = \frac{0.022}{1 + 0.862} = 0.0118$$

$$S_s = C'_\alpha H \log \left(\frac{t_2}{t_1} \right) = (0.0118)(5) \log \left(\frac{10}{3.5} \right) \approx 0.027 \text{ m}$$

Total consolidation settlement = primary consolidation settlement (S) + secondary consolidation settlement (S_s). From Example 6.4, $S = 101$ mm, so

total consolidation settlement $= 101 + 27 = $ **128 mm** ■

6.10 *Time Rate of Consolidation*

The total settlement caused by primary consolidation resulting from an increase in the stress on a soil layer can be calculated by using one of the three equations [(6.14), (6.16), or (6.17)] given in Section 6.6. However, the equations do not provide any information regarding the rate of primary consolidation. Terzaghi (1925) proposed the first theory to consider the rate of one-dimensional consolidation for saturated clay soils. The mathematical derivations are based on the following assumptions:

1. The clay–water system is homogeneous.
2. Saturation is complete.
3. Compressibility of water is negligible.
4. Compressibility of soil grains is negligible (but soil grains rearrange).
5. The flow of water is in one direction only (that is, in the direction of compression).
6. Darcy's law is valid.

Figure 6.18a shows a layer of clay of thickness $2H_{dr}$ located between two highly permeable sand layers. If the clay layer is subjected to an increased pressure of $\Delta\sigma$, the pore water pressure at any point A in the clay layer will increase. For one-dimensional consolidation, water will be squeezed out in the vertical direction toward the sand layers.

Figure 6.18b shows the flow of water through a prismatic element at A. For the soil element shown,

$$\begin{pmatrix} \text{rate of outflow} \\ \text{of water} \end{pmatrix} - \begin{pmatrix} \text{rate of inflow} \\ \text{of water} \end{pmatrix} = \begin{pmatrix} \text{rate of} \\ \text{volume change} \end{pmatrix}$$

Thus,

$$\left(v_z + \frac{\partial v_z}{\partial z} dz \right) dx\, dy - v_z\, dx\, dy = \frac{\partial V}{\partial t}$$

where V = volume of the soil element
v_z = velocity of flow in the z direction

or

$$\frac{\partial v_z}{\partial z} dx\, dy\, dz = \frac{\partial V}{\partial t} \tag{6.27}$$

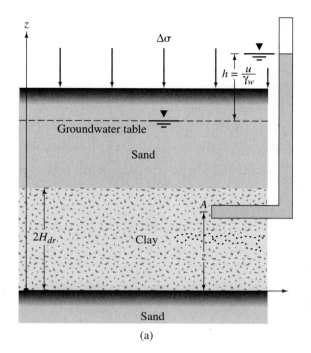

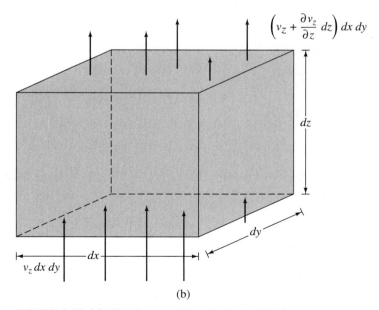

FIGURE 6.18 (a) Clay layer undergoing consolidation;
(b) flow of water at A during consolidation

Using Darcy's law, we have

$$v_z = ki = -k \frac{\partial h}{\partial z} = -\frac{k}{\gamma_w} \frac{\partial u}{\partial z} \tag{6.28}$$

where u = excess pore water pressure caused by the increase of stress. From Eqs. (6.27) and (6.28), we get

$$-\frac{k}{\gamma_w} \frac{\partial^2 u}{\partial z^2} = \frac{1}{dx\,dy\,dz} \frac{\partial V}{\partial t} \tag{6.29}$$

During consolidation, the rate of change in the volume of the soil element is equal to the rate of change in the volume of voids. So

$$\frac{\partial V}{\partial t} = \frac{\partial V_v}{\partial t} = \frac{\partial(V_s + eV_s)}{\partial t} = \frac{\partial V_s}{\partial t} + V_s \frac{\partial e}{\partial t} + e \frac{\partial V_s}{\partial t} \tag{6.30}$$

where V_s = volume of soil solids
V_v = volume of voids

But (assuming that soil solids are incompressible),

$$\frac{\partial V_s}{\partial t} = 0$$

and

$$V_s = \frac{V}{1 + e_0} = \frac{dx\,dy\,dz}{1 + e_0}$$

Substituting for $\partial V_s / \partial t$ and V_s in Eq. (6.30) yields

$$\frac{\partial V}{\partial t} = \frac{dx\,dy\,dz}{1 + e_0} \frac{\partial e}{\partial t} \tag{6.31}$$

where e_0 = initial void ratio. Combining Eqs. (6.29) and (6.30) gives

$$-\frac{k}{\gamma_w} \frac{\partial^2 u}{\partial z^2} = \frac{1}{1 + e_0} \frac{\partial e}{\partial t} \tag{6.32}$$

The change in the void ratio is caused by the increase in the effective stress (that is, decrease of excess pore water pressure). Assuming that those values are linearly related, we have

$$\partial e = a_v\, \partial(\Delta\sigma') = -a_v\, \partial u \tag{6.33}$$

where $\partial(\Delta\sigma')$ = change in effective pressure
a_v = coefficient of compressibility (a_v can be considered to be constant for a narrow range of pressure increases)

Combining Eqs. (6.32) and (6.33) gives

$$-\frac{k}{\gamma_w} \frac{\partial^2 u}{\partial z^2} = -\frac{a_v}{1 + e_0} \frac{\partial u}{\partial t} = -m_v \frac{\partial u}{\partial t}$$

where m_v = coefficient of volume compressibility = $a_v/(1 + e_0)$, or

$$\frac{\partial u}{\partial t} = c_v \frac{\partial^2 u}{\partial z^2} \tag{6.34}$$

where c_v = coefficient of consolidation = $k/(\gamma_w m_v)$.

Equation (6.34) is the basic differential equation of Terzaghi's consolidation theory and can be solved with the following boundary conditions:

$z = 0, \quad u = 0$

$z = 2H_{dr}, \quad u = 0$

$t = 0, \quad u = u_0$

The solution yields

$$u = \sum_{m=0}^{m=\infty} \left[\frac{2u_0}{M} \sin \left(\frac{Mz}{H_{dr}} \right) \right] e^{-M^2 T_v} \tag{6.35}$$

where m is an integer

$$M = \frac{\pi}{2}(2m + 1)$$

u_0 = initial excess pore water pressure

and

$$T_v = \frac{c_v t}{H_{dr}^2} = \text{time factor}$$

The time factor is a nondimensional number.

Because consolidation progresses by dissipation of excess pore water pressure, the degree of consolidation at a distance z at any time t is

$$U_z = \frac{u_0 - u_z}{u_0} = 1 - \frac{u_z}{u_0} \tag{6.36}$$

where u_z = excess pore water pressure at time t. Equations (6.35) and (6.36) can be combined to obtain the degree of consolidation at any depth z. This is shown in Figure 6.19.

The average degree of consolidation for the entire depth of the clay layer at any time t can be written from Eq. (6.36) as

$$U = \frac{S_t}{S} = 1 - \frac{\left(\dfrac{1}{2H_{dr}} \right) \displaystyle\int_0^{2H_{dr}} u_z \, dz}{u_0} \tag{6.37}$$

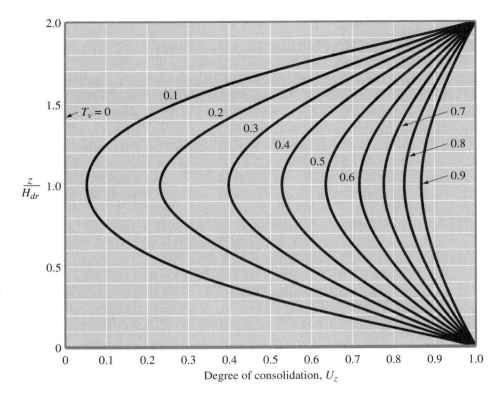

FIGURE 6.19 Variation of U_z with T_v and z/H_{dr}

where U = average degree of consolidation
       S_t = settlement of the layer at time t
      S = ultimate settlement of the layer from primary consolidation

Substituting the expression for excess pore water pressure, u_z, given in Eq. (6.36) into Eq. (6.37) gives

$$U = 1 - \sum_{m=0}^{m=\infty} \frac{2}{M^2} e^{-M^2 T_v} \tag{6.38}$$

The variation in the average degree of consolidation with the nondimensional time factor, T_v, is given in Table 6.2, which represents the case where u_0 is the same for the entire depth of the consolidating layer. The values of the time factor and their corresponding average degrees of consolidation may also be approximated by the following simple relationships:

$$\text{For } U = 0\% \text{ to } 60\%, \quad T_v = \frac{\pi}{4}\left(\frac{U\%}{100}\right)^2 \tag{6.39}$$

$$\text{For } U > 60\%, \quad T_v = 1.781 - 0.933 \log(100 - U\%) \tag{6.40}$$

Table 6.2 Variation of time factor with degree of consolidation*

U (%)	T_v	U (%)	T_v	U (%)	T_v
0	0	34	0.0907	68	0.377
1	0.00008	35	0.0962	69	0.390
2	0.0003	36	0.102	70	0.403
3	0.00071	37	0.107	71	0.417
4	0.00126	38	0.113	72	0.431
5	0.00196	39	0.119	73	0.446
6	0.00283	40	0.126	74	0.461
7	0.00385	41	0.132	75	0.477
8	0.00502	42	0.138	76	0.493
9	0.00636	43	0.145	77	0.511
10	0.00785	44	0.152	78	0.529
11	0.0095	45	0.159	79	0.547
12	0.0113	46	0.166	80	0.567
13	0.0133	47	0.173	81	0.588
14	0.0154	48	0.181	82	0.610
15	0.0177	49	0.188	83	0.633
16	0.0201	50	0.197	84	0.658
17	0.0227	51	0.204	85	0.684
18	0.0254	52	0.212	86	0.712
19	0.0283	53	0.221	87	0.742
20	0.0314	54	0.230	88	0.774
21	0.0346	55	0.239	89	0.809
22	0.0380	56	0.248	90	0.848
23	0.0415	57	0.257	91	0.891
24	0.0452	58	0.267	92	0.938
25	0.0491	59	0.276	93	0.993
26	0.0531	60	0.286	94	1.055
27	0.0572	61	0.297	95	1.129
28	0.0615	62	0.307	96	1.219
29	0.0660	63	0.318	97	1.336
30	0.0707	64	0.329	98	1.500
31	0.0754	65	0.340	99	1.781
32	0.0803	66	0.352	100	∞
33	0.0855	67	0.364		

*u_0 constant with depth.

Different types of drainage with u_0 constant

6.11   *Coefficient of Consolidation*

The coefficient of consolidation, c_v, generally decreases as the liquid limit of soil increases. The range of variation of c_v for a given liquid limit of soil is rather wide.

For a given load increment on a specimen, there are two commonly used graphic methods for determining c_v from laboratory one-dimensional consolidation tests. One of them is the *logarithm-of-time method* proposed by Casagrande and Fadum (1940), and the other is the *square-root-of-time method* suggested by Taylor (1942). The general procedures for obtaining c_v by the two methods are described next.

Logarithm-of-Time Method

For a given incremental loading of the laboratory test, the specimen deformation versus log-of-time plot is shown in Figure 6.20. The following constructions are needed to determine c_v:

1. Extend the straight-line portions of primary and secondary consolidations to intersect at A. The ordinate of A is represented by d_{100}—that is, the deformation at the end of 100% primary consolidation.
2. The initial curved portion of the plot of deformation versus log t is approximated to be a parabola on the natural scale. Select times t_1 and t_2 on the curved portion such that $t_2 = 4t_1$. Let the difference of the specimen deformation during time $(t_2 - t_1)$ be equal to x.

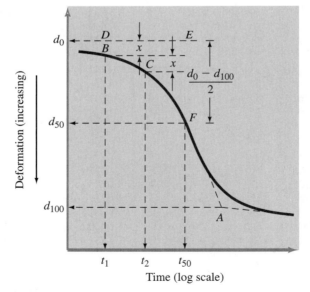

FIGURE 6.20 Logarithm-of-time method for determining coefficient of consolidation

3. Draw a horizontal line DE such that the vertical distance BD is equal to x. The deformation corresponding to the line DE is d_0 (that is, deformation at 0% consolidation).
4. The ordinate of point F on the consolidation curve represents the deformation at 50% primary consolidation, and its abscissa represents the corresponding time (t_{50}).
5. For 50% average degree of consolidation, $T_v = 0.197$ (Table 6.2);

$$T_{50} = \frac{c_v t_{50}}{H_{dr}^2}$$

or

$$c_v = \frac{0.197 H_{dr}^2}{t_{50}} \tag{6.41}$$

where H_{dr} = average longest drainage path during consolidation.

For specimens drained at both top and bottom, H_{dr} equals one-half of the average height of the specimen during consolidation. For specimens drained on only one side, H_{dr} equals the average height of the specimen during consolidation.

Square-Root-of-Time Method

In this method, a plot of deformation versus the square root of time is drawn for the incremental loading (Figure 6.21). Other graphic constructions required are

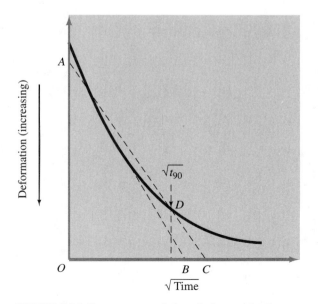

FIGURE 6.21 Square-root-of-time fitting method

as follows:

1. Draw a line AB through the early portion of the curve.
2. Draw a line AC such that $\overline{OC} = 1.15\,\overline{OB}$. The abscissa of point D, which is the intersection of AC and the consolidation curve, gives the square root of time for 90% consolidation ($\sqrt{t_{90}}$).
3. For 90% consolidation, $T_{90} = 0.848$ (Table 6.2), so

$$T_{90} = 0.848 = \frac{c_v t_{90}}{H_{dr}^2}$$

or

$$c_v = \frac{0.848 H_{dr}^2}{t_{90}} \tag{6.42}$$

H_{dr} in Eq. (6.42) is determined in a manner similar to the logarithm-of-time method.

EXAMPLE 6.7   A soil profile is shown in Figure 6.22. A surcharge load of 120 kN/m^2 = $\Delta\sigma$ is applied on the ground surface.

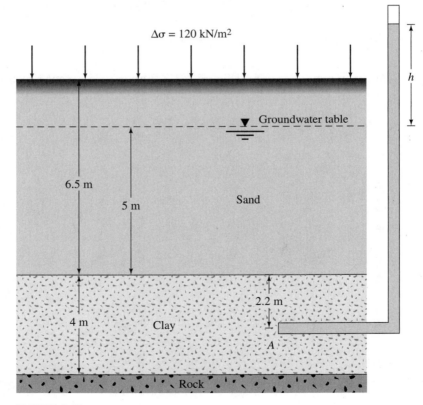

FIGURE 6.22

a. How high will the water rise in the piezometer immediately after the application of load?

b. What is the degree of consolidation at point A when $h = 6.5$ m?

c. Find h when the degree of consolidation at A is 60%.

Solution

a. Assuming a uniform increase in the initial excess pore water pressure through the 4 m depth of the clay layer, we have

$$u_0 = \Delta\sigma = 120 \text{ kN/m}^2$$

$$h = \frac{120}{9.81} = \mathbf{12.23 \text{ m}}$$

b. $U_A\% = \left(1 - \dfrac{u_A}{u_0}\right)100 = \left(1 - \dfrac{6.5 \times 9.81}{12.23 \times 9.81}\right)100 = \mathbf{46.8\%}$

c. $U_A = 0.6 = \left(1 - \dfrac{u_A}{u_0}\right)$

or

$$0.6 = \left(1 - \frac{u_A}{120}\right)$$

$$u_A = (1 - 0.6)(120) = 48 \text{ kN/m}^2$$

Hence,

$$h = \frac{48}{9.81} = \mathbf{4.89 \text{ m}} \qquad \blacksquare$$

EXAMPLE 6.8

The time required for 50% consolidation of a 25-mm-thick clay layer (drained at both top and bottom) in the laboratory is 2 min 20 sec. How long (in days) will it take for a 3-m-thick clay layer of the same clay in the field under the same pressure increment to reach 50% consolidation? In the field, there is a rock layer at the bottom of the clay.

Solution

$$T_{50} = \frac{c_v t_{\text{lab}}}{H_{dr(\text{lab})}^2} = \frac{c_v t_{\text{field}}}{H_{dr(\text{field})}^2}$$

or

$$\frac{t_{\text{lab}}}{H^2_{dr(\text{lab})}} = \frac{t_{\text{field}}}{H^2_{dr(\text{field})}}$$

$$\frac{140 \text{ sec}}{\left(\dfrac{0.025 \text{ m}}{2}\right)^2} = \frac{t_{\text{field}}}{(3 \text{ m})^2}$$

$$t_{\text{field}} = 8,064,000 \text{ sec} = \textbf{93.33 days} \qquad \blacksquare$$

**EXAMPLE
6.9**

Refer to Example 6.8. How long (in days) will it take in the field for 30% primary consolidation to occur? Use Eq. (6.39).

Solution From Eq. (6.39), we have

$$\frac{c_v t_{\text{field}}}{H^2_{dr(\text{field})}} = T_v \propto U^2$$

So

$$t \propto U^2$$

$$\frac{t_1}{t_2} = \frac{U_1^2}{U_2^2}$$

or

$$\frac{93.33 \text{ days}}{t_2} = \frac{50^2}{30^2}$$

$$t_2 = \textbf{33.6 days} \qquad \blacksquare$$

**EXAMPLE
6.10**

For a normally consolidated clay,

$$\sigma'_o = 200 \text{ kN/m}^2 \qquad e = e_0 = 1.22$$
$$\sigma'_o + \Delta\sigma' = 400 \text{ kN/m}^2 \qquad e = 0.98$$

The hydraulic conductivity, k, of the clay for the loading range is 0.61×10^{-4} m/day.

    a. How long (in days) will it take for a 4-m-thick clay layer (drained on one side) in the field to reach 60% consolidation?
    b. What is the settlement at that time (that is, at 60% consolidation)?

Solution

a. The coefficient of volume compressibility is

$$m_v = \frac{a_v}{1 + e_{av}} = \frac{(\Delta e / \Delta \sigma')}{1 + e_{av}}$$

$$\Delta e = 1.22 - 0.98 = 0.24$$

$$\Delta \sigma' = 400 - 200 = 200 \text{ kN/m}^2$$

$$e_{av} = \frac{1.22 + 0.98}{2} = 1.1$$

So

$$m_v = \frac{0.24/200}{1 + 1.1} = 5.7 \times 10^{-4} \text{ m}^2/\text{kN}$$

$$c_v = \frac{k}{m_v \gamma_w} = \frac{0.61 \times 10^{-4} \text{ m/day}}{(5.7 \times 10^{-4} \text{ m}^2/\text{kN})(9.81 \text{ kN/m}^2)} = 0.0109 \text{ m}^2/\text{day}$$

$$T_{60} = \frac{c_v t_{60}}{H_{dr}^2}$$

$$t_{60} = \frac{T_{60} H_{dr}^2}{c_v}$$

From Table 6.2, for $U = 60\%$, the value of T_{60} is 0.286, so

$$t_{60} = \frac{(0.286)(4)^2}{0.0109} = \textbf{419.8 days}$$

b. $C_c = \dfrac{e_1 - e_2}{\log(\sigma_2'/\sigma_1')} = \dfrac{1.22 - 0.98}{\log(400/200)} = 0.797$

From Eq. (6.14), we have

$$S = \frac{C_c H}{1 + e_0} \log\left(\frac{\sigma_o' + \Delta \sigma'}{\sigma_o'}\right)$$

$$= \frac{(0.797)(4)}{1 + 1.22} \log\left(\frac{400}{200}\right) = 0.432 \text{ m}$$

$$S \text{ at } 60\% = (0.6)(0.432 \text{ m}) \approx \textbf{0.259 m} \quad \blacksquare$$

EXAMPLE 6.11

A laboratory consolidation test on a soil specimen (drained on both sides) determined the following results:

thickness of the clay specimen = 25 mm

$\sigma'_1 = 50$ kN/m² $e_1 = 0.92$
$\sigma'_2 = 120$ kN/m² $e_2 = 0.78$

time for 50% consolidation = 2.5 min

Determine the hydraulic conductivity, *k*, of the clay for the loading range.

Solution

$$m_v = \frac{a_v}{1 + e_{av}} = \frac{(\Delta e / \Delta \sigma')}{1 + e_{av}}$$

$$= \frac{\dfrac{0.92 - 0.78}{120 - 50}}{1 + \dfrac{0.92 + 0.78}{2}} = 0.00108 \text{ m}^2/\text{kN}$$

$$c_v = \frac{T_{50} H_{dr}^2}{t_{50}}$$

From Table 6.2, for $U = 50\%$, the value of $T_v = 0.197$, so

$$c_v = \frac{(0.197)\left(\dfrac{0.025 \text{ m}}{2}\right)^2}{2.5 \text{ min}} = 1.23 \times 10^{-5} \text{ m}^2/\text{min}$$

$$k = c_v m_v \gamma_w = (1.23 \times 10^{-5})(0.00108)(9.81)$$

$$= \mathbf{1.303 \times 10^{-7} \text{ m/min}} \qquad \blacksquare$$

6.12   *Calculation of Primary Consolidation Settlement under a Foundation*

Chapter 5 showed that the increase in the vertical stress in soil caused by a load applied over a limited area decreases with depth *z* measured from the ground surface downward. Hence, to estimate the one-dimensional settlement of a foundation, we can use Eq. (6.14), (6.16), or (6.17). However, the increase of effective stress $\Delta \sigma'$ in these equations should be the average increase below the center of the foundation.

Assuming the pressure increase varies parabolically, we can estimate the value of $\Delta \sigma'_{av}$ as (Simpson's rule)

$$\Delta \sigma'_{av} = \frac{\Delta \sigma_t + 4 \Delta \sigma_m + \Delta \sigma_b}{6} \qquad (6.43)$$

where $\Delta\sigma_t$, $\Delta\sigma_m$, and $\Delta\sigma_b$ represent the increase in the pressure at the top, middle, and bottom of the layer, respectively.

The magnitudes of $\Delta\sigma_t$, $\Delta\sigma_m$, and $\Delta\sigma_b$ on a clay layer below the center of a foundation can be calculated using the principles given in Section 5.8. For ease in obtaining the pressure increase below the center of a rectangular area, we can also use the following relationship (based on Section 5.8):

$$\Delta\sigma_{(\text{below center})} = qI_c \qquad (6.44)$$

where q = net load per unit area on the foundation

$$I_c = f(m_1, n_1) \qquad (6.45)$$

$$m_1 = \frac{L}{B} \qquad (6.46)$$

$$n_1 = \frac{z}{(B/2)} \qquad (6.47)$$

L = length of the foundation
B = width of the foundation
z = distance measured from the bottom of the foundation

The variation of I_c with m_1 and n_1 is given in Table 6.3.

Table 6.3 Variation of I_c with m_1 and n_1

					m_1					
n_1	1	2	3	4	5	6	7	8	9	10
0.20	0.994	0.997	0.997	0.997	0.997	0.997	0.997	0.997	0.997	0.997
0.40	0.960	0.976	0.977	0.977	0.977	0.977	0.977	0.977	0.977	0.977
0.60	0.892	0.932	0.936	0.936	0.937	0.937	0.937	0.937	0.937	0.937
0.80	0.800	0.870	0.878	0.880	0.881	0.881	0.881	0.881	0.881	0.881
1.00	0.701	0.800	0.814	0.817	0.818	0.818	0.818	0.818	0.818	0.818
1.20	0.606	0.727	0.748	0.753	0.754	0.755	0.755	0.755	0.755	0.755
1.40	0.522	0.658	0.685	0.692	0.694	0.695	0.695	0.696	0.696	0.696
1.60	0.449	0.593	0.627	0.636	0.639	0.640	0.641	0.641	0.641	0.642
1.80	0.388	0.534	0.573	0.585	0.590	0.591	0.592	0.592	0.593	0.593
2.00	0.336	0.481	0.525	0.540	0.545	0.547	0.548	0.549	0.549	0.549
3.00	0.179	0.293	0.348	0.373	0.384	0.389	0.392	0.393	0.394	0.395
4.00	0.108	0.190	0.241	0.269	0.285	0.293	0.298	0.301	0.302	0.303
5.00	0.072	0.131	0.174	0.202	0.219	0.229	0.236	0.240	0.242	0.244
6.00	0.051	0.095	0.130	0.155	0.172	0.184	0.192	0.197	0.200	0.202
7.00	0.038	0.072	0.100	0.122	0.139	0.150	0.158	0.164	0.168	0.171
8.00	0.029	0.056	0.079	0.098	0.113	0.125	0.133	0.139	0.144	0.147
9.00	0.023	0.045	0.064	0.081	0.094	0.105	0.113	0.119	0.124	0.128
10.00	0.019	0.037	0.053	0.067	0.079	0.089	0.097	0.103	0.108	0.112

EXAMPLE 6.12

Calculate the primary consolidation settlement of the 3-m-thick clay layer (Figure 6.23) that will result from the load carried by a 1.5-m square footing. The clay is normally consolidated.

Solution For normally consolidated clay, from Eq. (6.14) we have

$$S = \frac{C_c H}{1 + e_0} \log \left(\frac{\sigma'_o + \Delta\sigma'}{\sigma'_o} \right)$$

where $C_c = 0.009(LL - 10) = 0.009(40 - 10) = 0.27$
$H = 3000$ mm
$e_0 = 1.0$
$\sigma'_o = 4.5 \times \gamma_{dry(sand)} + 1.5[\gamma_{sat(sand)} - 9.81] + \frac{3}{2} [\gamma_{sat(clay)} - 9.81]$
$\quad = 4.5 \times 15.7 + 1.5(18.9 - 9.81) + 1.5(17.3 - 9.81) = 95.53$ kN/m^2

In order to calculate $\Delta\sigma'$, we can prepare the following table:

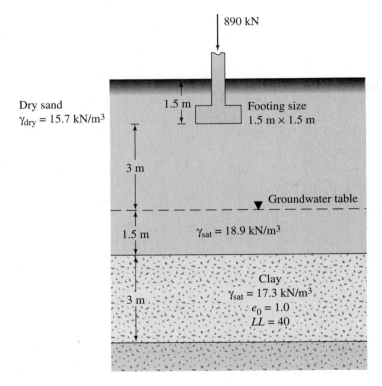

FIGURE 6.23

z (m)	$m_1 = \dfrac{L}{B}$	B (m)	$n_1 = \dfrac{z}{(B/2)}$	I_c (Table 6.3)	$\Delta\sigma = qI_c{}^*$ (kN/m²)
4.5	1	1.5	6	0.051	20.17
6.0	1	1.5	8	0.029	11.47
7.5	1	1.5	10	0.019	7.52

$$^*q = \frac{890}{2.25} = 395.6 \text{ kN/m}^2$$

We calculate

$$\Delta\sigma = \Delta\sigma' = \frac{20.17 + (4)(11.47) + 7.52}{6} = 12.26 \text{ kN/m}^2$$

Substituting these values into the settlement equation gives

$$S = \frac{(0.27)(3000)}{1+1} \log\left(\frac{95.53 + 12.26}{95.53}\right) = \textbf{21.2 mm}$$ ∎

6.13 *Precompression—General Considerations*

When highly compressible, normally consolidated clayey soil layers lie at a limited depth and large consolidation settlements are expected as a result of the construction of large buildings, highway embankments, or earth dams, precompression of soil may be used to minimize postconstruction settlement. The principles of precompression are best explained by referring to Figure 6.24. Here, the proposed structural load

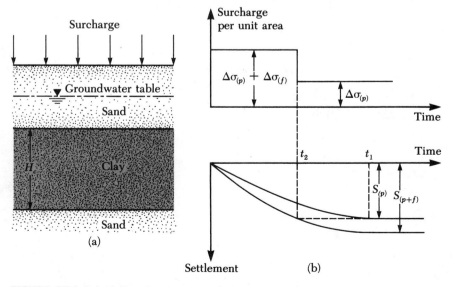

FIGURE 6.24 Principles of precompression

per unit area is $\Delta\sigma_{(p)}$ and the thickness of the clay layer undergoing consolidation is H_c. The maximum primary consolidation settlement caused by the structural load, $S_{(p)}$, then is

$$S_{(p)} = \frac{C_c H_c}{1 + e_0} \log \frac{\sigma_o' + \Delta\sigma_{(p)}}{\sigma_o'} \tag{6.48}$$

Note that at the end of consolidation, $\Delta\sigma' = \Delta\sigma_{(p)}$.

The settlement–time relationship under the structural load will be like that shown in Figure 6.24b. However, if a surcharge of $\Delta\sigma_{(p)} + \Delta\sigma_{(f)}$ is placed on the ground, then the primary consolidation settlement, $S_{(p+f)}$, will be

$$S_{(p+f)} = \frac{C_c H_c}{1 + e_0} \log \frac{\sigma_o' + [\Delta\sigma_{(p)} + \Delta\sigma_{(f)}]}{\sigma_o'} \tag{6.49}$$

Note that at the end of consolidation,

$$\Delta\sigma' = \Delta\sigma_{(p)} + \Delta\sigma_{(f)}$$

The settlement–time relationship under a surcharge of $\Delta\sigma_{(p)} + \Delta\sigma_{(f)}$ is also shown in Figure 6.24b. Note that a total settlement of $S_{(p)}$ would occur at a time t_2, which is much shorter than t_1. So, if a temporary total surcharge of $\Delta\sigma_{(f)} + \Delta\sigma_{(p)}$ is applied on the ground surface for time t_2, the settlement will equal $S_{(p)}$. At that time, if the surcharge is removed and a structure with a permanent load per unit area of $\Delta\sigma_{(p)}$ is built, no appreciable settlement will occur. The procedure just described is *precompression*. The total surcharge, $\Delta\sigma_{(p)} + \Delta\sigma_{(f)}$, can be applied by using temporary fills.

Derivation of Equations to Obtain $\Delta\sigma_{(f)}$ and t_2

Figure 6.24b shows that, under a surcharge of $\Delta\sigma_{(p)} + \Delta\sigma_{(f)}$, the degree of consolidation at time t_2 after load application is

$$U = \frac{S_{(p)}}{S_{(p+f)}} \tag{6.50}$$

Substitution of Eqs. (6.48) and (6.49) into Eq. (6.50) yields

$$U = \frac{\log\left[\dfrac{\sigma_o' + \Delta\sigma_{(p)}}{\sigma_o'}\right]}{\log\left[\dfrac{\sigma_o' + \Delta\sigma_{(p)} + \Delta\sigma_{(f)}}{\sigma_o'}\right]} = \frac{\log\left[1 + \dfrac{\Delta\sigma_{(p)}}{\sigma_o'}\right]}{\log\left\{1 + \dfrac{\Delta\sigma_{(p)}}{\sigma_o'}\left[1 + \dfrac{\Delta\sigma_{(f)}}{\Delta\sigma_{(p)}}\right]\right\}} \tag{6.51}$$

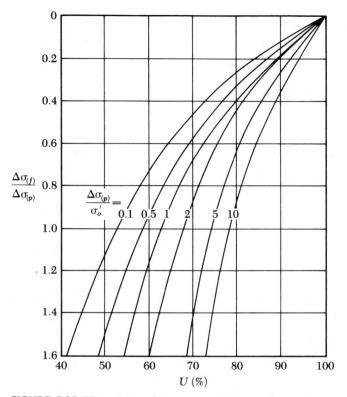

FIGURE 6.25 Plot of $\Delta\sigma_{(f)}/\Delta\sigma_{(p)}$ versus U for various values of $\Delta\sigma_{(p)}/\sigma'_o$—Eq. (6.51)

Figure 6.25 gives magnitudes of U for various combinations of $\Delta\sigma_{(p)}/\sigma'_o$ and $\Delta\sigma_{(f)}/\Delta\sigma_{(p)}$. The degree of consolidation referred to in Eq. (6.51) is actually the average degree of consolidation at time t_2, as shown in Figure 6.24. However, if the average degree of consolidation is used to determine time t_2, some construction problems might arise. The reason is that, after the removal of the surcharge and placement of the structural load, the portion of clay close to the drainage surface will continue to swell, and the soil close to the midplane will continue to settle (Figure 6.26). In some cases, net continuous settlement might result. A conservative approach may solve this problem; that is, assume that U in Eq. (6.51) is the midplane degree of consolidation (Johnson, 1970). Now, from Eq. (6.38), we have

$$U = f(T_v) \tag{6.52}$$

where T_v = time factor = $c_v t_2 / H_{dr}^2$
   c_v = coefficient of consolidation
   t_2 = time
   H_{dr} = maximum drainage path ($H/2$ for two-way drainage and H for one-way drainage)

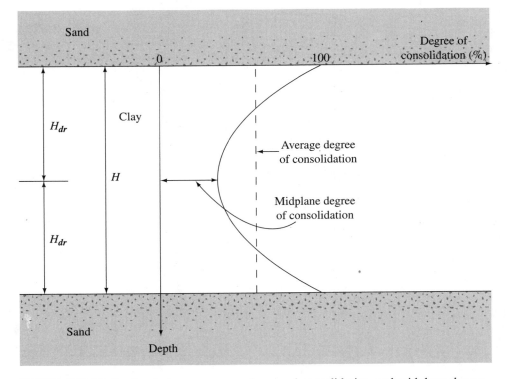

FIGURE 6.26 Distinction between average degree of consolidation and midplane degree of consolidation

The variation of U (midplane degree of consolidation) with T_v is shown in Figure 6.27.

Procedure for Obtaining Precompression Parameters

Engineers may encounter two problems during precompression work in the field:

1. The value of $\Delta\sigma_{(f)}$ is known, but t_2 must be obtained. In such case, obtain σ_o' and $\Delta\sigma_{(p)}$ and solve for U using Eq. (6.51) or Figure 6.25. For this value of U, obtain T_v from Figure 6.27. Then

$$t_2 = \frac{T_v H_{dr}^2}{c_v} \tag{6.53}$$

2. For a specified value of t_2, $\Delta\sigma_{(f)}$ must be obtained. In such case, calculate T_v. Then refer to Figure 6.27 to obtain the midplane degree of consolidation,

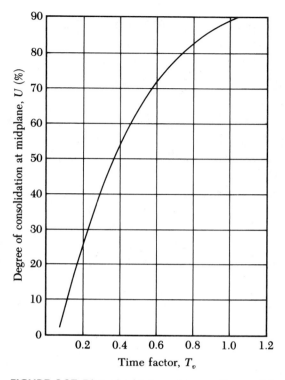

FIGURE 6.27 Plot of midplane degree of consolidation versus T_v

U. With the estimated value of *U*, go to Figure 6.25 to find the required $\Delta\sigma_{(f)}/\Delta\sigma_{(p)}$ and then calculate $\Delta\sigma_{(f)}$.

EXAMPLE 6.13

Refer to Figure 6.24. During the construction of a highway bridge, the average permanent load on the clay layer is expected to increase by about 115 kN/m². The average effective overburden pressure at the middle of the clay layer is 210 kN/m². Here, $H = 6$ m, $C_c = 0.28$, $e_0 = 0.9$, and $c_v = 0.36$ m²/mo. The clay is normally consolidated.

    a. Determine the total primary consolidation settlement of the bridge without precompression.

    b. What is the surcharge, $\Delta\sigma_{(f)}$, needed to eliminate by precompression the entire primary consolidation settlement in 9 months?

Solution

Part a The total primary consolidation settlement may be calculated from Eq. (6.48):

$$S_{(p)} = \frac{C_c H}{1 + e_0} \log\left[\frac{\sigma_o' + \Delta\sigma_{(p)}}{\sigma_o'}\right] = \frac{(0.28)(6)}{1 + 0.9} \log\left[\frac{210 + 115}{210}\right]$$

$$= 0.1677 \text{ m} = \textbf{167.7 mm}$$

Part b

$$T_v = \frac{c_v t_2}{H_{dr}^2}$$

$$c_v = 0.36 \text{ m}^2/\text{mo.}$$

$$H_{dr} = 3 \text{ m (two-way drainage)}$$

$$t_2 = 9 \text{ mo.}$$

Hence,

$$T_v = \frac{(0.36)(9)}{3^2} = 0.36$$

According to Figure 6.27, for $T_v = 0.36$, the value of U is 47%. Now

$$\Delta\sigma_{(p)} = 115 \text{ kN/m}^2$$

$$\sigma_o' = 210 \text{ kN/m}^2$$

So

$$\frac{\Delta\sigma_{(p)}}{\sigma_o'} = \frac{115}{210} = 0.548$$

According to Figure 6.25, for $U = 47\%$ and $\Delta\sigma_{(p)}/\sigma_o' = 0.548$, $\Delta\sigma_{(f)}/\Delta\sigma_{(p)} \approx 1.8$. So

$$\Delta\sigma_{(f)} = (1.8)(115) = \textbf{207 kN/m}^2 \qquad \blacksquare$$

6.14 Sand Drains

The use of *sand drains* is another way to accelerate the consolidation settlement of soft, normally consolidated clay layers and achieve precompression before foundation construction. Sand drains are constructed by drilling holes through the clay layer(s) in the field at regular intervals. The holes are backfilled with highly permeable sand (see Figure 6.28a), and then a surcharge is applied at the ground surface. This surcharge will increase the pore water pressure in the clay. The excess pore water pressure in the clay will be dissipated by drainage—both vertically and radially to the sand drains—which accelerates settlement of the clay layer.

Note that the radius of the sand drains is r_w (Figure 6.28a). Figure 6.28b also shows the plan of the layout of the sand drains. The effective zone from which the radial drainage will be directed toward a given sand drain is approximately cylindrical, with a diameter of d_e.

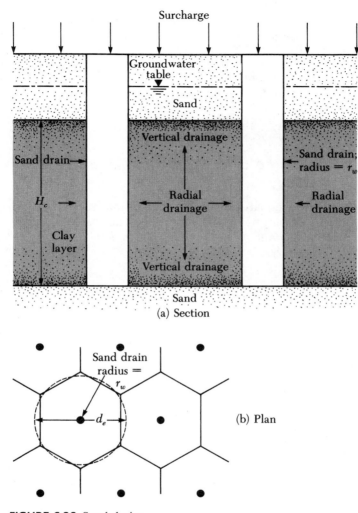

FIGURE 6.28 Sand drains

To determine the surcharge that needs to be applied at the ground surface and the length of time that it has to be maintained, refer to Figure 6.24 and use the corresponding equation, Eq. (6.51):

$$U_{v,r} = \dfrac{\log\left[1 + \dfrac{\Delta\sigma_{(p)}}{\sigma_o'}\right]}{\log\left\{1 + \dfrac{\Delta\sigma_{(p)}}{\sigma_o'}\left[1 + \dfrac{\Delta\sigma_{(f)}}{\Delta\sigma_{(p)}}\right]\right\}} \qquad (6.54)$$

The notations $\Delta\sigma_{(p)}$, σ'_o, and $\Delta\sigma_{(f)}$ are the same as those used in Eq. (6.51). However, unlike Eq. (6.51), the left-hand side of Eq. (6.54) is the *average degree* of consolidation instead of the degree of consolidation at midplane. Both *radial* and *vertical* drainage contribute to the average degree of consolidation. If $U_{v,r}$ can be determined for any time t_2 (see Figure 6.24b), then the total surcharge $\Delta\sigma_{(f)}$ + $\Delta\sigma_{(p)}$ may be easily obtained from Figure 6.25. The procedure for determining the average degree of consolidation $(U_{v,r})$ is given in the following sections.

Average Degree of Consolidation Due to Radial Drainage Only

The theory for equal-strain consolidation due to radial drainage only (with no smear) was developed by Barron (1948). The theory is based on the assumption that there is *no drainage in the vertical direction*. According to this theory,

$$U_r = 1 - \exp\left(\frac{-8T_r}{m}\right) \tag{6.55}$$

where U_r = average degree of consolidation due to radial drainage only

$$m = \left(\frac{n^2}{n^2 - 1}\right) \ln(n) - \frac{3n^2 - 1}{4n^2} \tag{6.56}$$

$$n = \frac{d_e}{2r_w} \tag{6.57}$$

T_r = nondimensional time factor for radial drainage only

$$= \frac{c_{vr}t_2}{d_e^2} \tag{6.58}$$

c_{vr} = coefficient of consolidation for radial drainage

$$= \frac{k_h}{\left[\dfrac{\Delta e}{\Delta\sigma'(1 + e_{av})}\right]\gamma_w} \tag{6.59}$$

Note that Eq. (6.59) is similar to that defined in Eq. (6.34). In Eq. (6.34), k is the hydraulic conductivity in the vertical direction of the clay layer. In Eq. (6.59) k is replaced by k_h, the hydraulic conductivity for flow in the horizontal direction. In some cases, k_h may be assumed to equal k; however, for soils like varved clay, $k_h > k$. Table 6.4 gives the variation of U_r with T_r for various values of n.

Table 6.4 Solution for radial drainage

Degree of consolidation, U_r (%)	Time factor, T_r, for values of n				
	5	10	15	20	25
0	0	0	0	0	0
1	0.0012	0.0020	0.0025	0.0028	0.0031
2	0.0024	0.0040	0.0050	0.0057	0.0063
3	0.0036	0.0060	0.0075	0.0086	0.0094
4	0.0048	0.0081	0.0101	0.0115	0.0126
5	0.0060	0.0101	0.0126	0.0145	0.0159
6	0.0072	0.0122	0.0153	0.0174	0.0191
7	0.0085	0.0143	0.0179	0.0205	0.0225
8	0.0098	0.0165	0.0206	0.0235	0.0258
9	0.0110	0.0186	0.0232	0.0266	0.0292
10	0.0123	0.0208	0.0260	0.0297	0.0326
11	0.0136	0.0230	0.0287	0.0328	0.0360
12	0.0150	0.0252	0.0315	0.0360	0.0395
13	0.0163	0.0275	0.0343	0.0392	0.0431
14	0.0177	0.0298	0.0372	0.0425	0.0467
15	0.0190	0.0321	0.0401	0.0458	0.0503
16	0.0204	0.0344	0.0430	0.0491	0.0539
17	0.0218	0.0368	0.0459	0.0525	0.0576
18	0.0232	0.0392	0.0489	0.0559	0.0614
19	0.0247	0.0416	0.0519	0.0594	0.0652
20	0.0261	0.0440	0.0550	0.0629	0.0690
21	0.0276	0.0465	0.0581	0.0664	0.0729
22	0.0291	0.0490	0.0612	0.0700	0.0769
23	0.0306	0.0516	0.0644	0.0736	0.0808
24	0.0321	0.0541	0.0676	0.0773	0.0849
25	0.0337	0.0568	0.0709	0.0811	0.0890
26	0.0353	0.0594	0.0742	0.0848	0.0931
27	0.0368	0.0621	0.0776	0.0887	0.0973
28	0.0385	0.0648	0.0810	0.0926	0.1016
29	0.0401	0.0676	0.0844	0.0965	0.1059
30	0.0418	0.0704	0.0879	0.1005	0.1103
31	0.0434	0.0732	0.0914	0.1045	0.1148
32	0.0452	0.0761	0.0950	0.1087	0.1193
33	0.0469	0.0790	0.0987	0.1128	0.1239
34	0.0486	0.0820	0.1024	0.1171	0.1285
35	0.0504	0.0850	0.1062	0.1214	0.1332
36	0.0522	0.0881	0.1100	0.1257	0.1380
37	0.0541	0.0912	0.1139	0.1302	0.1429
38	0.0560	0.0943	0.1178	0.1347	0.1479
39	0.0579	0.0975	0.1218	0.1393	0.1529
40	0.0598	0.1008	0.1259	0.1439	0.1580
41	0.0618	0.1041	0.1300	0.1487	0.1632
42	0.0638	0.1075	0.1342	0.1535	0.1685
43	0.0658	0.1109	0.1385	0.1584	0.1739

Table 6.4 (Continued)

Degree of consolidation, U_r (%)	Time factor, T_r, for values of n				
	5	10	15	20	25
44	0.0679	0.1144	0.1429	0.1634	0.1793
45	0.0700	0.1180	0.1473	0.1684	0.1849
46	0.0721	0.1216	0.1518	0.1736	0.1906
47	0.0743	0.1253	0.1564	0.1789	0.1964
48	0.0766	0.1290	0.1659	0.1897	0.2083
49	0.0788	0.1329	0.1659	0.1953	0.2144
50	0.0811	0.1368	0.1708	0.1953	0.2144
51	0.0835	0.1407	0.1758	0.2020	0.2206
52	0.0859	0.1448	0.1809	0.2068	0.2270
53	0.0884	0.1490	0.1860	0.2127	0.2335
54	0.0909	0.1532	0.1913	0.2188	0.2402
55	0.0935	0.1575	0.1968	0.2250	0.2470
56	0.0961	0.1620	0.2023	0.2313	0.2539
57	0.0988	0.1665	0.2080	0.2378	0.2610
58	0.1016	0.1712	0.2138	0.2444	0.2683
59	0.1044	0.1759	0.2197	0.2512	0.2758
60	0.1073	0.1808	0.2258	0.2582	0.2834
61	0.1102	0.1858	0.2320	0.2653	0.2912
62	0.1133	0.1909	0.2384	0.2726	0.2993
63	0.1164	0.1962	0.2450	0.2801	0.3075
64	0.1196	0.2016	0.2517	0.2878	0.3160
65	0.1229	0.2071	0.2587	0.2958	0.3247
66	0.1263	0.2128	0.2658	0.3039	0.3337
67	0.1298	0.2187	0.2732	0.3124	0.3429
68	0.1334	0.2248	0.2808	0.3210	0.3524
69	0.1371	0.2311	0.2886	0.3300	0.3623
70	0.1409	0.2375	0.2967	0.3392	0.3724
71	0.1449	0.2442	0.3050	0.3488	0.3829
72	0.1490	0.2512	0.3134	0.3586	0.3937
73	0.1533	0.2583	0.3226	0.3689	0.4050
74	0.1577	0.2658	0.3319	0.3795	0.4167
75	0.1623	0.2735	0.3416	0.3906	0.4288
76	0.1671	0.2816	0.3517	0.4021	0.4414
77	0.1720	0.2900	0.3621	0.4141	0.4546
78	0.1773	0.2988	0.3731	0.4266	0.4683
79	0.1827	0.3079	0.3846	0.4397	0.4827
80	0.1884	0.3175	0.3966	0.4534	0.4978
81	0.1944	0.3277	0.4090	0.4679	0.5137
82	0.2007	0.3383	0.4225	0.4831	0.5304
83	0.2074	0.3496	0.4366	0.4992	0.5481
84	0.2146	0.3616	0.4516	0.5163	0.5668
85	0.2221	0.3743	0.4675	0.5345	0.5868
86	0.2302	0.3879	0.4845	0.5539	0.6081
87	0.2388	0.4025	0.5027	0.5748	0.6311

Table 6.4 (Continued)

Degree of consolidation, U_r (%)	Time factor, T_r, for values of n				
	5	10	15	20	25
88	0.2482	0.4183	0.5225	0.5974	0.6558
89	0.2584	0.4355	0.5439	0.6219	0.6827
90	0.2696	0.4543	0.5674	0.6487	0.7122
91	0.2819	0.4751	0.5933	0.6784	0.7448
92	0.2957	0.4983	0.6224	0.7116	0.7812
93	0.3113	0.5247	0.6553	0.7492	0.8225
94	0.3293	0.5551	0.6932	0.7927	0.8702
95	0.3507	0.5910	0.7382	0.8440	0.9266
96	0.3768	0.6351	0.7932	0.9069	0.9956
97	0.4105	0.6918	0.8640	0.9879	1.0846
98	0.4580	0.7718	0.9640	1.1022	1.2100
99	0.5391	0.9086	1.1347	1.2974	1.4244

Average Degree of Consolidation Due to Vertical Drainage Only

The average degree of consolidation due to vertical drainage only may be obtained from Eqs. (6.39) and (6.40), (or Table 6.2):

$$T_v = \frac{\pi}{4}\left[\frac{U_v\%}{100}\right] \qquad \text{for } U_v = 0\% \text{ to } 60\% \tag{6.60}$$

and

$$T_v = 1.781 - 0.933 \log(100 - U_v\%) \qquad \text{for } U_v > 60\%$$

where U_v = average degree of consolidation due to vertical drainage only

$$T_v = \frac{c_v t_2}{H_{dr}^2} \tag{6.62}$$

c_v = coefficient of consolidation for vertical drainage

Average Degree of Consolidation Due to Vertical and Radial Drainage

For a given surcharge and duration t_2, the average degree of consolidation due to drainage in the vertical and radial directions is

$$U_{v,r} = 1 - (1 - U_r)(1 - U_v) \tag{6.63}$$

Wick Drains

The *wick drain* was recently developed as an alternative to the sand drain for inducing vertical drainage in saturated clay deposits. Wick drains appear to be better, faster, and more cost efficient. They essentially consist of paper or plastic strips that are held in a long tube. The tube is pushed into the soft clay deposit and then withdrawn, leaving behind the strips. These strips act as vertical drains and induce rapid consolidation. Wick drains can be placed at desired spacings like sand drains. The main advantage of wick drains over sand drains is that they do not require drilling, and thus installation is much faster.

EXAMPLE 6.14

Redo Example 6.13 with the addition of some sand drains. Assume that $r_w = 0.1$ m, $d_e = 3$ m, and $c_v = c_{vr}$.

Solution

Part a   The total primary consolidation settlement will be **167.7 mm** as before.

Part b   From Example 6.13, $T_v = 0.36$. The value of U_v from Table 6.2 is about 67%. From Eq. (6.57), we have

$$n = \frac{d_e}{2r_w} = \frac{3}{2 \times 0.1} = 15$$

Again,

$$T_r = \frac{c_{vr}t_2}{d_e^2} = \frac{(0.36)(9)}{(3)^2} = 0.36$$

From Table 6.4 for $n = 15$ and $T_r = 0.36$, the value of U_r is about 77%. Hence,

$$U_{v,r} = 1 - (1 - U_v)(1 - U_r) = 1 - (1 - 0.67)(1 - 0.77)$$
$$= 0.924 = 92.4\%$$

Now, from Figure 6.24 for $\Delta\sigma_{(p)}/\sigma_o' = 0.548$ and $U_{v,r} = 92.4\%$, the value of $\Delta\sigma_{(f)}/\Delta\sigma_{(p)} \approx 0.12$. Hence, we have

$$\Delta\sigma_{(f)} = (115)(0.12) = \mathbf{13.8 \ kN/m^2} \qquad \blacksquare$$

Problems

6.1   Figure 6.29 shows a soil profile. The uniformly distributed load on the ground surface is $\Delta\sigma$. Estimate the primary settlement of the clay layer given these values:

$$H_1 = 1.5 \text{ m}, H_2 = 2 \text{ m}, H_3 = 2.5 \text{ m}$$

Sand: $e = 0.62, G_s = 2.62$

Clay: $e = 0.98, G_s = 2.75, LL = 50$

$$\Delta\sigma = 110 \text{ kN/m}^2$$

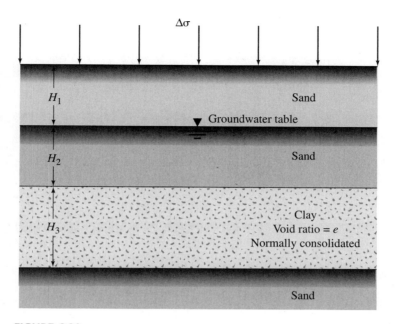

FIGURE 6.29

6.2 Repeat Problem 6.1 with the following values:

$H_1 = 1.5$ m, $H_2 = 2$ m, $H_3 = 2$ m

Sand: $e = 0.55$, $G_s = 2.67$

Clay: $e = 1.1$, $G_s = 2.73$, $LL = 45$

$\Delta\sigma = 120$ kN/m²

6.3 Repeat Problem 6.1 for these data:

$\Delta\sigma = 87$ kN/m²

$H_1 = 1$ m, $H_2 = 3$ m, $H_3 = 3.2$ m

Sand: $\gamma_{dry} = 14.6$ kN/m³, $\gamma_{sat} = 17.3$ kN/m³

Clay: $\gamma_{sat} = 19.3$ kN/m³, $LL = 38$, $e = 0.75$

6.4 If the clay layer in Problem 6.3 is preconsolidated and the average preconsolidation pressure is 80 kN/m², what will be the expected primary consolidation settlement if $C_s = \frac{1}{5}C_c$.

6.5 A soil profile is shown in Figure 6.30. The average preconsolidation pressure of the clay is 170 kN/m². Estimate the primary consolidation settlement that will take place as a result of a surcharge of 110 kN/m² if $C_s = \frac{1}{6}C_c$.

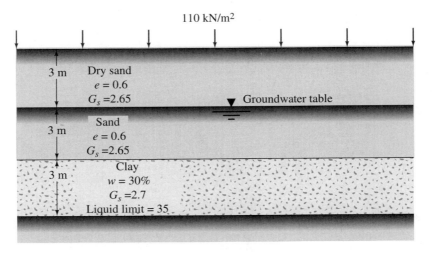

FIGURE 6.30

6.6 The results of a laboratory consolidation test on a clay specimen are given in the table.

Pressure, σ' (kN/m²)	Total height of specimen at end of consolidation (mm)
25	17.65
50	17.40
100	17.03
200	16.56
400	16.15
800	15.88

Also, initial height of specimen = 19 mm, G_s = 2.68, mass of dry specimen = 95.2 g, and area of specimen = 31.68 cm².

a. Draw the e–log σ' graph.
b. Determine the preconsolidation pressure.
c. Determine the compression index, C_c.

6.7 Following are the results of a consolidation test:

e	Pressure, σ' (kN/m²)
1.1	25
1.085	50
1.055	100
1.01	200
0.94	400
0.79	800
0.63	1600

a. Plot the e–log σ' curve.
b. Using Casagrande's method, determine the preconsolidation pressure.
c. Calculate the compression index, C_c.

6.8 The coordinates of two points on a virgin compression curve are given here:

$$\sigma_1' = 190 \text{ kN/m}^2 \qquad e_1 = 1.75$$

$$\sigma_2' = 385 \text{ kN/m}^2 \qquad e_2 = 1.49$$

Determine the void ratio that will correspond to an effective pressure of 600 kN/m².

6.9 The laboratory consolidation data for an undisturbed clay specimen are given:

$$\sigma_1' = 95 \text{ kN/m}^2 \qquad e_1 = 1.1$$

$$\sigma_2' = 475 \text{ kN/m}^2 \qquad e_2 = 0.9$$

What will be the void ratio for an effective pressure of 600 kN/m²? (*Note:* $\sigma_c' < 95 \text{ kN/m}^2$.)

6.10 Given are the relationships of e and σ' for a clay soil:

e	σ' (kN/m²)
1.0	20
0.97	50
0.85	180
0.75	320

For this clay soil in the field, the following values are given: $H = 2.5$ m, $\sigma_o' = 60$ kN/m², and $\sigma_o' + \Delta\sigma' = 210$ kN/m². Calculate the expected settlement caused by primary consolidation.

6.11 Consider the virgin compression curve described in Problem 6.8.
a. Find the coefficient of volume compressibility for the pressure range stated.
b. If the coefficient of consolidation for the pressure range is 0.0023 cm²/sec, find the hydraulic conductivity in (cm/sec) of the clay corresponding to the average void ratio.

6.12 Refer to Problem 6.1. Given $c_v = 0.003$ cm²/sec, how long will it take for 50% primary consolidation to take place?

6.13 Refer to Problem 6.2. Given $c_v = 2.62 \times 10^{-6}$ m²/min, how long will it take for 65% primary consolidation to take place?

6.14 Laboratory tests on a 25-mm-thick clay specimen drained at both the top and bottom show that 50% consolidation takes place in 8.5 min.
a. How long will it take for a similar clay layer in the field, 3.2 m thick and drained at the top only, to undergo 50% consolidation?
b. Find the time required for the clay layer in the field as described in part (a) to reach 65% consolidation.

6.15 A 3-m-thick layer (two-way drainage) of saturated clay under a surcharge loading underwent 90% primary consolidation in 75 days. Find the coefficient of consolidation of clay for the pressure range.

6.16 For a 30-mm-thick undisturbed clay specimen described in Problem 6.15, how long will it take to undergo 90% consolidation in the laboratory for a similar consolidation pressure range? The laboratory test's specimen will have two-way drainage.

6.17 A normally consolidated clay layer is 5 m thick (one-way drainage). From the application of a given pressure, the total anticipated primary consolidation settlement will be 160 mm.
   a. What is the average degree of consolidation for the clay layer when the settlement is 50 mm?
   b. If the average value of c_v for the pressure range is 0.003 cm²/sec, how long will it take for 50% settlement to occur?
   c. How long will it take for 50% consolidation to occur if the clay layer is drained at both top and bottom?

6.18 In laboratory consolidation tests on a clay specimen (drained on both sides), the following results were obtained:

   thickness of clay layer = 25 mm

   $\sigma_1' = 50 \text{ kN/m}^2$ $e_1 = 0.75$

   $\sigma_2' = 100 \text{ kN/m}^2$ $e_2 = 0.61$

   time for 50% consolidation (t_{50}) = 3.1 min

   Determine the hydraulic conductivity of the clay for the loading range.

6.19 A continuous footing is shown in Figure 6.31. Find the vertical stresses at A, B, and C caused by the load carried by the footing.

6.20 Use Eq. (6.43) to calculate the settlement of the footing described in Problem 6.19 from consolidation of the clay layer given:

   Sand: $e = 0.6$, $G_s = 2.65$; degree of saturation of sand above groundwater table is 30%

   Clay: $e = 0.85$, $G_s = 2.75$, $LL = 45$; the clay is normally consolidated

6.21 Refer to Figure 6.24. For the construction of an airport, a large fill operation is required. For the work, the average permanent load, $\Delta\sigma_{(p)}$, on the clay layer will increase by about 70 kN/m². The average effective overburden pressure on the clay layer before the fill operation is 95 kN/m². For the clay layer, which is normally consolidated and drained at top and bottom, $H = 5$ m, $C_c = 0.24$, $e_0 = 0.81$, and $c_v = 0.44$ m²/mo.
   a. Determine the primary consolidation settlement of the clay layer caused by the additional permanent load, $\Delta\sigma_{(p)}$.
   b. What is the time required for 90% of primary consolidation settlement under the additional permanent load only?

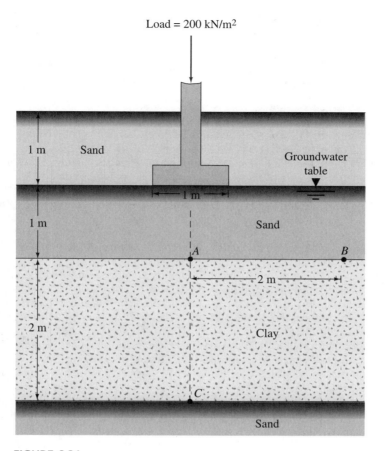

Load = 200 kN/m²

1 m

Sand

Groundwater
table

1 m

Sand

A

B

2 m

2 m

Clay

C

Sand

FIGURE 6.31

    c. What temporary surcharge, $\Delta\sigma_{(f)}$, will be required to eliminate the entire primary consolidation settlement in 6 months by the precompression technique?

6.22 Redo part (c) of Problem 6.21 for a time of elimination of primary consolidation settlement of 7 months.

6.23 Repeat Problem 6.21 with $\Delta\sigma_{(p)} = 30$ kN/m², the average effective overburden pressure on clay layer $= 50$ kN/m², $H = 5$ m, $C_c = 0.3$, $e_0 = 1.0$, and $c_v = 9.7 \times 10^{-2}$ cm²/min.

6.24 The diagram of a sand drain is shown in Figure 6.28. If $r_w = 0.25$ m, $d_e = 4$ m, $c_v = c_{vr} = 0.28$ m²/mo, and $H = 8.4$ m, determine the degree of consolidation caused only by the sand drain after 6 months of surcharge application.

6.25 Estimate the degree of consolidation for the clay layer described in Problem 6.24 that is caused by the combination of vertical drainage (drained on top and bottom) and radial drainage after 6 months of the application of surcharge.

6.26   A 4-m-thick clay layer is drained at top and bottom. Its characteristics are $c_{vr} = c_v$ (for vertical damage) $= 0.0039$ m^2/day, $r_w = 200$ mm, and $d_e = 2$ m. Estimate the degree of consolidation of the clay layer caused by the combination of vertical and radial drainage at $t = 0.2, 0.4, 0.8,$ and 1 yr.

References

Barron, R. A. (1948). "Consolidation of Fine-Grained Soils by Drain Wells," *Transactions,* American Society of Civil Engineers, Vol. 113, 718–754.

Casagrande, A. (1936). "Determination of the Preconsolidation Load and Its Practical Significance," *Proceedings,* 1st International Conference on Soil Mechanics and Foundation Engineering, Cambridge, MA, Vol. 3, 60–64.

Casagrande, A., and Fadum, R. E. (1940). "Notes on Soil Testing for Engineering Purposes," Harvard University Graduate School Engineering Publication No. 8.

Johnson, S. J. (1970). "Precompression for Improving Foundation Soils," *Journal of the Soil Mechanics and Foundations Division,* American Society of Civil Engineers, Vol. 96, No. SM1, 114–144.

Mesri, G. (1973). "Coefficient of Secondary Compression," *Journal of the Soil Mechanics and Foundations Division,* ASCE, Vol. 99, No. SM1, 122–137.

Nagaraj, T., and Murty, B. R. S. (1985). "Prediction of the Preconsolidation Pressure and Recompression Index of Soils," *Geotechnical Testing Journal,* ASTM, Vol. 8, No. 4, 199–202.

Rendon-Herrero, O. (1983). "Universal Compression Index Equation," *Discussion, Journal of Geotechnical Engineering,* ASCE, Vol. 109, No. 10, 1349.

Schmertmann, J. H. (1953). "Undistorted Consolidation Behavior of Clay," *Transactions,* ASCE, Vol. 120, 1201.

Taylor, D. W. (1942). "Research on Consolidation of Clays," *Serial No. 82,* Department of Civil and Sanitary Engineering, Massachusetts Institute of Technology, Cambridge, MA.

Terzaghi, K. (1925). *Erdbaumechanik auf Bodenphysikalischer Grundlage,* Deuticke, Vienna.

Terzaghi, K., and Peck, R. B. (1967). *Soil Mechanics in Engineering Practice,* 2nd ed., Wiley, New York.

Supplementary References for Further Study

Legget, R. F., and Peckover, F. L. (1973). "Foundation Performance of a 100-Year-Old Bridge," *Canadian Geotechnical Journal,* Vol. 10, No. 3, 504–519.

Leonards, G. A., and Altschaeffl, A. G. (1964). "Compressibility of Clay," *Journal of the Soil Mechanics and Foundations Division,* ASCE, Vol. 90, No. SM5, 133–156.

Leroueil, S. (1988). "Tenth Canadian Geotechnical Colloquium: Recent Developments in Consolidation of Natural Clays," *Canadian Geotechnical Journal,* Vol. 25, No. 1, 85–107.

Rendon-Herrero, O. (1980). "Universal Compression Index Equation," *Journal of the Geotechnical Engineering Division,* ASCE, Vol. 106, No. GT11, 1179–1200.

Robinson, R. G., and Allam, M. M. (1996). "Determination of Coefficient of Consolidation from Early Stage of log t Plot," *Geotechnical Testing Journal,* ASTM, Vol. 19, No. 3, 316–320.

7

Shear Strength of Soil

The *shear strength* of a soil mass is the internal resistance per unit area that the soil mass can offer to resist failure and sliding along any plane inside it. Engineers must understand the nature of shearing resistance in order to analyze soil stability problems such as bearing capacity, slope stability, and lateral pressure on earth-retaining structures.

7.1 *Mohr–Coulomb Failure Criteria*

Mohr (1900) presented a theory for rupture in materials. This theory contended that a material fails because of a critical combination of normal stress and shear stress, and not from either maximum normal or shear stress alone. Thus, the functional relationship between normal stress and shear stress on a failure plane can be expressed in the form (Figure 7.1a)

$$\tau_f = f(\sigma) \tag{7.1}$$

where τ_f = shear stress on the failure plane
σ = normal stress on the failure plane

The failure envelope defined by Eq. (7.1) is a curved line, as shown in Figure 7.1b. For most soil mechanics problems, it is sufficient to approximate the shear stress on the failure plane as a linear function of the normal stress (Coulomb, 1776). This relation can be written as

$$\tau_f = c + \sigma \tan \phi \tag{7.2}$$

where c = cohesion
ϕ = angle of internal friction

The preceding equation is called the *Mohr–Coulomb failure criteria*.

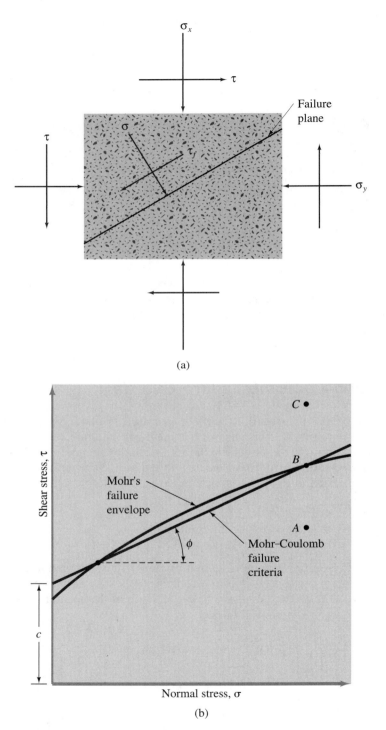

FIGURE 7.1 Mohr's failure envelope and the Mohr–Coulomb failure criteria

The significance of the failure envelope can be explained as follows: If the normal stress and the shear stress on a plane in a soil mass are such that they plot as point A in Figure 7.1b, then shear failure will not occur along that plane. If the normal stress and the shear stress on a plane plot as point B (which falls on the failure envelope), then shear failure will occur along that plane. A state of stress on a plane represented by point C cannot exist because it plots above the failure envelope, and shear failure in a soil would have occurred already.

7.2 *Inclination of the Plane of Failure Caused by Shear*

As stated by the Mohr–Coulomb failure criteria, failure from shear will occur when the shear stress on a plane reaches a value given by Eq. (7.2). To determine the inclination of the failure plane with the major principal plane, refer to Figure 7.2, where σ_1 and σ_3 are, respectively, the major and minor principal stresses. The failure plane EF makes an angle θ with the major principal plane. To determine the angle θ and the relationship between σ_1 and σ_3, refer to Figure 7.3, which is a plot of the Mohr's circle for the state of stress shown in Figure 7.2. In Figure 7.3, fgh is the failure envelope defined by the relationship $s = c + \sigma \tan \phi$. The radial line ab defines the major principal plane (CD in Figure 7.2), and the radial line ad defines the failure plane (EF in Figure 7.2). It can be shown that $\angle bad = 2\theta = 90 + \phi$, or

$$\theta = 45 + \frac{\phi}{2} \tag{7.3}$$

Again, from Figure 7.3, we have

$$\frac{\overline{ad}}{\overline{fa}} = \sin \phi \tag{7.4}$$

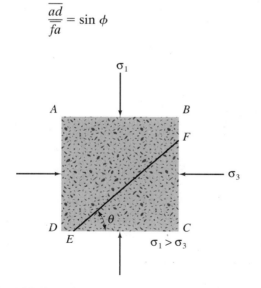

FIGURE 7.2 Inclination of failure plane in soil with major principal plane

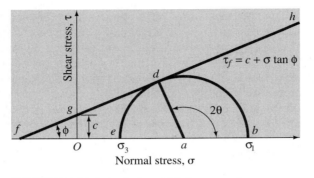

FIGURE 7.3 Mohr's circle and failure envelope

$$\overline{fa} = fO + Oa = c \cot \phi + \frac{\sigma_1 + \sigma_3}{2} \qquad (7.5)$$

Also,

$$\overline{ad} = \frac{\sigma_1 - \sigma_3}{2} \qquad (7.5b)$$

Substituting Eqs. (7.5a) and (7.5b) into Eq. (7.4), we obtain

$$\sin \phi = \frac{\dfrac{\sigma_1 - \sigma_3}{2}}{c \cot \phi + \dfrac{\sigma_1 + \sigma_3}{2}}$$

or

$$\sigma_1 = \sigma_3 \left(\frac{1 + \sin \phi}{1 - \sin \phi} \right) + 2c \left(\frac{\cos \phi}{1 - \sin \phi} \right) \qquad (7.6)$$

However,

$$\frac{1 + \sin \phi}{1 - \sin \phi} = \tan^2 \left(45 + \frac{\phi}{2} \right)$$

and

$$\frac{\cos \phi}{1 - \sin \phi} = \tan \left(45 + \frac{\phi}{2} \right)$$

Thus,

$$\sigma_1 = \sigma_3 \tan^2 \left(45 + \frac{\phi}{2} \right) + 2c \tan \left(45 + \frac{\phi}{2} \right) \qquad (7.7)$$

The preceding relationship is Mohr–Coulomb's failure criteria restated in terms of failure stresses.

7.3   *Shear Failure Law in Saturated Soil*

In saturated soil, the total normal stress at a point is the sum of the effective stress and the pore water pressure, or

$$\sigma = \sigma' + u$$

The effective stress, σ', is carried by the soil solids. So, to apply Eq. (7.2) to soil mechanics, we need to rewrite it as

$$\tau_f = c + (\sigma - u) \tan \phi = c + \sigma' \tan \phi \tag{7.8}$$

The value of c for sand and inorganic silt is 0. For normally consolidated clays, c can be approximated at 0. Overconsolidated clays have values of c that are greater than 0. The angle of friction, ϕ, is sometimes referred to as the *drained angle of friction*. Typical values of ϕ for some granular soils are given in Table 7.1.

For normally consolidated clays, the friction angle ϕ generally ranges from 20° to 30°. For overconsolidated clays, the magnitude of ϕ decreases. For natural noncemented, overconsolidated clays with preconsolidation pressure less than about 1000 kN/m², the magnitude of c is in the range of 5 to 15 kN/m².

LABORATORY DETERMINATION OF SHEAR STRENGTH PARAMETERS

The shear strength parameters of a soil are determined in the laboratory primarily with two types of tests: direct shear test and triaxial test. The procedures for conducting each of these tests are explained in some detail in the following sections.

Table 7.1 Typical values of drained angle of friction for sands and silts

Soil type	ϕ (deg)
Sand: Rounded grains	
Loose	27–30
Medium	30–35
Dense	35–38
Sand: Angular grains	
Loose	30–35
Medium	35–40
Dense	40–45
Gravel with some sand	34–48
Silts	26–35

7.4   *Direct Shear Test*

This is the oldest and simplest form of shear test arrangement. A diagram of the direct shear test apparatus is shown in Figure 7.4. The test equipment consists of a metal shear box in which the soil specimen is placed. The soil specimens may be square or circular. The size of the specimens generally used is about 20 to 25 cm² across and 25 to 30 mm high. The box is split horizontally into halves. Normal force on the specimen is applied from the top of the shear box. The normal stress on the specimens can be as great as 1000 kN/m². Shear force is applied by moving one half of the box relative to the other to cause failure in the soil specimen.

Depending on the equipment, the shear test can be either stress-controlled or strain-controlled. In stress-controlled tests, the shear force is applied in equal increments until the specimen fails. The failure takes place along the plane of split of the shear box. After the application of each incremental load, the shear displacement of the top half of the box is measured by a horizontal dial gauge. The change in the height of the specimen (and thus the volume change of the specimen) during the test can be obtained from the readings of a dial gauge that measures the vertical movement of the upper loading plate.

In strain-controlled tests, a constant rate of shear displacement is applied to one half of the box by a motor that acts through gears. The constant rate of shear displacement is measured by a horizontal dial gauge. The resisting shear force of the soil corresponding to any shear displacement can be measured by a horizontal proving ring or load cell. The volume change of the specimen during the test is

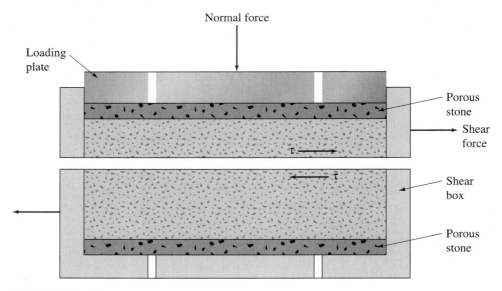

FIGURE 7.4 Diagram of direct shear test arrangement

obtained in a manner similar to the stress-controlled tests. Figure 7.5 is a photograph of strain-controlled direct shear test equipment.

The advantage of the strain-controlled tests is that, in the case of dense sand, peak shear resistance (that is, at failure) as well as lesser shear resistance (that is, at a point after failure called *ultimate strength*) can be observed and plotted. In stress-controlled tests, only peak shear resistance can be observed and plotted. Note that the peak shear resistance in stress-controlled tests can only be approximated. This is because failure occurs at a stress level somewhere between the prefailure load increment and the failure load increment. Nevertheless, stress-controlled tests probably simulate real field situations better than strain-controlled tests.

FIGURE 7.5 Direct shear test equipment (Courtesy of ELE International/Soiltest Products Division, Lake Bluff, Illinois)

For a given test, the normal stress can be calculated as

$$\sigma = \sigma' = \text{normal stress} = \frac{\text{normal force}}{\text{area of cross section of the specimen}} \tag{7.9}$$

The resisting shear stress for any shear displacement can be calculated as

$$\tau = \text{shear stress} = \frac{\text{resisting shear force}}{\text{area of cross section of the specimen}} \tag{7.10}$$

Figure 7.6 shows a typical plot of shear stress and change in the height of the specimen versus shear displacement for loose and dense sands. These observations were obtained from a strain-controlled test. The following generalizations can be made from Figure 7.6 regarding the variation of resisting shear stress with shear displacement:

1. In loose sand, the resisting shear stress increases with shear displacement until a failure shear stress of τ_f is reached. After that, the shear resistance

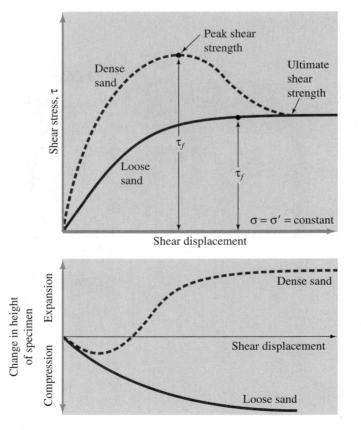

FIGURE 7.6 Plot of shear stress and change in height of specimen versus shear displacement for loose and dense dry sand (direct shear test)

remains approximately constant with any further increase in the shear displacement.

2. In dense sand, the resisting shear stress increases with shear displacement until it reaches a failure stress of τ_f. This τ_f is called the *peak shear strength*. After failure stress is attained, the resisting shear stress gradually decreases as shear displacement increases until it finally reaches a constant value called the *ultimate shear strength*.

Direct shear tests are repeated on similar specimens at various normal stresses. The normal stresses and the corresponding values of τ_f obtained from a number of tests are plotted on a graph, from which the shear strength parameters are determined. Figure 7.7 shows such a plot for tests on a dry sand. The equation for the average line obtained from experimental results is

$$\tau_f = \sigma' \tan \phi \tag{7.11}$$

(*Note:* $c = 0$ for sand and $\sigma = \sigma'$ for dry conditions.) So the friction angle

$$\phi = \tan^{-1}\left(\frac{\tau_f}{\sigma'}\right) \tag{7.12}$$

It is important to note that *in situ* cemented sands may show a c intercept.

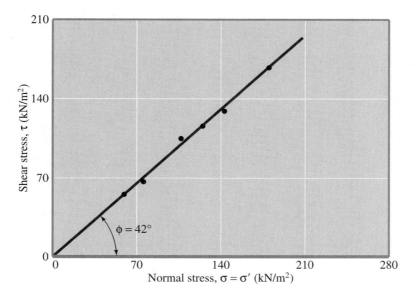

FIGURE 7.7 Determination of shear strength parameters for a dry sand using the results of direct shear tests

7.5 Drained Direct Shear Test on Saturated Sand and Clay

The shear box that contains the soil specimen is generally kept inside a container that can be filled with water to saturate the specimen. A *drained test* is made on a saturated soil specimen by keeping the rate of loading slow enough so that the excess pore water pressure generated in the soil completely dissipates by drainage. Pore water from the specimen is drained through two porous stones (see Figure 7.4).

Since the hydraulic conductivity of sand is high, the excess pore water pressure generated because of loading (normal and shear) is dissipated quickly. Hence, for an ordinary loading rate, essentially full drainage conditions exist. The friction angle ϕ obtained from a drained direct shear test of saturated sand will be the same as that for a similar specimen of dry sand.

The hydraulic conductivity of clay is very small compared with that of sand. When a normal load is applied to a clay soil specimen, a sufficient length of time must pass for full consolidation—that is, for dissipation of excess pore water pressure. For that reason, the shearing load has to be applied at a very slow rate. The test may last from 2 to 5 days. Figure 7.8 shows the results of a drained direct shear test on an overconsolidated clay. Figure 7.9 shows the plot of τ_f against σ' obtained from a number of drained direct shear tests on a normally consolidated and an overconsolidated clay. Note that $\sigma = \sigma'$ and the value of $c \approx 0$ for a normally consolidated clay.

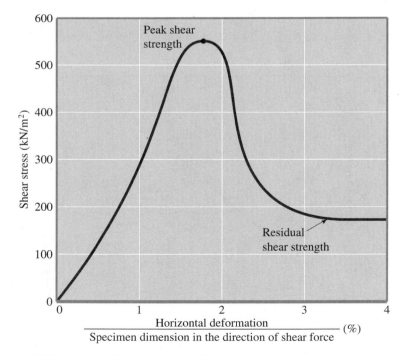

FIGURE 7.8 Results of a drained direct shear test on an overconsolidated clay

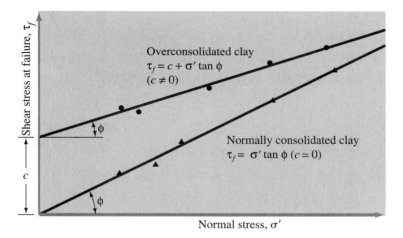

FIGURE 7.9 Failure envelope for clay obtained from drained direct
shear tests

**EXAMPLE
7.1**

Direct shear tests were performed on a dry, sandy soil. The size of the specimen
was 50 mm × 50 mm × 20 mm. Tests results were as given in the table.

Test no.	Normal force (N)	Normal stress,* $\sigma = \sigma'$ (kN/m²)	Shear force at failure (N)	Shear stress at failure,† τ_f (kN/m²)
1	90	36	54	21.6
2	135	54	82.35	32.9
3	315	126	189.5	75.8
4	450	180	270.5	108.2

$$* \; \sigma = \frac{\text{normal force}}{\text{area of specimen}} = \frac{\text{normal force} \times 10^{-3}\,\text{kN}}{50 \times 50 \times 10^{-6}\,\text{m}^2}$$

$$† \; \tau_f = \frac{\text{shear force}}{\text{area of specimen}} = \frac{\text{shear force} \times 10^{-3}\,\text{kN}}{50 \times 50 \times 10^{-6}\,\text{m}^2}$$

Find the shear stress parameters.

Solution The shear stresses, τ_f, obtained from the tests are plotted against the
normal stresses in Figure 7.10, from which we find $c = 0$, $\phi = 31°$. ∎

7.6 *Triaxial Shear Test*

The triaxial shear test is one of the most reliable methods available for determining
the shear strength parameters. It is widely used for both research and conventional

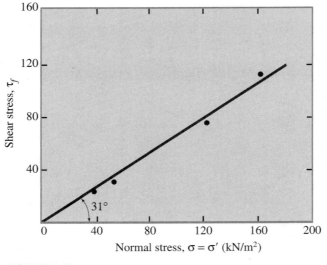

FIGURE 7.10

testing. The test is considered reliable for the following reasons:

1. It provides information on the stress–strain behavior of the soil that the direct shear test does not.
2. It provides more uniform stress conditions than the direct shear test does with its stress concentration along the failure plane.
3. It provides more flexibility in terms of loading path.

A diagram of the triaxial test layout is shown in Figure 7.11.

   In the triaxial shear test, a soil specimen about 36 mm in diameter and 76 mm long is generally used. The specimen is encased by a thin rubber membrane and placed inside a plastic cylindrical chamber that is usually filled with water or glycerine. The specimen is subjected to a confining pressure by compression of the fluid in the chamber. (Note that air is sometimes used as a compression medium.) To cause shear failure in the specimen, axial stress is applied through a vertical loading ram (sometimes called *deviator stress*). Stress is added in one of two ways:

1. Application of dead weights or hydraulic pressure in equal increments until the specimen fails. (Axial deformation of the specimen resulting from the load applied through the ram is measured by a dial gauge.)
2. Application of axial deformation at a constant rate by a geared or hydraulic loading press. This is a strain-controlled test. The axial load applied by the loading ram corresponding to a given axial deformation is measured by a proving ring or load cell attached to the ram.

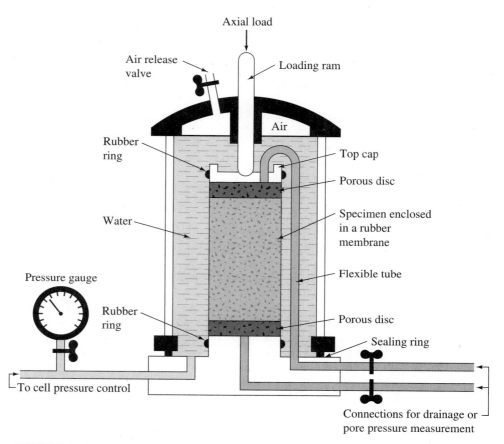

Axial load

Air release valve

Loading ram

Air

Rubber ring

Top cap

Porous disc

Water

Specimen enclosed in a rubber membrane

Pressure gauge

Flexible tube

Rubber ring

Porous disc

Sealing ring

To cell pressure control

Connections for drainage or pore pressure measurement

FIGURE 7.11 Diagram of triaxial test equipment (After Bishop and Bjerrum, 1960)

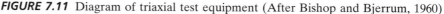

Connections to measure drainage into or out of the specimen, or to measure pressure in the pore water (as per the test conditions), are also provided. Three standard types of triaxial tests are generally conducted:

1. Consolidated-drained test or drained test (CD test)
2. Consolidated-undrained test (CU test)
3. Unconsolidated-undrained test or undrained test (UU test)

The general procedures and implications for each of the tests in *saturated soils* are described in the following sections.

7.7 Consolidated-Drained Test

In the consolidated-drained test, the specimen is first subjected to an all-around confining pressure, σ_3, by compression of the chamber fluid (Figure 7.12a). As

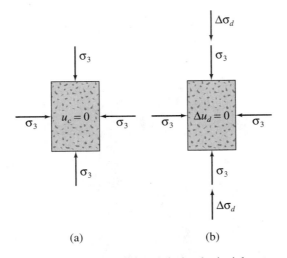

FIGURE 7.12 Consolidated-drained triaxial test:
(a) specimen under chamber confining pressure;
(b) deviator stress application

confining pressure is applied, the pore water pressure of the specimen increases by u_c. This increase in the pore water pressure can be expressed in the form of a nondimensional parameter:

$$B = \frac{u_c}{\sigma_3} \qquad (7.13)$$

where B = Skempton's pore pressure parameter (Skempton, 1954).

For saturated soft soils, B is approximately equal to 1; however, for saturated stiff soils, the magnitude of B can be less than 1. Black and Lee (1973) gave the theoretical values of B for various soils at complete saturation. These values are listed in Table 7.2.

When the connection to drainage is kept open, dissipation of the excess pore water pressure, and thus consolidation, will occur. With time, u_c will become equal to 0. In saturated soil, the change in the volume of the specimen (ΔV_c) that takes

Table 7.2 Theoretical values of B at complete saturation

Type of soil	Theoretical value
Normally consolidated soft clay	0.9998
Lightly overconsolidated soft clays and silts	0.9988
Overconsolidated stiff clays and sands	0.9877
Very dense sands and very stiff clays at high confining pressures	0.9130

place during consolidation can be obtained from the volume of pore water drained (Figure 7.13a). Then the deviator stress, $\Delta\sigma_d$, on the specimen is increased at a very slow rate (Figure 7.12b). The drainage connection is kept open, and the slow rate of deviator stress application allows complete dissipation of any pore water pressure that developed as a result ($\Delta u_d = 0$).

A typical plot of the variation of deviator stress against strain in loose sand and normally consolidated clay is shown in Figure 7.13b. Figure 7.13c shows a similar plot for dense sand and overconsolidated clay. The volume change, ΔV_d, of specimens that occurs because of the application of deviator stress in various soils is also shown in Figures 7.13d and e.

Since the pore water pressure developed during the test is completely dissipated, we have

total and effective confining stress $= \sigma_3 = \sigma_3'$

and

total and effective axial stress at failure $= \sigma_3 + (\Delta\sigma_d)_f = \sigma_1 = \sigma_1'$

In a triaxial test, σ_1' is the major principal effective stress at failure and σ_3' is the minor principal effective stress at failure.

Several tests on similar specimens can be conducted by varying the confining pressure. With the major and minor principal stresses at failure for each test, the Mohr's circles can be drawn and the failure envelopes can be obtained. Figure 7.14 shows the type of effective stress failure envelope obtained for tests in sand and normally consolidated clay. The coordinates of the point of tangency of the failure envelope with a Mohr's circle (that is, point *A*) give the stresses (normal and shear) on the failure plane of that test specimen.

Overconsolidation results when a clay is initially consolidated under an all-around chamber pressure of σ_c ($=\sigma_c'$) and is allowed to swell as the chamber pressure is reduced to σ_3 ($=\sigma_3'$). The failure envelope obtained from drained triaxial tests of such overconsolidated clay specimens shows two distinct branches (*ab* and *bc* in Figure 7.15). The portion *ab* has a flatter slope with a cohesion intercept, and the shear strength equation for this branch can be written as

$$\tau_f = c + \sigma' \tan \phi_1 \tag{7.14}$$

The portion *bc* of the failure envelope represents a normally consolidated stage of soil and follows the equation $\tau_f = \sigma' \tan \phi$.

A consolidated-drained triaxial test on a clayey soil may take several days to complete. The time is needed to apply deviator stress at a very slow rate to ensure full drainage from the soil specimen. For that reason, the CD type of triaxial test is not commonly used.

EXAMPLE 7.2

For a normally consolidated clay, these are the results of a drained triaxial test:

chamber confining pressure $= 112$ kN/m²

deviator stress at failure $= 175$ kN/m²

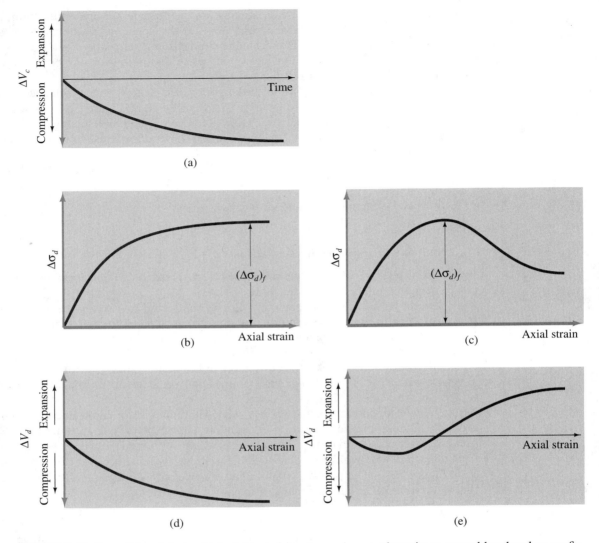

FIGURE 7.13 Consolidated-drained triaxial test: (a) volume change of specimen caused by chamber confining pressure; (b) plot of deviator stress against strain in the vertical direction for loose sand and normally consolidated clay; (c) plot of deviator stress against strain in the vertical direction for dense sand and overconsolidated clay; (d) volume change in loose sand and normally consolidated clay during deviator stress application; (e) volume change in dense sand and overconsolidated clay during deviator stress application

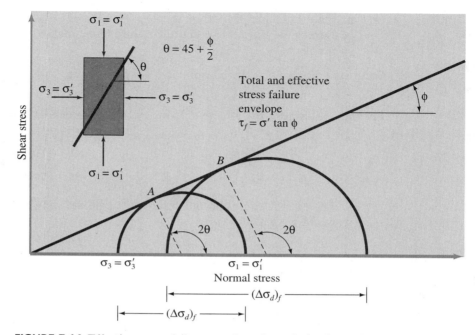

FIGURE 7.14 Effective stress failure envelope from drained tests in sand and normally consolidated clay

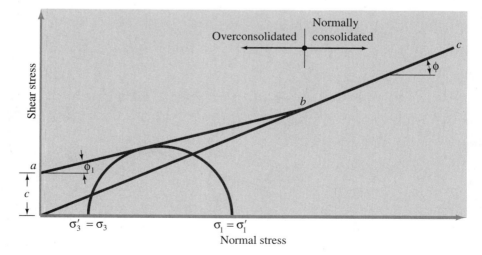

FIGURE 7.15 Effective stress failure envelope for overconsolidated clay

a. Find the angle of friction, ϕ.

b. Determine the angle θ that the failure plane makes with the major principal plane.

Solution   For a normally consolidated soil, the failure envelope equation is

$$\tau_f = \sigma' \tan \phi \qquad \text{(since } c = 0\text{)}$$

For the triaxial test, the effective major and minor principal stresses at failure are

$$\sigma_1' = \sigma_1 = \sigma_3 + (\Delta \sigma_d)_f = 112 + 175 = 287 \text{ kN/m}^2$$

and

$$\sigma_3' = \sigma_3 = 112 \text{ kN/m}^2$$

a. The Mohr's circle and the failure envelope are shown in Figure 7.16, from which we get

$$\sin \phi = \frac{AB}{OA} = \frac{\left(\dfrac{\sigma_1' - \sigma_3'}{2}\right)}{\left(\dfrac{\sigma_1' + \sigma_3'}{2}\right)}$$

or

$$\sin \phi = \frac{\sigma_1' - \sigma_3'}{\sigma_1' + \sigma_3'} = \frac{287 - 112}{287 + 112} = 0.438$$

$$\phi = \mathbf{26°}$$

b. $\theta = 45 + \dfrac{\phi}{2} = 45° + \dfrac{26}{2} = \mathbf{58°}$ ∎

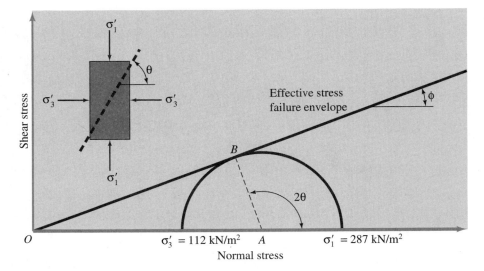

FIGURE 7.16

EXAMPLE 7.3   Refer to Example 7.2.

a. Find the normal stress, σ', and the shear stress, τ_f, on the failure plane.
b. Determine the effective normal stress on the plane of maximum shear stress.

Solution

a. From Figure 7.14, we can see that

$$\sigma' \text{ (on the failure plane)} = \frac{\sigma'_1 + \sigma'_3}{2} + \frac{\sigma'_1 - \sigma'_3}{2} \cos 2\theta \qquad \text{(a)}$$

and

$$\tau_f = \frac{\sigma'_1 - \sigma'_3}{2} \sin 2\theta \qquad \text{(b)}$$

Substituting the values of $\sigma'_1 = 287$ kN/m², $\sigma'_3 = 112$ kN/m², and $\theta = 58°$ into the preceding equations, we get

$$\sigma' = \frac{287 + 112}{2} + \frac{287 - 112}{2} \cos(2 \times 58) = \mathbf{161 \ kN/m^2}$$

and

$$\tau_f = \frac{287 - 112}{2} \sin(2 \times 58) = \mathbf{78.6 \ kN/m^2}$$

b. From Eq. (b), we can see that the maximum shear stress will occur on the plane with $\theta = 45°$. Substituting $\theta = 45°$ into Eq. (a) gives

$$\sigma' = \frac{287 + 112}{2} + \frac{287 - 112}{2} \cos 90 = \mathbf{199.5 \ kN/m^2} \qquad \blacksquare$$

EXAMPLE 7.4   The equation of the effective stress failure envelope for normally consolidated clayey soil is $\tau_f = \sigma' \tan 30°$. A drained triaxial test was conducted with the same soil at a chamber confining pressure of 70 kN/m². Calculate the deviator stress at failure.

Solution   For normally consolidated clay, $c = 0$. Thus, from Eq. (7.7), we have

$$\sigma'_1 = \sigma'_3 \tan^2 \left(45 + \frac{\phi}{2} \right)$$

$$\phi = 30°$$

$$\sigma_1' = 70 \tan^2 \left(45 + \frac{30}{2} \right) = 210 \text{ kN/m}^2$$

so

$$(\Delta\sigma_d)_f = \sigma_1' - \sigma_3' = 210 - 70 = \textbf{140 kN/m}^2 \qquad \blacksquare$$

EXAMPLE 7.5   We have the results of two drained triaxial tests on a saturated clay:

Specimen I: $\sigma_3 = 70 \text{ kN/m}^2$

$(\Delta\sigma_d)_f = 173 \text{ kN/m}^2$

Specimen II: $\sigma_3 = 105 \text{ kN/m}^2$

$(\Delta\sigma_d)_f = 235 \text{ kN/m}^2$

Determine the shear strength parameters.

Solution   Refer to Figure 7.17. For specimen I, the principal stresses at failure are

$$\sigma_3' = \sigma_3 = 70 \text{ kN/m}^2$$

and

$$\sigma_1' = \sigma_1 = \sigma_3 + (\Delta\sigma_d)_f = 70 + 173 = 243 \text{ kN/m}^2$$

Similarly, the principal stresses at failure for specimen II are

$$\sigma_3' = \sigma_3 = 105 \text{ kN/m}^2$$

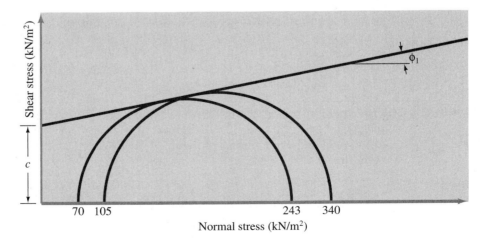

FIGURE 7.17

and

$$\sigma_1' = \sigma_1 = \sigma_3 + (\Delta\sigma_d)_f = 105 + 235 = 340 \text{ kN/m}^2$$

Using the relationship given by Eq. (7.7), we have

$$\sigma_1' = \sigma_3' \tan^2\left(45 + \frac{\phi_1}{2}\right) + 2c\tan\left(45 + \frac{\phi_1}{2}\right)$$

Thus, for specimen I,

$$243 = 70\tan^2\left(45 + \frac{\phi_1}{2}\right) + 2c\tan\left(45 + \frac{\phi_1}{2}\right)$$

and for specimen II,

$$340 = 105\tan^2\left(45 + \frac{\phi_1}{2}\right) + 2c\tan\left(45 + \frac{\phi_1}{2}\right)$$

Solving the two preceding equations, we obtain

$$\phi = 28° \qquad c = \mathbf{14.8 \text{ kN/m}^2} \qquad \blacksquare$$

7.8 *Consolidated-Undrained Test*

The consolidated-undrained test is the most common type of triaxial test. In this test, the saturated soil specimen is first consolidated by an all-round chamber fluid pressure, σ_3, that results in drainage. After the pore water pressure generated by the application of confining pressure is completely dissipated (that is, $u_c = B\sigma_3 = 0$), the deviator stress, $\Delta\sigma_d$, on the specimen is increased to cause shear failure. During this phase of the test, the drainage line from the specimen is kept closed. Since drainage is not permitted, the pore water pressure, Δu_d, will increase. During the test, measurements of $\Delta\sigma_d$ and Δu_d are made. The increase in the pore water pressure, Δu_d, can be expressed in a nondimensional form as

$$\overline{A} = \frac{\Delta u_d}{\Delta\sigma_d} \qquad\qquad (7.15)$$

where $\overline{A}$ = Skempton's pore pressure parameter (Skempton, 1954).

The general patterns of variation of $\Delta\sigma_d$ and Δu_d with axial strain for sand and clay soils are shown in Figures 7.18d, e, f, and g. In loose sand and normally consolidated clay, the pore water pressure increases with strain. In dense sand and overconsolidated clay, the pore water pressure increases with strain up to a certain limit, beyond which it decreases and becomes negative (with respect to the atmospheric pressure). This pattern is because the soil has a tendency to dilate.

Unlike in the consolidated-drained test, the total and effective principal stresses are not the same in the consolidated-undrained test. Since the pore water

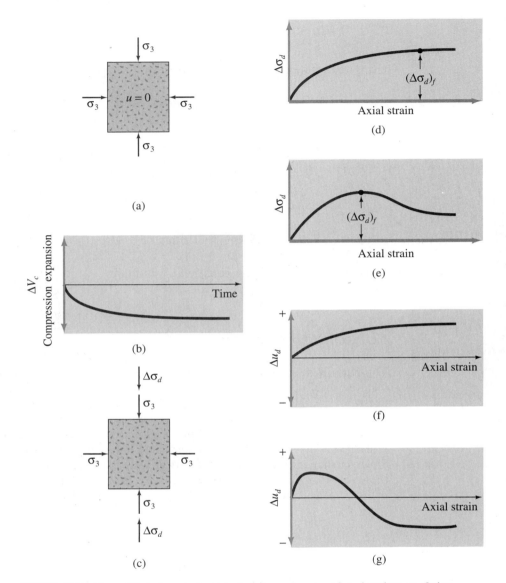

FIGURE 7.18 Consolidated-undrained test: (a) specimen under chamber confining pressure; (b) volume change in specimen caused by confining pressure; (c) deviator stress application; (d) deviator stress against axial strain for loose sand and normally consolidated clay; (e) deviator stress against axial strain for dense sand and overconsolidated clay; (f) variation of pore water pressure with axial strain for loose sand and normally consolidated clay; (g) variation of pore water pressure with axial strain for dense sand and overconsolidated clay

pressure at failure is measured in this test, the principal stresses may be analyzed as follows:

- Major principal stress at failure (total):

$$\sigma_3 + (\Delta\sigma_d)_f = \sigma_1$$

- Major principal stress at failure (effective):

$$\sigma_1 - (\Delta u_d)_f = \sigma_1'$$

- Minor principal stress at failure (total):

$$\sigma_3$$

- Minor principal stress at failure (effective):

$$\sigma_3 - (\Delta u_d)_f = \sigma_3'$$

where $(\Delta u_d)_f$ = pore water pressure at failure. The preceding derivations show that

$$\sigma_1 - \sigma_3 = \sigma_1' - \sigma_3'$$

Tests on several similar specimens with varying confining pressures may be done to determine the shear strength parameters. Figure 7.19 shows the total and effective stress Mohr's circles at failure obtained from consolidated-undrained triaxial tests in sand and normally consolidated clay. Note that A and B are two total

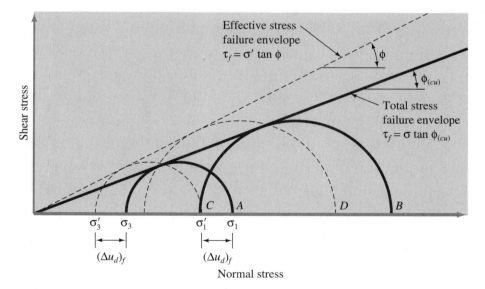

FIGURE 7.19 Total and effective stress failure envelopes for consolidated-undrained triaxial tests. (*Note:* The figure assumes that no back pressure is applied.)

stress Mohr's circles obtained from two tests. C and D are the effective stress Mohr's circles corresponding to total stress circles A and B, respectively. The diameters of circles A and C are the same; similarly, the diameters of circles B and D are the same.

In Figure 7.19, the total stress failure envelope can be obtained by drawing a line that touches all the total stress Mohr's circles. For sand and normally consolidated clays, this line will be approximately a straight line passing through the origin and may be expressed by the equation

$$\tau_f = \sigma \tan \phi_{(cu)} \tag{7.16}$$

where σ = total stress

   $\phi_{(cu)}$ = the angle that the total stress failure envelope makes with the normal stress axis, also known as the consolidated-undrained angle of shearing resistance

Equation (7.16) is seldom used for practical considerations.

Again referring to Figure 7.19, we see that the failure envelope that is tangent to all the effective stress Mohr's circles can be represented by the equation $\tau_f = \sigma' \tan \phi$, which is the same as the failure envelope obtained from consolidated-drained tests (see Figure 7.14).

In overconsolidated clays, the total stress failure envelope obtained from consolidated-undrained tests takes the shape shown in Figure 7.20. The straight line $a'b'$ is represented by the equation

$$\tau_f = c_{(cu)} + \sigma \tan \phi_{1(cu)} \tag{7.17}$$

and the straight line $b'c'$ follows the relationship given by Eq. (7.16). The effective stress failure envelope drawn from the effective stress Mohr's circles is similar to that shown in Figure 7.20.

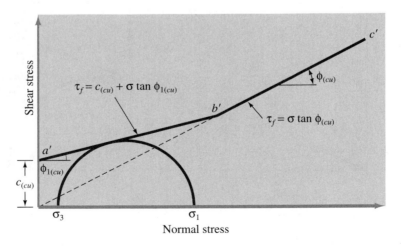

FIGURE 7.20 Total stress failure envelope obtained from consolidated-undrained tests in overconsolidated clay

Consolidated-drained tests on clay soils take considerable time. For that reason, consolidated-undrained tests can be conducted on such soils with pore pressure measurements to obtain the drained shear strength parameters. Since drainage is not allowed in these tests during the application of deviator stress, the tests can be performed rather quickly.

Skempton's pore water pressure parameter $\overline{A}$ was defined in Eq. (7.15). At failure, the parameter $\overline{A}$ can be written as

$$\overline{A} = \overline{A}_f = \frac{(\Delta u_d)_f}{(\Delta \sigma_d)_f} \qquad (7.18)$$

The general range of $\overline{A}_f$ values in most clay soils is as follows:

- Normally consolidated clays: 0.5 to 1
- Overconsolidated clays: −0.5 to 0

EXAMPLE 7.6

A consolidated-undrained test on a normally consolidated clay yielded the following results:

$\sigma_3 = 84 \ kN/m^2$

deviator stress, $(\Delta\sigma_d)_f = 63.7 \ kN/m^2$

pore pressure, $(\Delta u_d)_f = 47.6 \ kN/m^2$

Calculate the consolidated-undrained friction angle and the drained friction angle.

Solution   Refer to Figure 7.21.

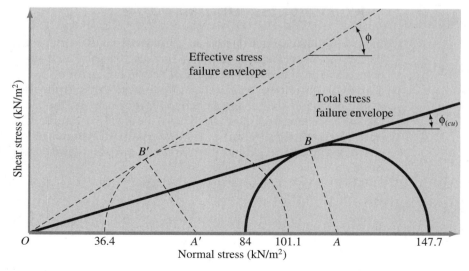

FIGURE 7.21

$$\sigma_3 = 84 \text{ kN/m}^2$$

$$\sigma_1 = \sigma_3 + (\Delta\sigma_d)_f = 84 + 63.7 = 147.7 \text{ kN/m}^2$$

$$\sigma_1 = \sigma_3 \tan^2\left(45 + \frac{\phi_{(cu)}}{2}\right)$$

$$147.7 = 84 \tan^2\left(45 + \frac{\phi_{(cu)}}{2}\right)$$

$$\phi_{(cu)} = 2\left[\tan^{-1}\left(\frac{147.7}{84}\right)^{0.5} - 45\right] = \mathbf{16°}$$

Again,

$$\sigma_3' = \sigma_3 - (\Delta u_d)_f = 84 - 47.6 = 36.4 \text{ kN/m}^2$$

$$\sigma_1' = \sigma_1 - (\Delta u_d)_f = 147.7 - 47.6 = 100.1 \text{ kN/m}^2$$

$$\sigma_1' = \sigma_3' \tan^2\left(45 + \frac{\phi}{2}\right)$$

$$100.1 = 36.4 \tan^2\left(45 + \frac{\phi}{2}\right)$$

$$\phi = 2\left[\tan^{-1}\left(\frac{100.1}{36.4}\right)^{0.5} - 45\right] = \mathbf{27.8°}$$ ∎

7.9   *Unconsolidated-Undrained Test*

In unconsolidated-undrained tests, drainage from the soil specimen is not permitted during the application of chamber pressure, σ_3. The test specimen is sheared to failure by the application of deviator stress, $\Delta\sigma_d$, with no drainage allowed. Since drainage is not allowed at any stage, the test can be performed very quickly. Because of the application of chamber confining pressure, σ_3, the pore water pressure in the soil specimen will increase by u_c. There will be a further increase in the pore water pressure, Δu_d, because of the deviator stress application. Hence, the total pore water pressure, u, in the specimen at any stage of deviator stress application can be given as

$$u = u_c + \Delta u_d \tag{7.19}$$

From Eqs. (7.13) and (7.15), we have $u_c = B\sigma_3$ and $\Delta u_d = \overline{A} \, \Delta\sigma_d$, so

$$u = B\sigma_3 + \overline{A} \, \Delta\sigma_d = B\sigma_3 + \overline{A}(\sigma_1 - \sigma_3) \tag{7.20}$$

The unconsolidated-undrained test is usually conducted on clay specimens and depends on a very important strength concept for saturated cohesive soils. The added axial stress at failure $(\Delta\sigma_d)_f$ is practically the same regardless of the chamber confining pressure. This result is shown in Figure 7.22. The failure envelope for the total stress Mohr's circles becomes a horizontal line and hence is called a $\phi = 0$ condition, and

$$\tau_f = c_u \tag{7.21}$$

where c_u is the undrained shear strength and is equal to the radius of the Mohr's circles.

The reason for obtaining the same added axial stress $(\Delta\sigma_d)_f$ regardless of the confining pressure is as follows: If a clay specimen (no. 1) is consolidated at a chamber pressure σ_3 and then sheared to failure with no drainage allowed, then the total stress conditions at failure can be represented by the Mohr's circle P in Figure 7.23. The pore pressure developed in the specimen at failure is equal to $(\Delta u_d)_f$. Thus, the major and minor principal effective stresses at failure are

$$\sigma'_1 = [\sigma_3 + (\Delta\sigma_d)_f] - (\Delta u_d)_f = \sigma_1 - (\Delta u_d)_f$$

and

$$\sigma'_3 = \sigma_3 - (\Delta u_d)_f$$

Q is the effective stress Mohr's circle drawn with the preceding principal stresses. Note that the diameters of circles P and Q are the same.

Now let us consider another similar clay specimen (no. 2) that is consolidated at a chamber pressure σ_3. If the chamber pressure is increased by $\Delta\sigma_3$ with no drainage allowed, then the pore water pressure increases by an amount Δu_c. For

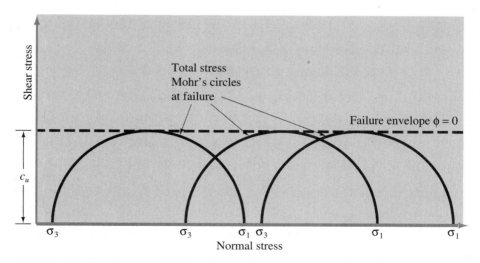

FIGURE 7.22 Total stress Mohr's circles and failure envelope ($\phi = 0$) obtained from unconsolidated-undrained triaxial tests

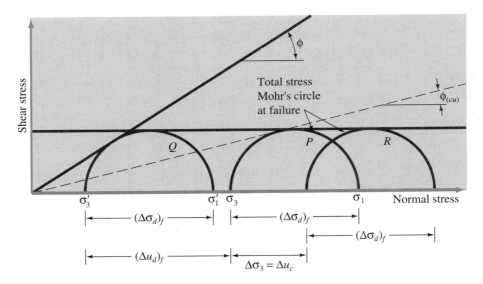

FIGURE 7.23 The $\phi = 0$ concept

saturated soils under isotropic stresses, the pore water pressure increase is equal to the total stress increase, so $\Delta u_c = \Delta \sigma_3$. At this time, the effective confining pressure is equal to $\sigma_3 + \Delta \sigma_3 - \Delta u_c = \sigma_3 + \Delta \sigma_3 - \Delta \sigma_3 = \sigma_3$. This is the same as the effective confining pressure of specimen no. 1 before the application of deviator stress. Hence, if specimen no. 2 is sheared to failure by increasing the axial stress, it should fail at the same deviator stress $(\Delta \sigma_d)_f$ that was obtained for specimen no. 1. The total stress Mohr's circle at failure will be R (Figure 7.23). The added pore pressure increase caused by the application of $(\Delta \sigma_d)_f$ will be $(\Delta u_d)_f$.

At failure, the minor principal effective stress is

$$[\sigma_3 + \Delta \sigma_3] - [\Delta u_c + (\Delta u_d)_f] = \sigma_3 - (\Delta u_d)_f = \sigma_3'$$

and the major principal effective stress is

$$[\sigma_3 + \Delta \sigma_3 + (\Delta \sigma_d)_f] - [\Delta u_c + (\Delta u_d)_f] = [\sigma_3 + (\Delta \sigma_d)_f] - (\Delta u_d)_f$$

$$= \sigma_1 - (\Delta u_d)_f = \sigma_1'$$

Thus, the effective stress Mohr's circle will still be Q because strength is a function of effective stress. Note that the diameters of circles P, Q, and R are all the same.

Any value of $\Delta \sigma_3$ could have been chosen for testing specimen no. 2. In any case, the deviator stress $(\Delta \sigma_d)_f$ to cause failure would have been the same.

7.10   *Unconfined Compression Test on Saturated Clay*

The unconfined compression test is a special type of unconsolidated-undrained test that is commonly used for clay specimens. In this test, the confining pressure σ_3 is

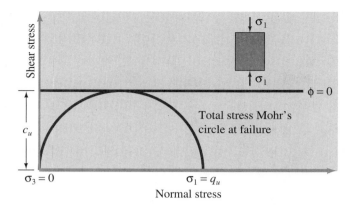

FIGURE 7.24 Unconfined compression test

0. An axial load is rapidly applied to the specimen to cause failure. At failure, the total minor principal stress is 0 and the total major principal stress is σ_1 (Figure 7.24). Since the undrained shear strength is independent of the confining pressure, we have

$$\tau_f = \frac{\sigma_1}{2} = \frac{q_u}{2} = c_u \tag{7.22}$$

where q_u is the *unconfined compression strength*. Table 7.3 gives the approximate consistencies of clays based on their unconfined compression strengths. A photograph of unconfined compression test equipment is shown in Figure 7.25.

Theoretically, for similar saturated clay specimens, the unconfined compression tests and the unconsolidated-undrained triaxial tests should yield the same values of c_u. In practice, however, unconfined compression tests on saturated clays

Table 7.3 General relationship of consistency and unconfined compression strength of clays

Consistency	q_u (kN/m²)
Very soft	0–25
Soft	25–50
Medium	50–100
Stiff	100–200
Very stiff	200–400
Hard	>400

FIGURE 7.25 Unconfined compression test equipment
(Courtesy of ELE International/Soiltest Products
Division, Lake Bluff, Illinois)

yield slightly lower values of c_u than those obtained from unconsolidated-undrained tests. This fact is demonstrated in Figure 7.26.

7.11   *Sensitivity and Thixotropy of Clay*

For many naturally deposited clay soils, the unconfined compression strength is greatly reduced when the soils are tested after remolding without any change in the moisture content, as shown in Figure 7.27. This property of clay soils is called *sensitivity*. The degree of sensitivity may be defined as the ratio of the unconfined compression strength in an undisturbed state to that in a remolded state, or

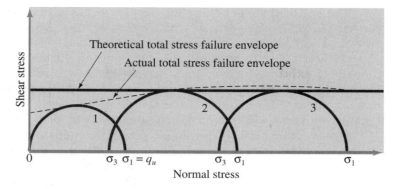

FIGURE 7.26 Comparison of results of unconfined compression tests and unconsolidated-undrained tests for a saturated clay soil. (*Note:* Mohr's circle no. 1 is for unconfined compression test; Mohr's circles no. 2 and 3 are for unconsolidated-undrained triaxial tests.)

$$S_t = \frac{q_{u(\text{undisturbed})}}{q_{u(\text{remolded})}} \tag{7.23}$$

The sensitivity ratio of most clays ranges from about 1 to 8; however, highly flocculent marine clay deposits may have sensitivity ratios ranging from about 10 to 80. There are also some clays that turn to viscous fluids upon remolding. These clays are found mostly in the previously glaciated areas of North America and Scandinavia and are referred to as "quick" clays.

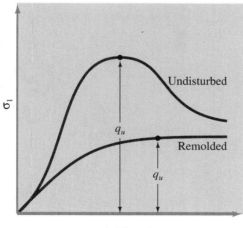

FIGURE 7.27 Unconfined compression strength for undisturbed and remolded clay

The loss of strength of clay soils from remolding is primarily caused by the destruction of the clay particle structure that was developed during the original process of sedimentation. If, however, after remolding, a soil specimen is kept in an undisturbed state (that is, without any change in the moisture content), it will continue to gain strength with time. This phenomenon is referred to as *thixotropy*. Thixotropy is a time-dependent reversible process in which materials under constant composition and volume soften when remolded. This loss of strength is gradually regained with time when the materials are allowed to rest.

Most soils are partially thixotropic; part of the strength loss caused by remolding is never regained with time. For soils, the difference between the undisturbed strength and the strength after thixotropic hardening can be attributed to the destruction of the clay-particle structure that was developed during the original process of sedimentation.

7.12   *Vane Shear Test*

Fairly reliable results for the *in situ* undrained shear strength c_u ($\phi = 0$ concept) of very plastic cohesive soils may be obtained directly from vane shear tests. The shear vane usually consists of four thin, equal-sized steel plates welded to a steel torque rod (Figure 7.28). First, the vane is pushed into the soil. Then torque is applied at the top of the torque rod to rotate the vane at a uniform speed. A cylinder of soil of height h and diameter d will resist the torque until the soil fails. The undrained shear strength of the soil can be calculated as follows.

If T is the maximum torque applied at the head of the torque rod to cause

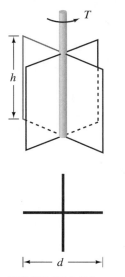

FIGURE 7.28 Diagram of vane shear equipment

failure, it should be equal to the sum of the resisting moment of the shear force along the side surface of the soil cylinder (M_s) and the resisting moment of the shear force at each end (M_e) (Figure 7.29):

$$T = M_s + \underbrace{M_e + M_e}_{\text{Two ends}} \tag{7.24}$$

The resisting moment M_s can be given as

$$M_s = \underbrace{(\pi dh)c_u}_{\substack{\text{Surface} \\ \text{area}}} \ \underbrace{(d/2)}_{\substack{\text{Moment} \\ \text{arm}}} \tag{7.25}$$

where d = diameter of the shear vane
 h = height of the shear vane

For calculating M_e, investigators assume three types of distribution of shear strength mobilization at the ends of the soil cylinder:

1. *Triangular:* Shear strength mobilization is c_u at the periphery of the soil cylinder and decreases linearly to 0 at the center.
2. *Uniform:* Shear strength mobilization is constant (that is, c_u) from the periphery to the center of the soil cylinder.
3. *Parabolic:* Shear strength mobilization is c_u at the periphery of the soil cylinder and decreases parabolically to 0 at the center.

These variations in shear strength mobilization are shown in Figure 7.29b. In general, the torque, T, at failure can be expressed as

$$T = \pi c_u \left[\frac{d^2 h}{2} + \beta \frac{d^3}{4} \right] \tag{7.26}$$

or

$$c_u = \frac{T}{\pi \left[\dfrac{d^2 h}{2} + \beta \dfrac{d^3}{4} \right]} \tag{7.27}$$

where $\beta = \frac{1}{2}$ for triangular mobilization of undrained shear strength
 $\beta = \frac{2}{3}$ for uniform mobilization of undrained shear strength
 $\beta = \frac{3}{5}$ for parabolic mobilization of undrained shear strength

[Eq. (7.27) is usually referred to as Calding's equation.]

Vane shear tests can be conducted in the laboratory and in the field during soil exploration. The laboratory shear vane has dimensions of about 12.7 mm (diameter) and 25.4 mm (height). Figure 7.30 is a photograph of laboratory vane shear equipment. Field shear vanes with the following dimensions are used by the

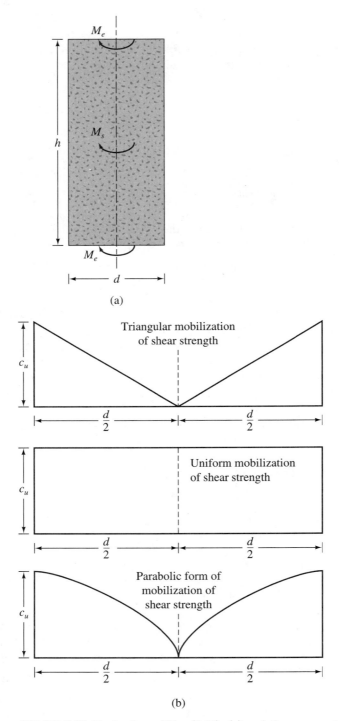

(a)

(b)

FIGURE 7.29 Derivation of Eq. (7.26): (a) resisting moment of shear force; (b) variations in shear strength mobilization

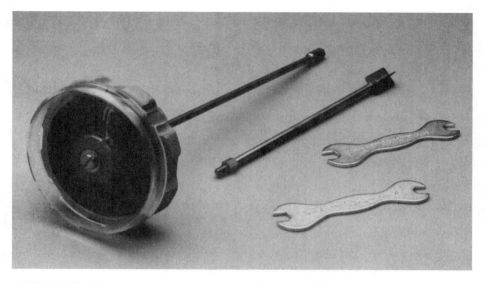

FIGURE 7.30 Laboratory vane shear device (Courtesy of ELE International/Soiltest Products Division, Lake Bluff, Illinois)

U.S. Bureau of Reclamation:

$$d = 50.8 \text{ mm}; \quad h = 101.6 \text{ mm}$$

$$d = 76.2 \text{ mm}; \quad h = 152.4 \text{ mm}$$

$$d = 101.6 \text{ mm}; \quad h = 203.2 \text{ mm}$$

In the field, where the undrained shear strength varies considerably with depth, vane shear tests are extremely useful. In a short period of time, it is possible to establish a reasonable pattern of the change of c_u with depth. However, if the clay deposit at a given site is more or less uniform, a few unconsolidated-undrained triaxial tests on undisturbed samples will give a reasonable estimation of soil parameters for design work. Vane shear tests are also limited by the strength of soils in which they can be used. The undrained shear strength obtained from a vane shear test is also dependent on the rate of application of torque T.

Bjerrum (1974) showed that as the plasticity of soils increases, c_u obtained from vane shear tests may give unsafe results for foundation design. For that reason, he suggested the following correction:

$$c_{u(\text{design})} = \lambda c_{u(\text{vane shear})} \tag{7.28}$$

where λ = correction factor = $1.7 - 0.54 \log(PI)$ $\qquad$ (7.29)
PI = plasticity index

More recently, Morris and Williams (1994) gave the following correlations of λ:

$$\lambda = 1.18e^{-0.08(PI)} + 0.57 \qquad \text{(for } PI > 5\text{)} \tag{7.30}$$

and

$$\lambda = 7.01e^{-0.08(LL)} + 0.57 \qquad \text{(for } LL > 20\text{)} \tag{7.31}$$

where LL = liquid limit (%).

7.13 *Empirical Relationships Between Undrained Cohesion (c_u) and Effective Overburden Pressure (σ'_o)*

Several empirical relationships can be observed between c_u and the effective overburden pressure, σ'_o, in the field. Some of these relationships are summarized in Table 7.4.

The overconsolidation ratio was defined in Chapter 6 as

$$OCR = \frac{\sigma'_c}{\sigma'_o} \tag{7.32}$$

where σ'_c = preconsolidation pressure.

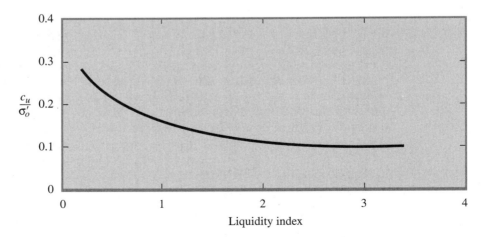

FIGURE 7.31 Variation of c_u/σ'_o with liquidity index (Based on Bjerrum and Simons, 1960)

Table 7.4 Empirical equations related to c_u and σ'_o

Reference	Relationship	Remarks
Skempton (1957)	$\dfrac{c_{u(\text{VST})}}{\sigma'_o} = 0.11 + 0.0037(PI)$ PI = plasticity index (%) $c_{u(\text{VST})}$ = undrained shear strength from vane shear test	For normally consolidated clay
Chandler (1988)	$\dfrac{c_{u(\text{VST})}}{\sigma'_c} = 0.11 + 0.0037(PI)$ σ'_c = preconsolidation pressure	Can be used in overconsolidated soil; accuracy ±25%; not valid for sensitive and fissured clays
Jamiolkowski et al. (1985)	$\dfrac{c_u}{\sigma'_c} = 0.23 \pm 0.04$	For lightly overconsolidated clays
Mesri (1989)	$\dfrac{c_u}{\sigma'_o} = 0.22$	
Bjerrum and Simons (1960)	$\dfrac{c_u}{\sigma'_o} = f(LI)$ LI = liquidity index	See Figure 7.31; for normally consolidated clays
Ladd et al. (1977)	$\dfrac{\left(\dfrac{c_u}{\sigma'_o}\right)_{\text{overconsolidated}}}{\left(\dfrac{c_u}{\sigma'_o}\right)_{\text{normally consolidated}}} = (OCR)^{0.8}$ OCR = overconsolidation ratio	

Problems

7.1 A direct shear test was conducted on a specimen of dry sand with a normal stress of 140 kN/m². Failure occurred at a shear stress of 94.5 kN/m². The size of the specimen tested was 50 mm × 50 mm × 25 mm (height). Determine the angle of friction, ϕ. For a normal stress of 84 kN/m², what shear force would be required to cause failure in the specimen?

7.2 The size of a sand specimen in a direct shear test was 50 mm × 50 mm × 30 mm (height). It is known that, for the sand, tan $\phi = 0.65/e$ (where e = void ratio) and the specific gravity of solids $G_s = 2.65$. During the test, a normal stress of 140 kN/m² was applied. Failure occurred at a shear stress of 105 kN/m². What was the weight of the sand specimen?

7.3 The angle of friction of a compacted dry sand is 38°. In a direct shear test on the sand, a normal stress of 84 kN/m² was applied. The size of the specimen was 50 mm × 50 mm × 30 mm (height). What shear force (in kN) will cause failure?

7.4 Repeat Problem 7.3 with the following changes:

friction angle = 37°

normal stress = 150 kN/m²

7.5 Following are the results of four drained direct shear tests on a normally consolidated clay:

diameter of specimen = 50 mm

height of specimen = 25 mm

Test no.	Normal force (N)	Shear force at failure (N)
1	271	120.6
2	406.25	170.64
3	474	204.1
4	541.65	244.3

Draw a graph of the shear stress at failure versus the normal stress. Determine the drained angle of friction from the graph.

7.6 The relationship between the relative density, D_r, and the angle of friction, ϕ, of a sand can be given as $\phi° = 25 + 0.18D_r$ (D_r in %). A drained triaxial test on the same sand was conducted with a chamber confining pressure of 105 kN/m². The relative density of compaction was 45%. Calculate the major principal stress at failure.

7.7 Consider the triaxial test described in Problem 7.6.
   a. Estimate the angle that the failure plane makes with the major principal plane.
   b. Determine the normal and shear stresses (when the specimen failed) on a plane that makes an angle of 30° with the major principal plane.

7.8 The effective stress failure envelope of a sand can be given as $\tau_f = \sigma'\tan 41°$. A drained triaxial test was conducted on the same sand. The specimen failed when the deviator stress was 400.5 kN/m². What was the chamber confining pressure during the test?

7.9 Refer to Problem 7.8.
   a. Estimate the angle that the failure plane makes with the minor principal plane.
   b. Determine the normal stress and the shear stress on a plane that makes an angle of 35° with the minor principal plane.

7.10 For a normally consolidated clay, the results of a drained triaxial test are as follows:

chamber confining pressure = 150 kN/m²

deviator stress at failure = 275 kN/m²

Determine the soil friction angle, ϕ.

7.11 For a normally consolidated clay, we are given $\phi = 25°$. In a drained triaxial test, the specimen failed at a deviator stress of 154 kN/m^2. What was the chamber confining pressure, σ_3?

7.12 A consolidated-drained triaxial test was conducted on a normally consolidated clay. The results were as follows:

$$\sigma_3 = 276 \text{ kN/m}^2$$

$$(\Delta\sigma_d)_f = 276 \text{ kN/m}^2$$

   a. Find the angle of friction, ϕ.
   b. What is the angle θ that the failure plane makes with the major principal stress?
   c. Determine the normal stress σ' and the shear stress τ_f on the failure plane.

7.13 Refer to Problem 7.12.
   a. Determine the effective normal stress on the plane of maximum shear stress.
   b. Explain why the shear failure took place along the plane as determined in part (b) and not along the plane of maximum shear stress.

7.14 The results of two drained triaxial tests on a saturated clay are given here:
   Specimen I: chamber confining pressure = 69 kN/m^2
              deviator stress at failure = 213 kN/m^2
   Specimen II: chamber confining pressure = 120 kN/m^2
              deviator stress at failure = 258.7 kN/m^2
   Calculate the shear strength parameters of the soil.

7.15 A sandy soil has a drained angle of friction of 36°. In a drained triaxial test on the same soil, the deviator stress at failure is 268 kN/m^2. What is the chamber confining pressure?

7.16 A consolidated-undrained test was conducted on a normally consolidated specimen with a chamber confining pressure of 140 kN/m^2. The specimen failed while the deviator stress was 126 kN/m^2. The pore water pressure in the specimen at that time was 76.3 kN/m^2. Determine the consolidated-undrained and the drained friction angles.

7.17 Repeat Problem 7.16 with the following values:

$$\sigma_3 = 84 \text{ kN/m}^2$$

$$(\Delta\sigma_d)_f = 58.7 \text{ kN/m}^2$$

$$(\Delta u_d)_f = 39.2 \text{ kN/m}^2$$

7.18 The shear strength of a normally consolidated clay can be given by the equation $\tau_f = \sigma' \tan 28°$. A consolidated-undrained triaxial test was conducted on the clay. Following are the results of the test:

   chamber confining pressure = 105 kN/m^2

   deviator stress at failure = 97 kN/m^2

    a. Determine the consolidated-undrained friction angle, $\phi_{(cu)}$.

    b. What is the pore water pressure developed in the clay specimen at failure?

7.19 For the clay specimen described in Problem 7.18, what would have been the deviator stress at failure if a drained test had been conducted with the same chamber confining pressure (that is, $\sigma_3 = 105 \text{ kN/m}^2$)?

7.20 For a clay soil, we are given $\phi = 28°$ and $\phi_{(cu)} = 18°$. A consolidated-undrained triaxial test was conducted on this clay soil with a chamber confining pressure of 105 kN/m^2. Determine the deviator stress and the pore water pressure at failure.

7.21 During a consolidated-undrained triaxial test on a clayey soil specimen, the minor and major principal stresses at failure were 96 kN/m^2 and 187 kN/m^2, respectively. What will be the axial stress at failure if a similar specimen is subjected to an unconfined compression test?

7.22 The friction angle, ϕ, of a normally consolidated clay specimen collected during field exploration was determined from drained triaxial tests to be $22°$. The unconfined compression strength, q_u, of a similar specimen was found to be 120 kN/m^2. Determine the pore water pressure at failure for the unconfined compression test.

7.23 Repeat Problem 7.22 with $\phi = 25°$ and $q_u = 121.5 \text{ kN/m}^2$.

References

Bishop, A. W., and Bjerrum, L. (1960). "The Relevance of the Triaxial Test to the Solution of Stability Problems," *Proceedings,* Research Conference on Shear Strength of Cohesive Soils, ASCE, 437–501.

Bjerrum, L. (1974). "Problems of Soil Mechanics and Construction on Soft Clays," Norwegian Geotechnical Institute, *Publications No. 110,* Oslo.

Bjerrum, L., and Simons, N. E. (1960). "Compression of Shear Strength Characteristics of Normally Consolidated Clay," *Proceedings,* Research Conference on Shear Strength of Cohesive Soils, ASCE, 711–726.

Black, D. K., and Lee, K. L. (1973). "Saturating Laboratory Samples by Back Pressure," *Journal of the Soil Mechanics and Foundations Division,* ASCE, Vol. 99, No. SM1, 75–93.

Chandler, R. J. (1988). "The *in situ* Measurement of the Undrained Shear Strength of Clays Using the Field Vane," *STP 1014, Vane Shear Strength Testing in Soils: Field and Laboratory Studies,* ASTM, 13–44.

Coulomb, C. A. (1776). "Essai sur une application des regles de Maximums et Minimis á quelques Problèmes de Statique, relatifs á l'Architecture," *Memoires de Mathematique et de Physique,* Présentés, á l'Academie Royale des Sciences, Paris, Vol. 3, 38.

Jamiolkowski, M., Ladd, C. C., Germaine, J. T., and Lancellotta, R. (1985). "New Developments in Field and Laboratory Testing of Soils," *Proceedings,* XIth International Conference on Soil Mechanics and Foundation Engineering, San Francisco, Vol. 1, 57–153.

Ladd, C. C., Foote, R., Ishihara, K., Schlosser, F., and Poulos, H. G. (1977). "Stress Deformation and Strength Characteristics," *Proceedings,* 9th International Conference on Soil Mechanics and Foundation Engineering, Tokyo, Vol. 2, 421–494.

Mesri, G. (1989). "A Re-evaluation of $s_{u(mob)} \approx 0.22\sigma_p$ Using Laboratory Shear Tests," *Canadian Geotechnical Journal,* Vol. 26, No. 1, 162–164.

Mohr, O. (1900). "Welche Umstände Bedingen die Elastizitätsgrenze und den Bruch eines Materiales?" *Zeitschrift des Vereines Deutscher Ingenieure,* Vol. 44, 1524–1530, 1572–1577.

Morris, P. M., and Williams, D. J. (1994). "Effective Stress Vane Shear Strength Correction Factor Correlations," *Canadian Geotechnical Journal,* Vol. 31, No. 3, 335–342.

Skempton, A. W. (1954). "The Pore Water Coefficients A and B," *Geotechnique,* Vol. 4, 143–147.

Skempton, A. W. (1957). "Discussion: The Planning and Design of New Hong Kong Airport," *Proceedings,* Institute of Civil Engineers, London, Vol. 7, 305–307.

8

Subsurface Exploration

The process of identifying the layers of deposits that underlie a proposed structure and their physical characteristics is generally referred to as *subsurface exploration*. The purpose of subsurface exploration is to obtain information that will aid the geotechnical engineer in these tasks:

1. Selecting the type and depth of foundation suitable for a given structure
2. Evaluating the load-bearing capacity of the foundation
3. Estimating the probable settlement of a structure
4. Determining potential foundation problems (for example, expansive soil, collapsible soil, sanitary landfill, and so on)
5. Determining the location of the water table
6. Predicting lateral earth pressure for structures such as retaining walls, sheet pile bulkheads, and braced cuts
7. Establishing construction methods for changing subsoil conditions

Subsurface exploration is also necessary for underground construction and excavation. It may be required when additions or alterations to existing structures are contemplated.

8.1 Subsurface Exploration Program

Subsurface exploration comprises several steps, including collection of preliminary information, reconnaissance, and site investigation.

Collection of Preliminary Information

Information must be obtained regarding the type of structure to be built and its general use. For the construction of buildings, the approximate column loads and their spacing and the local building-code and basement requirements should be known. The construction of bridges requires determining span length and the loading on piers and abutments.

A general idea of the topography and the type of soil to be encountered near and around the proposed site can be obtained from the following sources:

1. U.S. Geological Survey maps
2. State government geological survey maps
3. U.S. Department of Agriculture's Soil Conservation Service county soil reports
4. Agronomy maps published by the agriculture departments of various states
5. Hydrological information published by the U.S. Corps of Engineers, including the records of stream flow, high flood levels, tidal records, and so on
6. Highway department soils manuals published by several states

The information collected from these sources can be extremely helpful to those planning a site investigation. In some cases, substantial savings are realized by anticipating problems that may be encountered later in the exploration program.

Reconnaissance

The engineer should always make a visual inspection of the site to obtain information about these features:

1. The general topography of the site and the possible existence of drainage ditches, abandoned dumps of debris, or other materials. Also, evidence of creep of slopes and deep, wide shrinkage cracks at regularly spaced intervals may be indicative of expansive soils.
2. Soil stratification from deep cuts, such as those made for construction of nearby highways and railroads.
3. Type of vegetation at the site, which may indicate the nature of the soil. For example, a mesquite cover in central Texas may indicate the existence of expansive clays that can cause possible foundation problems.
4. High-water marks on nearby buildings and bridge abutments.
5. Groundwater levels, which can be determined by checking nearby wells.
6. Types of construction nearby and existence of any cracks in walls or other problems.

The nature of stratification and physical properties of the soil nearby can also be obtained from any available soil-exploration reports for nearby existing structures.

Site Investigation

The site investigation phase of the exploration program consists of planning, making test boreholes, and collecting soil samples at desired intervals for subsequent observation and laboratory tests. The approximate required minimum depth of the borings should be predetermined; however, the depth can be changed during the drilling

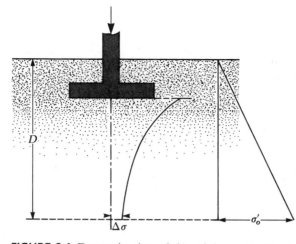

FIGURE 8.1 Determination of the minimum depth of boring

operation, depending on the subsoil encountered. To determine the approximate minimum depth of boring for foundations, engineers may use the rules established by the American Society of Civil Engineers (1972):

1. Determine the net increase of stress, $\Delta\sigma$, under a foundation with depth as shown in Figure 8.1. (The general equations for estimating stress increase are given in Chapter 5.)
2. Estimate the variation of the vertical effective stress, σ_o', with depth.
3. Determine the depth, $D = D_1$, at which the stress increase $\Delta\sigma$ is equal to $\left(\frac{1}{10}\right) q$ (q = estimated net stress on the foundation).
4. Determine the depth, $D = D_2$, at which $\Delta\sigma/\sigma_o' = 0.05$.
5. Unless bedrock is encountered, the smaller of the two depths, D_1 and D_2, just determined is the approximate minimum depth of boring required.

If the preceding rules are used, the depths of boring for a building with a width of 30 m will be approximately as listed in Table 8.1, according to Sowers and

Table 8.1 Approximate depths of borings for buildings with a width of 30 m

No. of stories	Boring depth (m)
1	3.5
2	6
3	10
4	16
5	24

Table 8.2 Approximate spacing of boreholes

Type of project	Spacing (m)
Multistory building	10–30
One-story industrial plants	20–60
Highways	250–500
Residential subdivision	250–500
Dams and dikes	40–80

Sowers (1970). For hospitals and office buildings, they also use the following rule to determine boring depth:

$$D_b = 3S^{0.7} \qquad \text{(for light steel or narrow concrete buildings)} \qquad (8.1)$$

$$D_b = 6S^{0.7} \qquad \text{(for heavy steel or wide concrete buildings)} \qquad (8.2)$$

where D_b = depth of boring (m)
S = number of stories

When deep excavations are anticipated, the depth of boring should be at least 1.5 times the depth of excavation.

Sometimes subsoil conditions require that the foundation load be transmitted to bedrock. The minimum depth of core boring into the bedrock is about 3 m. If the bedrock is irregular or weathered, the core borings may have to be deeper.

There are no hard and fast rules for borehole spacing. Table 8.2 gives some general guidelines. The spacing can be increased or decreased, depending on the subsoil condition. If various soil strata are more or less uniform and predictable, fewer boreholes are needed than in nonhomogeneous soil strata.

The engineer should also take into account the ultimate cost of the structure when making decisions regarding the extent of field exploration. The exploration cost generally should be 0.1% to 0.5% of the cost of the structure.

8.2   *Exploratory Borings in the Field*

Soil borings can be made by several methods, including auger boring, wash boring, percussion drilling, and rotary drilling.

Auger boring is the simplest method of making exploratory boreholes. Figure 8.2 shows two types of hand auger: the *post hole auger* and the *helical auger*. Hand augers cannot be used for advancing holes to depths exceeding 3–5 m; however, they can be used for soil exploration work for some highways and small structures. *Portable power-driven helical augers* (30 to 75 mm in diameter) are available for making deeper boreholes. The soil samples obtained from such borings are highly disturbed. In some noncohesive soils or soils that have low cohesion, the walls of

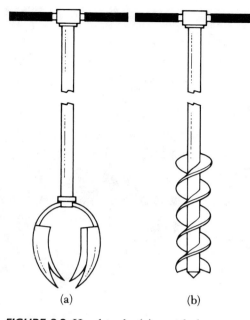

FIGURE 8.2 Hand tools: (a) post hole auger;
(b) helical auger

the boreholes will not stand unsupported. In such circumstances, a metal pipe is used as a *casing* to prevent the soil from caving in.

When power is available, *continuous-flight augers* are probably the most common method used for advancing a borehole. The power for drilling is delivered by truck- or tractor-mounted drilling rigs. Boreholes up to about 60–70 m can be made easily by this method. Continuous-flight augers are available in sections of about 1–2 m with either a solid or hollow stem. Some of the commonly used solid stem augers have outside diameters of 67 mm, 83 mm, 102 mm, and 114 mm. Hollow stem augers commercially available have dimensions of 64 mm inside diameter (ID) and 158 mm outside diameter (OD), 70 mm ID and 178 mm OD, 76 mm ID and 203 mm OD, and 83 mm ID and 229 mm OD.

The tip of the auger is attached to a cutter head. During the drilling operation (Figure 8.3), section after section of auger can be added and the hole extended downward. The flights of the augers bring the loose soil from the bottom of the hole to the surface. The driller can detect changes in soil type by noting changes in the speed and sound of drilling. When solid stem augers are used, the auger must be withdrawn at regular intervals to obtain soil samples and also to conduct other operations such as standard penetration tests. Hollow stem augers have a distinct advantage over solid stem augers in that they do not have to be removed frequently for sampling or other tests. The outside of the hollow stem auger acts like a casing. A removable plug is attached to the bottom of the auger with a center rod. During the drilling, the plug can be pulled out with the auger in place, and soil

FIGURE 8.3 Drilling with continuous-flight augers (Courtesy of Danny R. Anderson, Danny R. Anderson Consultants, El Paso, Texas)

sampling and standard penetration tests can be performed. When hollow stem augers are used in sandy soils below the water table, the sand may be pushed several meters into the stem of the auger by excess hydrostatic pressure immediately after the plug is removed. Under such conditions, the plug should not be used. Instead, water inside the hollow stem should be maintained at a level higher than the water table.

Wash boring is another method of advancing boreholes. In this method, a casing about 2–3 m long is driven into the ground. The soil inside the casing is then removed using a chopping bit attached to a drilling rod. Water is forced through the drilling rod and exits at a very high velocity through the holes at the bottom of the chopping bit (Figure 8.4). The water and the chopped soil particles rise in the drill hole and overflow at the top of the casing through a T connection. The washwater is collected in a container. The casing can be extended with additional pieces as the borehole progresses; however, that is not required if the borehole will stay open and not cave in.

Rotary drilling is a procedure by which rapidly rotating drilling bits attached to the bottom of drilling rods cut and grind the soil and advance the borehole. Rotary drilling can be used in sand, clay, and rocks (unless badly fissured). Water, or *drilling mud,* is forced down the drilling rods to the bits, and the return flow forces the cuttings to the surface. Boreholes with diameters of 50–200 mm can be

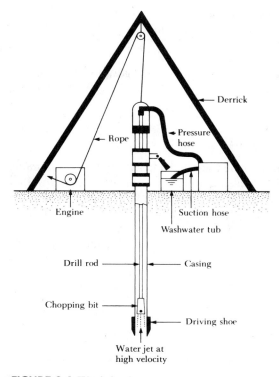

FIGURE 8.4 Wash boring

made easily by this technique. The drilling mud is a slurry of water and bentonite. Generally, rotary drilling is used when the soil encountered is likely to cave in. When soil samples are needed, the drilling rod is raised and the drilling bit is replaced by a sampler.

Percussion drilling is an alternative method of advancing a borehole, particularly through hard soil and rock. A heavy drilling bit is raised and lowered to chop the hard soil. The chopped soil particles are brought up by the circulation of water. Percussion drilling may require casing.

8.3 *Procedures for Sampling Soil*

Two types of soil samples can be obtained during subsurface exploration: *disturbed* and *undisturbed*. Disturbed, but representative, samples can generally be used for the following types of laboratory test:

1. Grain-size analysis
2. Determination of liquid and plastic limits

   3. Specific gravity of soil solids
   4. Organic content determination
   5. Classification of soil

Disturbed soil samples, however, cannot be used for consolidation, hydraulic conductivity, or shear strength tests. Undisturbed soil samples must be obtained for these laboratory tests.

Split-Spoon Sampling

Split-spoon samplers can be used in the field to obtain soil samples that are generally disturbed but still representative. A section of a *standard split-spoon sampler* is shown in Figure 8.5a. It consists of a tool-steel driving shoe, a steel tube that is split longitudinally in half, and a coupling at the top. The coupling connects the sampler to the drill rod. The standard split tube has an inside diameter of 34.93 mm and an outside diameter of 50.8 mm; however, samplers that have inside and outside diameters up to 63.5 mm and 76.2 mm, respectively, are also available.

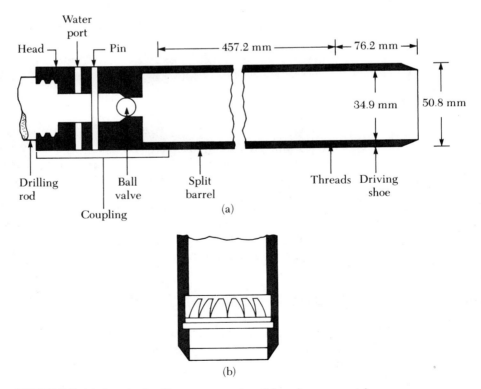

FIGURE 8.5 (a) Standard split-spoon sampler; (b) spring core catcher

When a borehole is extended to a predetermined depth, the drill tools are removed and the sampler is lowered to the bottom of the borehole. The sampler is driven into the soil by hammer blows to the top of the drill rod. The standard weight of the hammer is 62.3 N and, for each blow, the hammer drops a distance of 762 mm. The number of blows required for spoon penetration of three 152.4-mm intervals is recorded. The numbers of blows required for the last two intervals are added to give the *standard penetration number* at that depth. This number is generally referred to as the *N value* (American Society for Testing and Materials, 1997, Designation D-1586). The sampler is then withdrawn, and the shoe and coupling are removed. The soil sample recovered from the tube is then placed in a glass bottle and transported to the laboratory.

The degree of disturbance for a soil sample is usually expressed as

$$A_R(\%) = \frac{D_o^2 - D_i^2}{D_i^2}(100)$$

(8.3)

where A_R = area ratio
D_o = outside diameter of the sampling tube
D_i = inside diameter of the sampling tube

When the area ratio is 10% or less, the sample is generally considered to be undisturbed.

Split-spoon samples generally are taken at intervals of about 1.5 m. When the material encountered in the field is sand (particularly fine sand below the water table), sample recovery by a split-spoon sampler may be difficult. In that case, a device such as a *spring core catcher* (Figure 8.5b) may have to be placed inside the split spoon.

Besides obtaining soil samples, standard penetration tests provide several useful correlations. For example, the consistency of clayey soils can often be estimated from the standard penetration number, N, as shown in Table 8.3. However,

Table 8.3 Consistency of clays and approximate correlation to the standard penetration number, N

Standard penetration number, N	Consistency	Unconfined compression strength, q_u (kN/m²)
0–2	Very soft	0–25
2–5	Soft	25–50
5–10	Medium stiff	50–100
10–20	Stiff	100–200
20–30	Very stiff	200–400
>30	Hard	>400

correlations for clays require tests to verify that the relationships are valid for the clay deposit being examined.

The literature contains many correlations between the standard penetration number and the undrained shear strength of clay, c_u. Based on the results of undrained triaxial tests conducted on insensitive clays, Stroud (1974) suggested that

$$c_u = KN \tag{8.4}$$

where K = constant = 3.5–6.5 kN/m²
N = standard penetration number obtained from the field

The average value of K is about 4.4 kN/m². Hara et al. (1971) also suggested that

$$c_u \, (\text{kN/m}^2) = 29N^{0.72} \tag{8.5}$$

The overconsolidation ratio, OCR, of a natural clay deposit can also be correlated with the standard penetration number. Based on the regression analysis of 110 data points, Mayne and Kemper (1988) obtained the relationship

$$OCR = 0.193 \left(\frac{N}{\sigma_o'} \right)^{0.689} \tag{8.6}$$

where σ_o' = effective vertical stress (MN/m²).

It is important to point out that any correlation between c_u and N is only approximate. The sensitivity, S_t, of clay soils also plays an important role in the actual N value obtained from the field. Figure 8.6 shows a plot of $N_{(\text{measured})}/N_{(\text{at } S_t=1)}$ versus S_t, as predicted by Schmertmann (1975).

In granular soils, the N value is affected by the effective overburden pressure, σ_o'. For that reason, the N value obtained from field exploration under different effective overburden pressures should be changed to correspond to a standard value of σ_o'; that is,

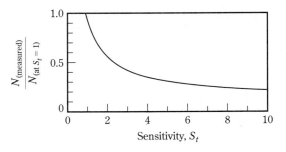

FIGURE 8.6 Variation of $N_{(\text{measured})}/N_{(\text{at } S_t=1)}$ with S_t of clays (Based on Schmertmann, 1975)

Table 8.4 Empirical relationships for C_N (*Note:* σ'_o is in kN/m²)

Source	C_N
Liao and Whitman (1986)	$9.78\sqrt{\dfrac{1}{\sigma'_o}}$
Skempton (1986)	$\dfrac{2}{1 + 0.01\sigma'_o}$
Seed et al. (1975)	$1 - 1.25\log\left(\dfrac{\sigma'_o}{95.6}\right)$
Peck et al. (1974)	$0.77\log\left(\dfrac{1912}{\sigma'_o}\right)$ for $\sigma'_o \geq 25$ kN/m²

$$N_{\text{cor}} = C_N N_F \tag{8.7}$$

where N_{cor} = corrected N value to a standard value of σ'_o (95.6 kN/m²)
C_N = correction factor
N_F = N value obtained from the field

A number of empirical relationships have been proposed for C_N. Some of the relationships are given in Table 8.4. The most commonly cited relationships are those given by Liao and Whitman (1986) and Skempton (1986).

An approximate relationship between the corrected standard penetration number and the relative density of sand is given in Table 8.5. These values are approximate, primarily because the effective overburden pressure and the stress history of the soil significantly influence the N_F values of sand. An extensive study

Table 8.5 Relation between the corrected N values and the relative density in sands

Standard penetration number, N_{cor}	Approximate relative density, D_r (%)
0–5	0–5
5–10	5–30
10–30	30–60
30–50	60–95

conducted by Marcuson and Bieganousky (1977) produced the empirical relationship

$$D_r\,(\%) = 11.7 + 0.76(222N_F + 1600 - 7.68\sigma_o' - 50C_u^2)^{0.5} \qquad (8.8)$$

where D_r = relative density
N_F = standard penetration number in the field
σ_o' = effective overburden pressure (kN/m^2)
C_u = uniformity coefficient of the sand

The *peak* angle of friction of granular soils, ϕ, was correlated to the corrected standard penetration number by Peck, Hanson, and Thornburn (1974). They gave a correlation between N_{cor} and ϕ in a graphical form, which can be approximated as (Wolff, 1989)

$$\phi\,(\text{deg}) = 27.1 + 0.3N_{cor} - 0.00054N_{cor}^2 \qquad (8.9)$$

Schmertmann (1975) provided a correlation among N_F, σ_o', and ϕ, which is shown in Figure 8.7. The correlation can be approximated as (Kulhawy and Mayne, 1990)

$$\phi = \tan^{-1}\left[\frac{N_F}{12.2 + 20.3\left(\dfrac{\sigma_o'}{p_a}\right)}\right]^{0.34} \qquad (8.10)$$

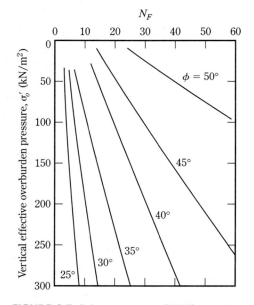

FIGURE 8.7 Schmertmann's (1975) correlation among N_F, σ_o', and ϕ for granular soils

where N_F = field standard penetration number

σ'_o = effective overburden pressure

p_a = atmospheric pressure in the same unit as σ'_o

ϕ = soil friction angle

More recently, Hatanaka and Uchida (1996) provided a simple correlation between ϕ and N_{cor} (Figure 8.8), which can be expressed as

$$\phi = \sqrt{20N_{cor}} + 20 \tag{8.11}$$

When the standard penetration resistance values are used in the preceding correlations to estimate soil parameters, the following qualifications should be noted:

1. The equations are approximate and largely empirical.
2. Because the soil is not homogeneous, the N_F values obtained from a given borehole vary widely.
3. In soil deposits that contain large boulders and gravel, standard penetration numbers may be erratic and unreliable.

Although the correlations are approximate, with correct interpretation the standard penetration test provides a good evaluation of soil properties. The primary sources of errors in standard penetration tests are inadequate cleaning of the borehole, careless measurement of the blow count, eccentric hammer strikes on the drill rod, and inadequate maintenance of water head in the borehole.

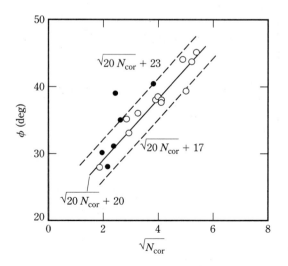

FIGURE 8.8 Laboratory test result of Hatanaka and Uchida (1996) for correlation between ϕ and $\sqrt{N_{cor}}$

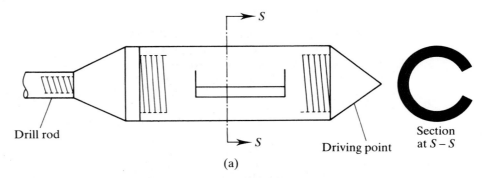

FIGURE 8.9 Sampling devices: (a) scraper bucket; (b) thin wall tube; (c) and (d) piston sampler

Scraper Bucket

When the soil deposit is sand mixed with pebbles, obtaining samples by split spoon with a spring core catcher may not be possible because the pebbles may prevent the springs from closing. In such cases, a scraper bucket may be used to obtain disturbed representative samples (Figure 8.9a). The scraper bucket has a driving point and can be attached to a drilling rod. The sampler is driven down into the soil and rotated, and the scrapings from the side fall into the bucket.

Thin Wall Tube

Thin wall tubes are sometimes called *Shelby tubes*. They are made of seamless steel and are commonly used to obtain undisturbed clayey soils. The commonly used thin wall tube samplers have outside diameters of 50.8 mm and 76.2 mm. The bottom end of the tube is sharpened. The tubes can be attached to drilling rods (Figure 8.9b). The drilling rod with the sampler attached is lowered to the bottom of the borehole, and the sampler is pushed into the soil. The soil sample inside the tube is then pulled out. The two ends of the sampler are sealed, and it is sent to the laboratory for testing.

Samples obtained in this manner may be used for consolidation or shear tests. A thin wall tube with a 50.8-mm outside diameter has an inside diameter of about 47.63 mm. The area ratio is

$$A_R\,(\%) = \frac{D_o^2 - D_i^2}{D_i^2}\,(100) = \frac{(50.8)^2 - (47.63)^2}{(47.63)^2}\,(100) = 13.75\%$$

Increasing the diameters of samples increases the cost of obtaining them.

Piston Sampler

When undisturbed soil samples are very soft or larger than 76.2 mm in diameter, they tend to fall out of the sampler. Piston samplers are particularly useful under

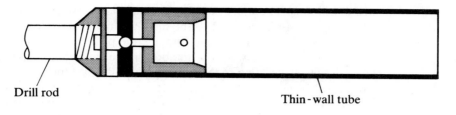

Drill rod

Thin-wall tube

(b)

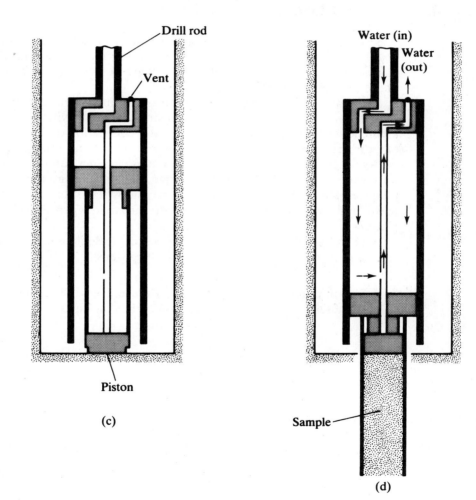

(c)

(d)

FIGURE 8.9 (Continued)

such conditions. There are several types of piston sampler; however, the sampler proposed by Osterberg (1952) is the most useful (see Figure 8.9c and d). It consists of a thin wall tube with a piston. Initially, the piston closes the end of the thin wall tube. The sampler is lowered to the bottom of the borehole (Figure 8.9c), and the thin wall tube is pushed into the soil hydraulically, past the piston. Then the pressure is released through a hole in the piston rod (Figure 8.9d). To a large extent, the presence of the piston prevents distortion in the sample by not letting the soil squeeze into the sampling tube very fast and by not admitting excess soil. Consequently, samples obtained in this manner are less disturbed than those obtained by Shelby tubes.

8.4   *Observation of Water Levels*

The presence of a water table near a foundation significantly affects a foundation's load-bearing capacity and settlement. The water level will change seasonally. In many cases, establishing the highest and lowest possible levels of water during the life of a project may become necessary.

In soils with high hydraulic conductivity, the level of water in a borehole will stabilize about 24 hours after completion of the boring. The depth of the water table can then be recorded by lowering a chain or tape into the borehole.

If water is encountered in a borehole during a field exploration, that fact should be recorded. In highly impermeable layers, the water level in a borehole may not stabilize for several weeks. In such cases, if accurate water level measurements are required, a *piezometer* can be used. A piezometer basically consists of a porous stone or a perforated pipe with a plastic standpipe attached to it. Figure 8.10 shows the general placement of a piezometer in a borehole.

For silty soils, Hvorslev (1949) proposed a technique to determine the water level (see Figure 8.11) involving the following steps:

1. Bail water out of the borehole to a level below the estimated water table.
2. Observe the water levels in the borehole at times

$$t = 0$$

$$t = t_1$$

$$t = t_2$$

$$t = t_3$$

Note that $t_1 - 0 = t_1 - t_2 = t_2 - t_3 = \Delta t$.
3. Calculate Δh_1, Δh_2, and Δh_3 (see Figure 8.11).
4. Calculate

$$h_0 = \frac{\Delta h_1^2}{\Delta h_1 - \Delta h_2} \tag{8.12a}$$

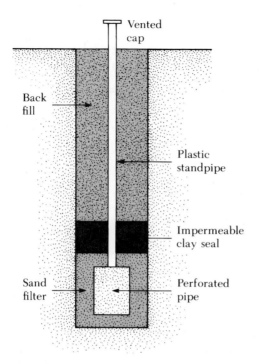

FIGURE 8.10 Casagrande-type porous stone piezometer

$$h_2 = \frac{\Delta h_2^2}{\Delta h_1 - \Delta h_2} \tag{8.12b}$$

$$h_3 = \frac{\Delta h_3^2}{\Delta h_2 - \Delta h_3} \tag{8.12c}$$

5. Plot h_0, h_2, and h_3 above the water levels observed at times $t = 0$, t_2, and t_3, respectively, to determine the final water level in the borehole.

EXAMPLE 8.1

Refer to Figure 8.11. For a borehole, $h_w + h_0 = 9.5$ m,

$$\Delta t = 24 \text{ hr}$$

$$\Delta h_1 = 0.9 \text{ m}$$

$$\Delta h_2 = 0.70 \text{ m}$$

$$\Delta h_3 = 0.54 \text{ m}$$

Make the necessary calculations and locate the water level.

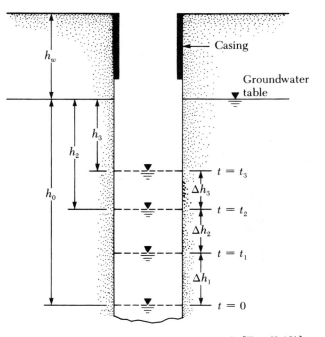

FIGURE 8.11 Determination of water levels [Eq. (8.12)]

Solution   Using Eq. (8.12), we have

$$h_0 = \frac{\Delta h_1^2}{\Delta h_1 - \Delta h_2} = \frac{0.9^2}{0.9 - 0.70} = 4.05 \text{ m}$$

$$h_2 = \frac{\Delta h_2^2}{\Delta h_1 - \Delta h_2} = \frac{0.70^2}{0.9 - 0.70} = 2.45 \text{ m}$$

$$h_3 = \frac{\Delta h_3^2}{\Delta h_2 - \Delta h_3} = \frac{0.54^2}{0.70 - 0.54} = 1.82 \text{ m}$$

Figure 8.12 shows a plot of the preceding calculations and the estimated water levels. Note that $h_w = 5.5$ m.

8.5   *Vane Shear Test*

The *vane shear test* (ASTM Test Designation D-2573) may be used during the drilling operation to determine the *in situ* undrained shear strength, c_u, of clay soils—particularly soft clays. The vane shear apparatus consists of four blades on the end of a rod, as shown in Figure 8.13. The height, h, of the vane is twice the diameter, d. The vane can be either rectangular or tapered (see Figure 8.13). The dimensions of vanes used in the field are given in Table 8.6. The vanes of the

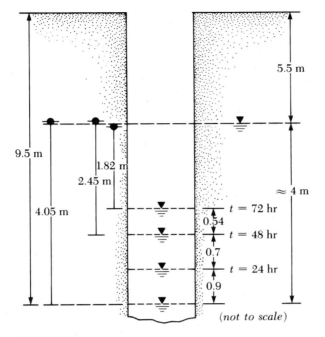

FIGURE 8.12

apparatus are pushed into the soil at the bottom of a borehole without disturbing the soil appreciably. Torque is applied at the top of the rod to rotate the vanes at a standard rate of 0.1°/sec. This rotation will induce failure in a soil of cylindrical shape surrounding the vanes. The maximum torque, T, applied to cause failure is measured. Note that

$$T = f(c_u, h, \text{ and } d) \tag{8.13}$$

or

$$c_u = \frac{T}{K} \tag{8.14}$$

where T is in N · m, and c_u is in kN/m²

$K = $ a constant with a magnitude depending on the dimension and shape of the vane

$$K = \left(\frac{\pi}{10^6}\right)\left(\frac{d^2 h}{2}\right)\left(1 + \frac{d}{3h}\right) \tag{8.15}$$

where $d = $ diameter of vane (cm)

$h = $ measured height of vane (cm)

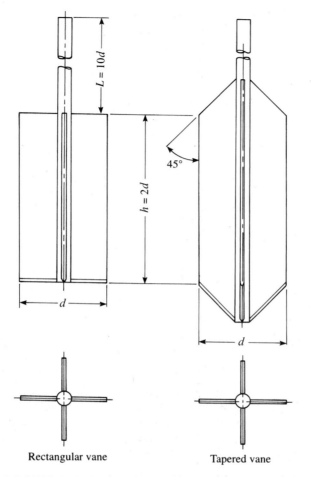

Rectangular vane   Tapered vane

FIGURE 8.13 Geometry of field vane (After ASTM, 1997)

Table 8.6 ASTM recommended dimensions of field vanes*

Casting size	Diameter, d (mm)	Height, h (mm)	Thickness of blade (mm)	Diameter of rod (mm)
AX	38.1	76.2	1.6	12.7
BX	50.8	101.6	1.6	12.7
NX	63.5	127.0	3.2	12.7
101.6 mm[†]	92.1	184.1	3.2	12.7

*Selection of the vane size is directly related to the consistency of the soil being tested; that is, the softer the soil, the larger the vane diameter should be.
[†]Inside diameter.

If $h/d = 2$, Eq. (8.15) yields

$$K = 366 \times 10^{-8}d^3$$
$$\uparrow$$
$$\text{(cm)}$$

(8.16)

Field vane shear tests are moderately rapid and economical and are used extensively in field soil-exploration programs. The test gives good results in soft and medium stiff clays, and it is also an excellent test to determine the properties of sensitive clays.

Sources of significant error in the field vane shear test are poor calibration of torque measurement and damaged vanes. Other errors may be introduced if the rate of vane rotation is not properly controlled.

The field vane shear strength can also be correlated to preconsolidation pressure and the overconsolidation ratio of the clay. Using 343 data points, Mayne and Mitchell (1988) derived the following empirical relationship for estimating the preconsolidation pressure of a natural clay deposit:

$$\sigma'_c = 7.04[c_{u(\text{field})}]^{0.83}$$

(8.17)

where σ'_c = preconsolidation pressure (kN/m²)
$c_{u(\text{field})}$ = field vane shear strength (kN/m²)

Mayne and Mitchell also showed that the overconsolidation ratio (OCR) can be correlated to $c_{u(\text{field})}$ as

$$OCR = \beta \frac{c_{u(\text{field})}}{\sigma'_o}$$

(8.18a)

where σ'_o = effective overburden pressure

$$\beta = 22(PI)^{-0.48}$$

(8.18b)

PI = plasticity index

8.6 *Cone Penetration Test*

The cone penetration test (CPT), originally known as the Dutch cone penetration test, is a versatile sounding method that can be used to determine the materials in a soil profile and estimate their engineering properties. This test is also called the *static penetration test*, and no boreholes are necessary to perform it. In the original version, a 60° cone with a base area of 10 cm² was pushed into the ground at a steady rate of about 20 mm/sec, and the resistance to penetration (called the point resistance) was measured.

The cone penetrometers in use at present measure (a) the *cone resistance*, q_c, to penetration developed by the cone, which is equal to the vertical force applied to the cone divided by its horizontally projected area, and (b) the *frictional resistance*, f_c, which is the resistance measured by a sleeve located above the cone with the local soil surrounding it. The frictional resistance is equal to the vertical force applied to the sleeve divided by its surface area—actually, the sum of friction and adhesion.

Generally, two types of penetrometers are used to measure q_c and f_c:

1. *Mechanical friction-cone penetrometer* (Figure 8.14). In this case, the penetrometer tip is connected to an inner set of rods. The tip is first advanced about 40 mm, thus giving the cone resistance. With further thrusting, the tip engages the friction sleeve. As the inner rod advances, the rod force is

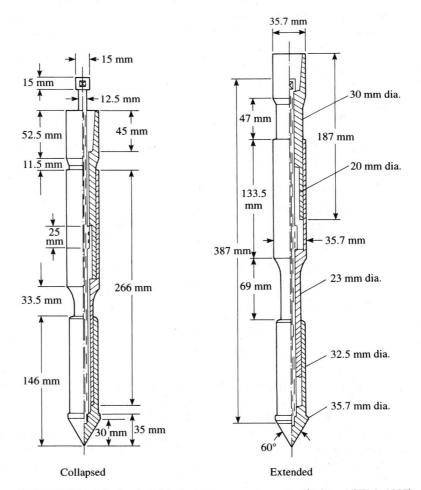

FIGURE 8.14 Mechanical friction-cone penetrometer (After ASTM, 1997)

equal to the sum of the vertical forces on the cone and the sleeve. Subtracting the force on the cone gives the side resistance.

2. *Electric friction-cone penetrometer* (Figure 8.15). In this case, the tip is attached to a string of steel rods. The tip is pushed into the ground at the rate of 20 mm/sec. Wires from the transducers are threaded through the center of the rods and continuously give the cone and side resistances.

Figure 8.16 shows the results of penetrometer tests in a soil profile with friction measurement by a mechanical friction-cone penetrometer and an electric friction-cone penetrometer.

Several correlations that are useful in estimating the properties of soils encountered during an exploration program have been developed for the cone resistance, q_c, and the friction ratio, F_r, obtained from the cone penetration tests. The friction ratio, F_r, is defined as

$$F_r = \frac{\text{frictional resistance}}{\text{cone resistance}} = \frac{f_c}{q_c} \tag{8.19}$$

Lancellotta (1983) and Jamiolkowski et al. (1985) showed that the relative density of normally consolidated sand, D_r, and q_c can be correlated as

$$D_r(\%) = A + B \log_{10}\left(\frac{q_c}{\sqrt{\sigma_o'}}\right) \tag{8.20}$$

where A, B = constants
σ_o' = vertical effective stress

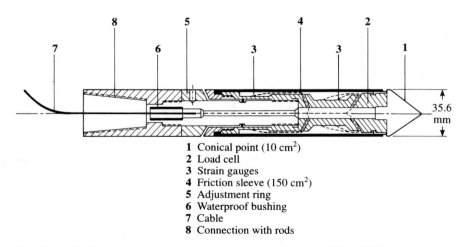

1 Conical point (10 cm²)
2 Load cell
3 Strain gauges
4 Friction sleeve (150 cm²)
5 Adjustment ring
6 Waterproof bushing
7 Cable
8 Connection with rods

FIGURE 8.15 Electric friction-cone penetrometer (After ASTM, 1997)

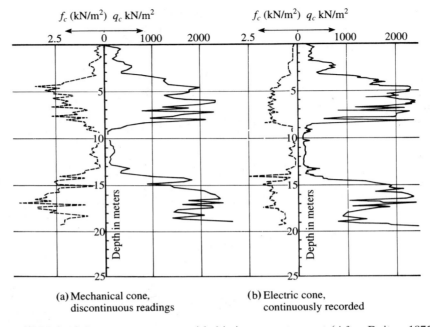

(a) Mechanical cone,
discontinuous readings

(b) Electric cone,
continuously recorded

FIGURE 8.16 Penetrometer tests with friction measurement (After Ruiter, 1971)

The values of A and B are -98 and 66, respectively, and q_c and σ'_o are in metric ton/m^2.

Baldi et al. (1982) and Robertson and Campanella (1983) also recommended an empirical relationship among vertical effective stress (σ'_o), relative density (D_r), and q_c for *normally consolidated sand*. This is shown in Figure 8.17.

Robertson and Campanella (1983) provided a graphical relationship among σ'_o, q_c, and the peak friction angle for normally consolidated quartz sand. The correlation can be expressed as (Kulhawy and Mayne, 1990)

$$\phi = \tan^{-1}\left[0.1 + 0.38\log\left(\frac{q_c}{\sigma'_o}\right)\right] \tag{8.21}$$

Robertson and Campanella (1983) also found a general correlation among q_c, friction ratio F_r, and the type of soil encountered in the field (Figure 8.18).

According to Mayne and Kemper (1988), in clayey soil the undrained shear strength c_u, the preconsolidation pressure σ'_c, and the overconsolidation ratio can be correlated as

$$\frac{c_u}{\sigma'_o} = \left(\frac{q_c - \sigma_o}{\sigma'_o}\right)\frac{1}{N_K} \tag{8.22}$$

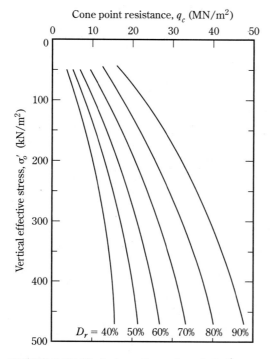

FIGURE 8.17 Variation of q_c, σ'_o, and D_r for normally consolidated quartz sand (Based on Baldi et al., 1982, and Robertson and Campanella, 1983)

or

$$c_u = \frac{q_c - \sigma_o}{N_K} \tag{8.22a}$$

where N_K = bearing capacity factor (N_K = 15 for electric cone and N_K = 20 for mechanical cone)

   σ_o = *total* vertical stress

   σ'_o = effective vertical stress

Consistent units of c_u, σ_o, σ'_o, and q_c should be used with Eq. (8.22):

$$\sigma'_c = 0.243(q_c)^{0.96} \tag{8.23}$$

$$\uparrow \qquad \uparrow$$

$$\text{(MN/m}^2) \qquad \text{(MN/m}^2)$$

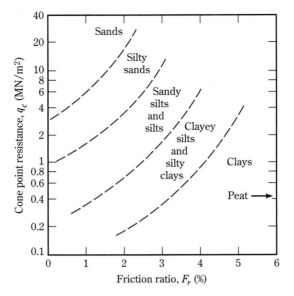

FIGURE 8.18 Robertson and Campanella correlation (1983) of q_c, F_r, and the soil type

and

$$OCR = 0.37 \left(\frac{q_c - \sigma_o}{\sigma_o'} \right)^{1.01}$$

(8.24)

where σ_o and σ_o' are the total and effective stress, respectively.

8.7 Pressuremeter Test (PMT)

The pressuremeter test is an *in situ* test conducted in a borehole. It was originally developed by Menard (1956) to measure the strength and deformability of soil. It has also been adopted by ASTM as Test Designation 4719. The Menard-type PMT essentially consists of a probe with three cells. The top and bottom ones are *guard cells* and the middle one is the *measuring cell,* as shown schematically in Figure 8.19a. The test is conducted in a pre-bored hole. The pre-bored hole should have a diameter that is between 1.03 and 1.2 times the nominal diameter of the probe. The probe that is most commonly used has a diameter of 58 mm and a length of 420 mm. The probe cells can be expanded by either liquid or gas. The guard cells are expanded to reduce the end-condition effect on the measuring cell. The measuring cell has a volume, V_o, of 535 cm^3. Table 8.7 lists the probe diameters and the diameters of the boreholes as recommended by ASTM.

To conduct a test, the measuring cell volume, V_o, is measured and the probe

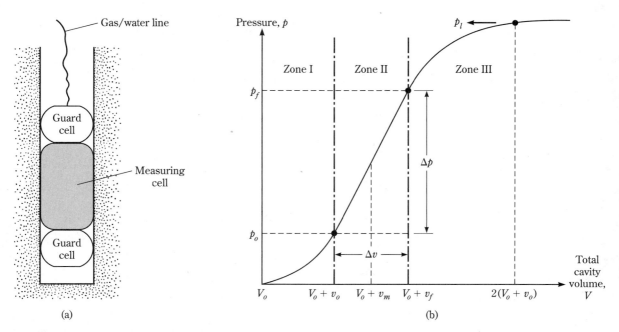

FIGURE 8.19 (a) Pressuremeter; (b) plot of pressure versus total cavity volume

is inserted into the borehole. Pressure is applied in increments, and the volumatic expansion of the cell is measured. This process is continued until the soil fails or until the pressure limit of the device is reached. The soil is considered to have failed when the total volume of the expanded cavity, V, is about twice the volume of the original cavity. After the completion of the test, the probe is deflated and advanced for testing at another depth.

The results of the pressuremeter test are expressed in a graphical form of pressure versus volume in Figure 8.19b. In this figure, Zone I represents the reloading portion during which the soil around the borehole is pushed back into the initial state (that is, the state it was in before drilling). The pressure, p_o, represents the *in situ* total horizontal stress. Zone II represents a pseudo-elastic zone in which the

Table 8.7 Probe and borehole diameter for pressuremeter test

Probe diameter (mm)	Borehole diameter	
	Nominal (mm)	Maximum (mm)
44	45	53
58	60	70
74	76	89

relation of cell volume to cell pressure is practically linear. The pressure, p_f, represents the creep, or yield, pressure. The zone marked III is the plastic zone. The pressure, p_l, represents the limit pressure.

The pressuremeter modulus, E_p, of the soil is determined using the theory of expansion of an infinitely thick cylinder. Thus,

$$E_p = 2(1 + \mu)(V_o + v_m)\left(\frac{\Delta p}{\Delta v}\right) \tag{8.25}$$

where $v_m = \dfrac{v_o + v_f}{2}$

$$\Delta p = p_f - p_o$$
$$\Delta v = v_f - v_o$$
$$\mu = \text{Poisson's ratio (which may be assumed to be 0.33)}$$

The limit pressure, p_l, is usually obtained by extrapolation and not by direct measurement.

To overcome the difficulty of preparing the borehole to the proper size, self-boring pressuremeters (SBPMT) have also been developed. The details concerning SBPMTs can be found in the work of Baguelin et al. (1978).

Correlations between various soil parameters and the results obtained from the pressuremeter tests were developed by various investigators. Kulhawy and Mayne (1990) proposed that

$$\sigma_c' = 0.45p_l \tag{8.26}$$

where $\sigma_c' =$ preconsolidation pressure.

Based on the cavity expansion theory, Baguelin et al. (1978) proposed that

$$c_u = \frac{p_l - p_o}{N_p} \tag{8.27}$$

where $c_u =$ undrained shear strength of a clay and

$$N_p = 1 + \ln\left(\frac{E_p}{3c_u}\right)$$

Typical values of N_p vary between 5 and 12, with an average of about 8.5. Ohya et al. (1982) (see also Kulhawy and Mayne, 1990) correlated E_p with field standard penetration numbers, N_F, for sand and clay as follows:

clay: $E_p \text{ (kN/m}^2) = 1930N_F^{0.63}$ \hfill (8.28)

sand: $E_p \text{ (kN/m}^2) = 908N_F^{0.66}$ \hfill (8.29)

8.8   Dilatometer Test

The use of the flat-plate dilatometer test (DMT) is relatively recent (Marchetti, 1980; Schmertmann, 1986). The equipment essentially consists of a flat plate measuring 220 mm (length) $\times$ 95 mm (width) $\times$ 14 mm (thickness). A thin, flat, circular expandable steel membrane with a diameter of 60 mm is located flush at the center on one side of the plate (Figure 8.20a). The dilatometer probe is inserted into the ground using a cone penetrometer testing rig (Figure 8.20b). Gas and electric lines extend from the surface control box through the penetrometer rod into the blade. At the required depth, high-pressure nitrogen gas is used to inflate the membrane. Two pressure readings are taken:

1. The pressure A to "lift off" the membrane
2. The pressure B at which the membrane expands 1.1 mm into the surrounding soil

The A and B readings are corrected as follows (Schmertmann, 1986):

$$\text{contact stress, } p_o = 1.05(A + \Delta A - Z_m) - 0.05(B - \Delta B - Z_m) \tag{8.30}$$

$$\text{expansion stress, } p_1 = B - Z_m - \Delta B \tag{8.31}$$

where ΔA = vacuum pressure required to keep the membrane in contact with its seating
ΔB = air pressure required inside the membrane to deflect it outward to a center expansion of 1.1 mm
Z_m = gauge pressure deviation from 0 when vented to atmospheric pressure

The test is normally conducted at depths 200 to 300 mm apart. The result of a given test is used to determine three parameters:

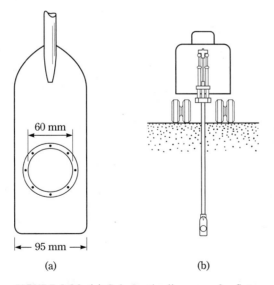

60 mm

95 mm

(a)   (b)

FIGURE 8.20 (a) Schematic diagram of a flat-plate dilatometer; (b) dilatometer probe inserted into ground

1. Material index, $I_D = \dfrac{p_1 - p_o}{p_o - u_o}$

2. Horizontal stress index, $K_D = \dfrac{p_o - u_o}{\sigma'_o}$

3. Dilatometer modulus, E_D (kN/m²) $= 34.7(p_1 \text{ kN/m}^2 - p_o \text{ kN/m}^2)$

where u_o = pore water pressure

$\quad\quad\sigma'_o$ = *in situ* vertical effective stress

Figure 8.21 shows the results of a dilatometer test conducted in Porto Tolle, Italy (Marchetti, 1980). The subsoil consisted of recent, normally consolidated delta deposits of the Po River. A thick layer of silty clay was found below a depth of about 3m ($c = 0$; $\phi \approx 28°$). The results obtained from the dilatometer tests were correlated with several soil properties (Marchetti, 1980). Some of these correlations are given here:

$$K_o = \left(\frac{K_D}{1.5}\right)^{0.47} - 0.6 \tag{8.32}$$

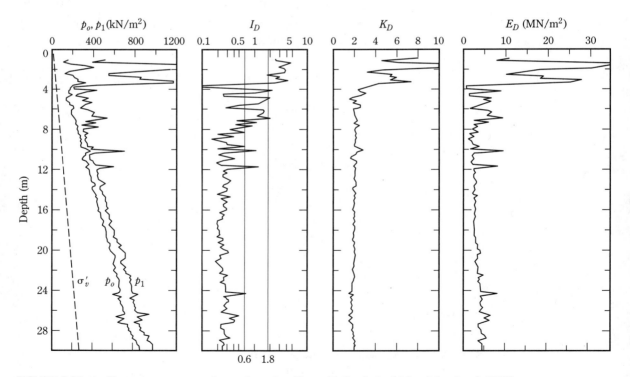

FIGURE 8.21 A dilatometer test result conducted at Porto Tolle, Italy (After Marchetti, 1980)

$$OCR = (0.5K_D)^{1.6} \tag{8.33}$$

$$\frac{c_u}{\sigma_o'} = 0.22 \qquad \text{(for normally consolidated clay)} \tag{8.34}$$

$$\left(\frac{c_u}{\sigma_o'}\right)_{OC} = \left(\frac{c_u}{\sigma_o'}\right)_{NC} (0.5K_D)^{1.25} \tag{8.35}$$

$$E = (1 - \mu^2)E_D \tag{8.36}$$

where K_o = coefficient of at-rest earth pressure
OCR = overconsolidation ratio
OC = overconsolidated soil
NC = normally consolidated soil
E = modulus of elasticity

Schmertmann (1986) also provided a correlation between the material index, I_D, and the dilatometer modulus, E_D, for determination of soil description and unit weight, γ. This relationship is shown in Figure 8.22.

8.9 *Coring of Rocks*

When a rock layer is encountered during a drilling operation, rock coring may be necessary. For coring of rocks, a *core barrel* is attached to a drilling rod. A *coring bit* is attached to the bottom of the core barrel (Fig. 8.23). The cutting elements may be diamond, tungsten, carbide, or others. Table 8.8 summarizes the various types of core barrel and their sizes, as well as the compatible drill rods commonly used for foundation exploration. The coring is advanced by rotary drilling. Water is circulated through the drilling rod during coring, and the cutting is washed out.

Two types of core barrel are available: the *single-tube core barrel* (Figure 8.23a) and the *double-tube core barrel* (Figure 8.23b). Rock cores obtained by single-tube core barrels can be highly disturbed and fractured because of torsion. Rock cores smaller than the BX size tend to fracture during the coring process.

Table 8.8 Standard size and designation of casing, core barrel, and compatible drill rod

Casing and core barrel designation	Outside diameter of core barrel bit (mm)	Drill rod designation	Outside diameter of drill rod (mm)	Diameter of borehole (mm)	Diameter of core sample (mm)
EX	36.51	E	33.34	38.1	22.23
AX	47.63	A	41.28	50.8	28.58
BX	58.74	B	47.63	63.5	41.28
NX	74.61	N	60.33	76.2	53.98

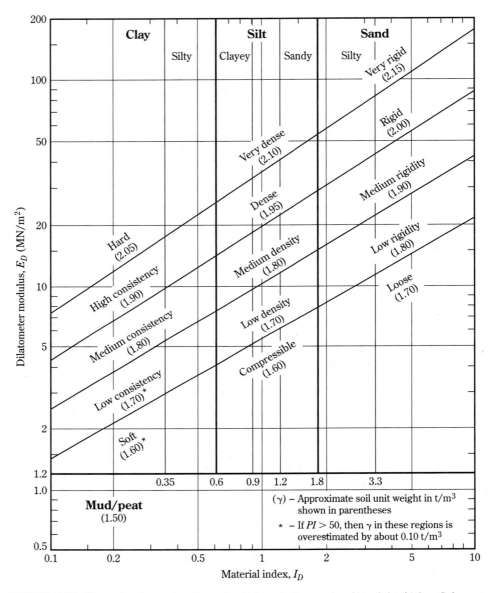

FIGURE 8.22 Chart for determination of soil description and unit weight (After Schmertmann, 1986). *Note:* 1 t/m³ = 9.81 kN/m³.

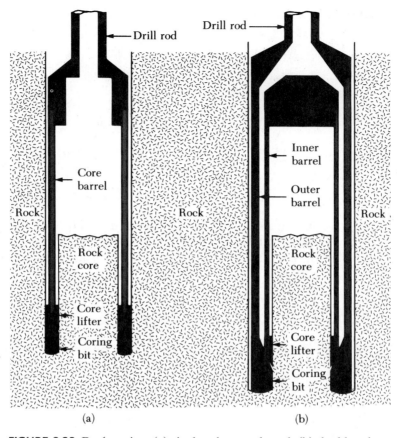

FIGURE 8.23 Rock coring: (a) single-tube core barrel; (b) double-tube core barrel

When the core samples are recovered, the depth of recovery should be properly recorded for further evaluation in the laboratory. Based on the length of the rock core recovered from each run, the following quantities may be calculated for a general evaluation of the rock quality encountered:

$$\text{recovery ratio} = \frac{\text{length of core recovered}}{\text{theoretical length of rock cored}} \qquad (8.37)$$

$$\begin{aligned}
&\text{rock quality designation } (RQD) \\
&= \frac{\Sigma \text{ length of recovered pieces equal to or larger than } 101.6 \text{ mm}}{\text{theoretical length of rock cored}}
\end{aligned} \qquad (8.38)$$

Table 8.9 Relationship between *in situ* rock quality and *RQD*

RQD	Rock quality
0–0.25	Very poor
0.25–0.5	Poor
0.5–0.75	Fair
0.75–0.9	Good
0.9–1	Excellent

A recovery ratio of 1 indicates the presence of intact rock; for highly fractured rocks, the recovery ratio may be 0.5 or less. Table 8.9 presents the general relationship (Deere, 1963) between the *RQD* and the *in situ* rock quality.

8.10   *Preparation of Boring Logs*

The detailed information gathered from each borehole is presented in a graphical form called the *boring log*. As a borehole is advanced downward, the driller generally should record the following information in a standard log:

1. Name and address of the drilling company
2. Driller's name
3. Job description and number
4. Number and type of boring and boring location
5. Date of boring
6. Subsurface stratification, which can be obtained by visual observation of the soil brought out by auger, split-spoon sampler, and thin wall Shelby tube sampler
7. Elevation of water table and date observed, use of casing and mud losses, and so on
8. Standard penetration resistance and the depth
9. Number, type, and depth of soil sample collected
10. In case of rock coring, type of core barrel used and, for each run, the actual length of coring, length of core recovery, and the *RQD*

This information should never be left to memory because not recording the data often results in erroneous boring logs.

After completing all the necessary laboratory tests, the geotechnical engineer prepares a finished log that includes notes from the driller's field log and the results of tests conducted in the laboratory. Figure 8.24 shows a typical boring log. These logs should be attached to the final soil exploration report submitted to the client. Note that Figure 8.24 also lists the classifications of the soils in the left-hand column, along with the description of each soil (based on the Unified Soil Classification System).

Boring Log

Name of the Project   Two-story apartment building

Location   Johnson & Olive St.   Date of Boring   March 2, 1999

Boring No.   3   Type of   Hollow stem auger   Ground Elevation   60.8 m
Boring

Soil description	Depth (m)	Soil sample type and number	N_F	w_n (%)	Comments
Light brown clay (fill)					
Silty sand (SM)	1 — 2	SS-1	9	8.2	
°G.W.T. ▽ 3.5 m	3 — 4	SS-2	12	17.6	$LL = 38$ $PI = 11$
Light gray clayey silt (ML)	5	ST-1		20.4	$LL = 36$ $q_u = 112 \, kN/m^2$
	6	SS-3	11	20.6	
Sand with some gravel (SP)	7				
End of boring @ 8 m	8	SS-4	27	9	

N_F = standard penetration number (blows/304.8 mm)
w_n = natural moisture content
LL = liquid limit; PI = plasticity index
q_u = unconfined compression strength
SS = split-spoon sample; ST = Shelby tube sample

°Groundwater table observed after 1 week of drilling

FIGURE 8.24 A typical boring log

8.11   *Soil Exploration Report*

At the end of all soil exploration programs, the soil and/or rock specimens collected in the field are subject to visual observation and appropriate laboratory testing. After all the required information has been compiled, a soil exploration report is prepared for use by the design office and for reference during future construction work. Although the details and sequence of information in the report may vary to some degree, depending on the structure under consideration and the person compiling the report, each report should include the following items:

1. The scope of the investigation
2. A description of the proposed structure for which the subsoil exploration has been conducted
3. A description of the location of the site, including structure(s) nearby, drainage conditions of the site, nature of vegetation on the site and surrounding it, and any other feature(s) unique to the site
4. Geological setting of the site
5. Details of the field exploration—that is, number of borings, depths of borings, type of boring, and so on
6. General description of the subsoil conditions as determined from soil specimens and from related laboratory tests, standard penetration resistance and cone penetration resistance, and so on
7. Water table conditions
8. Foundation recommendations, including the type of foundation recommended, allowable bearing pressure, and any special construction procedure that may be needed; alternative foundation design procedures should also be discussed in this portion of the report
9. Conclusions and limitations of the investigations

The following graphical presentations should be attached to the report:

1. Site location map
2. A plan view of the location of the borings with respect to the proposed structures and those existing nearby
3. Boring logs
4. Laboratory test results
5. Other special graphical presentations

The exploration reports should be well planned and documented. They will help in answering questions and solving foundation problems that may arise later during design and construction.

Problems

8.1   A Shelby tube has an outside diameter of 50.8 mm and an inside diameter of 47.6 mm.

  a. What is the area ratio of the tube?
  b. If the outside diameter remains the same, what should be the inside diameter of the tube to give an area ratio of 10%?

8.2 A soil profile is shown in Figure 8.25 along with the standard penetration numbers in the clay layer. Use Eqs. (8.5) and (8.6) to determine and plot the variation of c_u and OCR with depth.

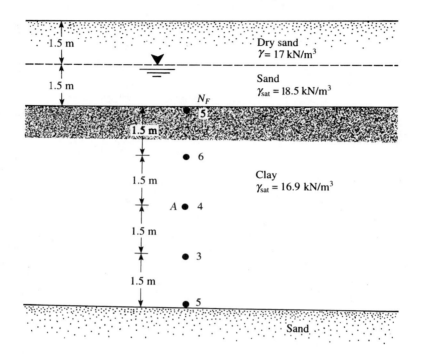

FIGURE 8.25

8.3 The average value of the field standard penetration number in a saturated clay layer is 6. Estimate the unconfined compression strength of the clay. Use Eq. (8.4) ($K \approx 4.2$ kN/m²).

8.4 The table gives the variation of the field standard penetration number, N_F, in a sand deposit:

Depth (m)	N_F
1.5	5
3	7
4.5	9
6	8
7.5	12
9	11

The groundwater table is located at a depth of 5.5 m. The dry unit weight of sand from 0 to a depth of 5.5 m is 18.08 kN/m³, and the saturated unit weight of sand for depths of 5.5 to 10.5 m is 19.34 kN/m³. Use the relationship of Liao and Whitman given in Table 8.4 to calculate the corrected penetration numbers.

8.5   For the soil profile described in Problem 8.4, estimate an average peak soil friction angle. Use Eq. (8.9).

8.6   The table gives the standard penetration numbers determined from a sandy soil deposit in the field:

Depth (m)	Unit weight of soil (kN/m³)	N_F
3.0	16.7	7
4.5	16.7	9
6.0	16.7	11
7.5	18.6	16
9.0	18.6	18
10.5	18.6	20
12	18.6	22

Using Eq. (8.10), determine the variation of the peak soil friction angle, ϕ. Estimate an average value of ϕ for the design of a shallow foundation. *Note:* For depth greater than 6 m, the unit weight of soil is 18.6 kN/m³.

8.7   Redo Problem 8.6 using Skempton's relationship given in Table 8.4 and Eq. (8.11).

8.8   The details for a soil deposit in sand are given in the table:

Depth (m)	Effective overburden pressure (kN/m²)	Field standard penetration number, N_F
3.0	55.1	9
4.5	82.7	11
6.0	97.3	12

Assume the uniformity coefficient, C_u, of the sand to be 3.2. Estimate the average relative density of the sand between the depths of 3 m and 6 m. Use Eq. (8.8).

8.9   Refer to Figure 8.11. For a borehole in a silty clay soil, the following values are given:

$$h_w + h_o = 7.6 \text{ m}$$

$$t_1 = 24 \text{ hr} \qquad \Delta h_1 = 0.73 \text{ m}$$
$$t_2 = 48 \text{ hr} \qquad \Delta h_2 = 0.52 \text{ m}$$
$$t_3 = 72 \text{ hr} \qquad \Delta h_3 = 0.37 \text{ m}$$

Determine the depth of the water table measured from the ground surface.

8.10 Repeat Problem 8.9 for

$$h_w + h_o = 12.8 \text{ m}$$

$$t_1 = 24 \text{ hr} \qquad \Delta h_1 = 1.83 \text{ m}$$
$$t_2 = 48 \text{ hr} \qquad \Delta h_2 = 1.46 \text{ m}$$
$$t_3 = 72 \text{ hr} \qquad \Delta h_3 = 1.16 \text{ m}$$

8.11 Refer to Figure 8.25. Vane shear tests were conducted in the clay layer. The vane dimensions were 63.5 mm (d) × 127 mm (h). For the test at A, the torque required to cause failure was 0.051 N·m. For the clay, liquid limit was 46 and plastic limit was 21. Estimate the undrained cohesion of the clay for use in the design by using each equation:

a. Bjerrum's λ relationship (Eq. 7.29)

b. Morris and Williams' λ and PI relationship (Eq. 7.30)

c. Morris and Williams' λ and LL relationship (Eq. 7.31)

8.12 **a.** A vane shear test was conducted in a saturated clay. The height and diameter of the vane were 101.6 mm and 50.8 mm, respectively. During the test, the maximum torque applied was 0.0168 N·m. Determine the undrained shear strength of the clay.

b. The clay soil described in part (a) has a liquid limit of 64 and a plastic limit of 29. What would be the corrected undrained shear strength of the clay for design purposes? Use Bjerrum's relationship for λ (Eq. 7.29).

8.13 Refer to Problem 8.11. Determine the overconsolidation ratio for the clay. Use Eqs. (8.18a) and (8.18b).

8.14 In a deposit of normally consolidated dry sand, a cone penetration test was conducted. The table gives the results:

Depth (m)	Point resistance of cone, q_c (MN/m^2)
1.5	2.05
3.0	4.23
4.5	6.01
6.0	8.18
7.5	9.97
9.0	12.42

Assume the dry unit weight of sand is 15.5 kN/m^3.

a. Estimate the average peak friction angle, ϕ, of the sand. Use Eq. (8.21).

b. Estimate the average relative density of the sand. Use Figure 8.17.

8.15   Refer to the soil profile shown in Figure 8.26. Assume the cone penetration resistance, q_c, at A as determined by an electric friction-cone penetrometer is 0.6 MN/m².
   a. Determine the undrained cohesion, c_u.
   b. Find the overconsolidation ratio, OCR.

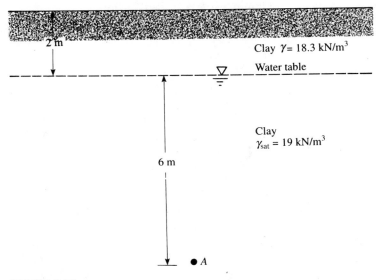

FIGURE 8.26

8.16   Consider a pressuremeter test in a soft saturated clay.

   measuring cell volume, $V_o = 535$ cm³

   $p_o = 42.4$ kN/m² $v_o = 46$ cm³
   $p_f = 326.5$ kN/m² $v_f = 180$ cm³

   Assuming Poisson's ratio, μ, to be 0.5 and referring to Figure 8.19, calculate the pressuremeter modulus, E_p.

8.17   A dilatometer test was conducted in a clay deposit. The groundwater table was located at a depth of 3 m below the ground surface. At a depth of 8 m below the ground surface, the contact pressure, p_o, was 280 kN/m² and the expansion stress, p_1, was 350 kN/m².
   a. Determine the coefficient of at-rest earth pressure, K_o.
   b. Find the overconsolidation ratio, OCR.
   c. What is the modulus of elasticity, E?
   Assume σ_o' at a depth of 8 m to be 95 kN/m² and $\mu = 0.35$.

8.18   During a field exploration, coring of rock was required. The core barrel was advanced 1.5 m during the coring. The length of the core recovered was 0.98 m. What was the recovery ratio?

References

American Society for Testing and Materials (1997). *Annual Book of ASTM Standards,* Vol. 04.08, West Conshohocken, PA.

American Society of Civil Engineers (1972). "Subsurface Investigation for Design and Construction of Foundations of Buildings," *Journal of the Soil Mechanics and Foundations Division,* American Society of Civil Engineers, Vol. 98. No. SM5, 481–490.

Baguelin, F., Jézéquel, J. F., and Shields, D. H. (1978). *The Pressuremeter and Foundation Engineering,* Trans Tech Publications, Clausthal.

Baldi, G., Bellotti, R., Ghionna, V., and Jamiolkowski, M. (1982). "Design Parameters for Sands from CPT," *Proceedings,* Second European Symposium on Penetration Testing, Amsterdam, Vol. 2, 425–438.

Deere, D. U. (1963). "Technical Description of Rock Cores for Engineering Purposes," *Felsmechanik und Ingenieurgeologie,* Vol. 1, No. 1, 16–22.

Hara, A., Ohata, T., and Niwa, M. (1971). "Shear Modulus and Shear Strength of Cohesive Soils," *Soils and Foundations,* Vol. 14, No. 3, 1–12.

Hatanaka, M., and Uchida, A. (1996). "Empirical Correlation Between Penetration Resistance and Internal Friction Angle of Sandy Soils," *Soils and Foundations,* Vol. 36, No. 4, 1–10.

Hvorslev, M. J. (1949). *Subsurface Exploration and Sampling of Soils for Civil Engineering Purposes,* Waterways Experiment Station, Vicksburg, MS.

Jamiolkowski, M., Ladd, C. C., Germaine, J. T., and Lancellotta, R. (1985). "New Developments in Field and Laboratory Testing of Soils," *Proceedings,* 11th International Conference on Soil Mechanics and Foundation Engineering, Vol. 1, 57–153.

Kulhawy, F. H., and Mayne, P. W. (1990). *Manual on Estimating Soil Properties for Foundation Design,* Electric Power Research Institute, Palo Alto, CA.

Lancellotta, R. (1983). *Analisi di Affidabilità in Ingegneria Geotecnia,* Atti Istituto Sciencza Costruzioni, No. 625, Politecnico di Torino.

Liao, S. S. C., and Whitman, R. V. (1986). "Overburden Correction Factors for SPT in Sand," *Journal of Geotechnical Engineering,* American Society of Civil Engineers, Vol. 112, No. 3, 373–377.

Marchetti, S. (1980). "*In Situ* Test by Flat Dilatometer," *Journal of Geotechnical Engineering Division,* ASCE, Vol. 106, GT3, 299–321.

Marcuson, W. F. III, and Bieganousky, W. A. (1977). "SPT and Relative Density in Coarse Sands," *Journal of Geotechnical Engineering Division,* American Society of Civil Engineers, Vol. 103, No. 11, 1295–1309.

Mayne, P. W., and Kemper, J. B. (1988). "Profiling OCR in Stiff Clays by CPT and SPT," *Geotechnical Testing Journal,* ASTM, Vol. 11, No. 2, 139–147.

Mayne, P. W., and Mitchell, J. K. (1988). "Profiling of Overconsolidation Ratio in Clays by Field Vane," *Canadian Geotechnical Journal,* Vol. 25, No. 1, 150–158.

Menard, L. (1956). *An Apparatus for Measuring the Strength of Soils in Place,* M.S. Thesis, University of Illinois, Urbana, IL.

Ohya, S., Imai, T., and Matsubara, M. (1982). "Relationships Between N Value by SPT and LLT Pressuremeter Results," *Proceedings,* 2nd European Symposium on Penetration Testing, Amsterdam, Vol. 1, 125–130.

Osterberg, J. O. (1952). "New Piston-Type Soil Sampler," *Engineering News-Record,* April 24.

Peck, R. B., Hanson, W. E., and Thornburn, T. H. (1974). *Foundation Engineering,* 2nd ed., Wiley, New York.

Robertson, P. K., and Campanella, R. G. (1983). "Interpretation of Cone Penetration Tests. Part I: Sand," *Canadian Geotechnical Journal,* Vol. 20, No. 4, 718–733.

Ruiter, J. (1971). "Electric Penetrometer for Site Investigations," *Journal of the Soil Mechanics and Foundations Division,* American Society of Civil Engineers, Vol. 97, No. 2, 457–472.

Schmertmann, J. H. (1975). "Measurement of *In Situ* Shear Strength," *Proceedings,* Specialty Conference on *In Situ* Measurement of Soil Properties, ASCE, Vol. 2, 57–138.

Schmertmann, J. H. (1986). "Suggested Method for Performing the Flat Dilatometer Test," *Geotechnical Testing Journal,* ASTM, Vol. 9, No. 2, 93–101.

Seed, H. B., Arango, I., and Chan, C. K. (1975). "Evaluation of Soil Liquefaction Potential During Earthquakes,"*Report No. EERC 75-28,* Earthquake Engineering Research Center, University of California, Berkeley.

Skempton, A. W. (1986). "Standard Penetration Test Procedures and the Effect in Sands of Overburden Pressure, Relative Density, Particle Size, Aging and Overconsolidation," *Geotechnique,* Vol. 36, No. 3, 425–447.

Sowers, G. B., and Sowers, G. F. (1970). *Introductory Soil Mechanics and Foundations,* 3rd ed., Macmillan, New York.

Stroud, M. (1974). "SPT in Insensitive Clays," *Proceedings,* European Symposium on Penetration Testing, Vol. 2.2, 367–375.

Wolff, T. F. (1989). "Pile Capacity Prediction Using Parameter Functions," in *Predicted and Observed Axial Behavior of Piles, Results of a Pile Prediction Symposium,* sponsored by Geotechnical Engineering Division, ASCE, Evanston, IL, June 1989, ASCE Geotechnical Special Publication No. 23, 96–106.

9

Lateral Earth Pressure

Retaining structures, such as retaining walls, basement walls, and bulkheads, are commonly encountered in foundation engineering, and they may support slopes of earth masses. Proper design and construction of these structures require a thorough knowledge of the lateral forces that act between the retaining structures and the soil masses being retained. These lateral forces are caused by lateral earth pressure. This chapter is devoted to the study of various earth pressure theories.

9.1 Earth Pressure at Rest

Let us consider the mass of soil shown in Figure 9.1. The mass is bounded by a frictionless wall AB that extends to an infinite depth. A soil element located at a depth z is subjected to *effective* vertical and horizontal pressures of σ'_o and σ'_h, respectively. For this case, since the soil is dry, we have

$$\sigma'_o = \sigma_o$$

and

$$\sigma'_h = \sigma_h$$

where σ_o and σ_h = *total* vertical and horizontal pressures, respectively. Also, note that there are no shear stresses on the vertical and horizontal planes.

If the wall AB is static—that is, if it does not move either to the right or to the left of its initial position—the soil mass will be in a state of *elastic equilibrium;* that is, the horizontal strain is 0. The ratio of the effective horizontal stress to the vertical stress is called the *coefficient of earth pressure at rest, K_o,* or

$$K_o = \frac{\sigma'_h}{\sigma'_o} \tag{9.1}$$

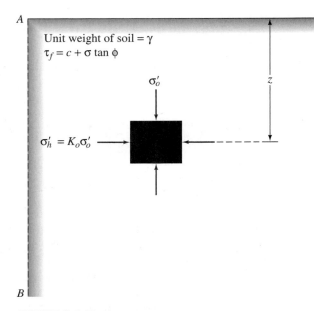

FIGURE 9.1 Earth pressure at rest

Since $\sigma'_o = \gamma z$, we have

$$\sigma'_h = K_o(\gamma z) \tag{9.2}$$

For coarse-grained soils, the coefficient of earth pressure at rest can be estimated by the empirical relationship (Jaky, 1944)

$$K_o = 1 - \sin \phi \tag{9.3}$$

where ϕ = drained friction angle. For fine-grained, normally consolidated soils, Massarsch (1979) suggested the following equation for K_o:

$$K_o = 0.44 + 0.42 \left[\frac{PI\,(\%)}{100} \right] \tag{9.4}$$

For overconsolidated clays, the coefficient of earth pressure at rest can be approximated as

$$K_{o(\text{overconsolidated})} = K_{o(\text{normally consolidated})} \sqrt{OCR} \tag{9.5}$$

where OCR = overconsolidation ratio. The overconsolidation ratio was defined in Chapter 5 as

$$OCR = \frac{\text{preconsolidation pressure}}{\text{present effective overburden pressure}} \tag{9.6}$$

The magnitude of K_o in most soils ranges between 0.5 and 1.0, with perhaps higher values for heavily overconsolidated clays.

Figure 9.2 shows the distribution of earth pressure at rest on a wall of height H. The total force per unit length of the wall, P_o, is equal to the area of the pressure diagram, so

$$P_o = \frac{1}{2} K_o \gamma H^2 \tag{9.7}$$

Earth Pressure at Rest for Partially Submerged Soil

Figure 9.3a shows a wall of height H. The groundwater table is located at a depth H_1 below the ground surface, and there is no compensating water on the other side of the wall. For $z \le H_1$, the total lateral earth pressure at rest can be given as

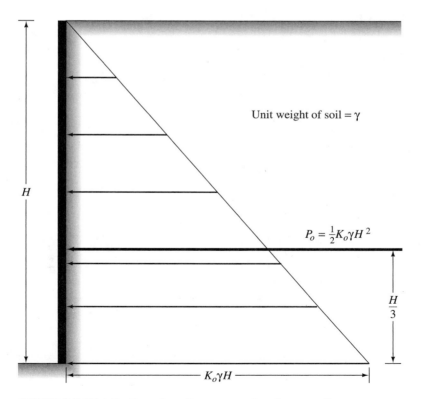

FIGURE 9.2 Distribution of earth pressure at rest on a wall

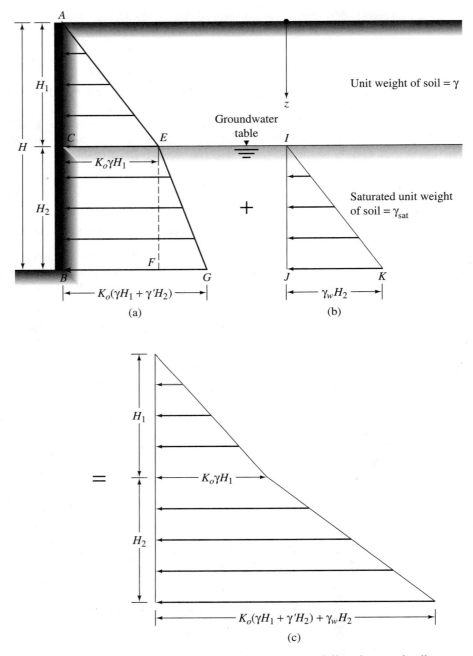

FIGURE 9.3 Distribution of earth pressure at rest for partially submerged soil

$\sigma_h' = K_o\gamma z$. The variation of σ_h' with depth is shown by triangle ACE in Figure 9.3a. However, for $z \geq H_1$ (that is, below the groundwater table), the pressure on the wall is found from the effective stress and pore water pressure components in the following manner:

$$\text{effective vertical pressure} = \sigma_o' = \gamma H_1 + \gamma'(z - H_1) \tag{9.8}$$

where $\gamma' = \gamma_{sat} - \gamma_w$ = effective unit weight of soil. So, the effective lateral pressure at rest is

$$\sigma_h' = K_o\sigma_o' = K_o[\gamma H_1 + \gamma'(z - H_1)] \tag{9.9}$$

The variation of σ_h' with depth is shown by $CEGB$ in Figure 9.3a. Again, the lateral pressure from pore water is

$$u = \gamma_w(z - H_1) \tag{9.10}$$

The variation of u with depth is shown in Figure 9.3b.

Hence, the total lateral pressure from earth and water at any depth $z \geq H_1$ is equal to

$$\sigma_h = \sigma_h' + u$$
$$= K_o[\gamma H_1 + \gamma'(z - H_1)] + \gamma_w(z - H_1) \tag{9.11}$$

The force per unit width of the wall can be found from the sum of the areas of the pressure diagrams in Figures 9.3a and b and is equal to

$$P_o = \underbrace{\frac{1}{2}K_o\gamma H_1^2}_{\substack{\text{Area} \\ ACE}} + \underbrace{K_o\gamma H_1 H_2}_{\substack{\text{Area} \\ CEFB}} + \underbrace{\frac{1}{2}(K_o\gamma' + \gamma_w)H_2^2}_{\substack{\text{Areas} \\ EFG \text{ and } IJK}} \tag{9.12}$$

or

$$P_o = \frac{1}{2}K_o[\gamma H_1^2 + 2\gamma H_1 H_2 + \gamma'H_2^2] + \gamma_w H_2^2 \tag{9.13}$$

9.2 Rankine's Theory of Active and Passive Earth Pressures

The term *plastic equilibrium* in soil refers to the condition in which every point in a soil mass is on the verge of failure. Rankine (1857) investigated the stress conditions in soil at a state of plastic equilibrium. This section deals with Rankine's theory of earth pressure.

Rankine's Active State

Figure 9.4a shows the same soil mass that was illustrated in Figure 9.1. It is bounded by a frictionless wall AB that extends to an infinite depth. The vertical and horizontal effective principal stresses on a soil element at a depth z are σ_o', and σ_h', respectively. As we saw in Section 9.1, if the wall AB is not allowed to move at all, then $\sigma_h' = K_o\sigma_o'$. The stress condition in the soil element can be represented by the Mohr's circle a in Figure 9.4b. However, if the wall AB is allowed to move away from the soil mass gradually, then the horizontal effective principal stress will decrease. Ultimately a state will be reached at which the stress condition in the soil element can be represented by the Mohr's circle b, the state of plastic equilibrium, and failure of the soil will occur. This state is *Rankine's active state,* and the pressure σ_a' on the vertical plane (which is a principal plane) is *Rankine's active earth pressure.* Following is the derivation for expressing σ_a' in terms of γ, z, c, and ϕ. From Figure 9.4b, we have

$$\sin \phi = \frac{CD}{AC} = \frac{CD}{AO + OC}$$

but

$$CD = \text{radius of the failure circle} = \frac{\sigma_o' - \sigma_a'}{2}$$

$$AO = c \cot \phi$$

and

$$OC = \frac{\sigma_o' + \sigma_a'}{2}$$

so

$$\sin \phi = \frac{\dfrac{\sigma_o' - \sigma_a'}{2}}{c \cot \phi + \dfrac{\sigma_o' + \sigma_a'}{2}}$$

or

$$c \cos \phi + \frac{\sigma_o' + \sigma_a'}{2} \sin \phi = \frac{\sigma_o' - \sigma_a'}{2}$$

or

$$\sigma_a' = \sigma_o' \frac{1 - \sin \phi}{1 + \sin \phi} - 2c \frac{\cos \phi}{1 + \sin \phi} \tag{9.14}$$

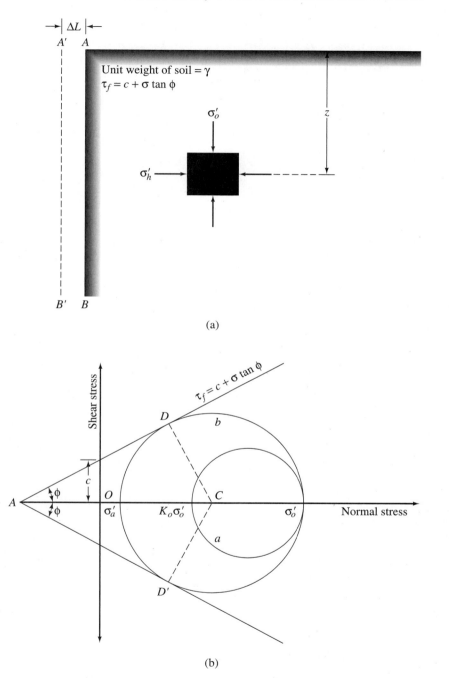

(a)

(b)

FIGURE 9.4 Rankine's active earth pressure

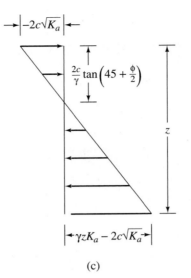

$$-2c\sqrt{K_a}$$

$$\frac{2c}{\gamma}\tan\left(45+\frac{\phi}{2}\right)$$

z

$$\gamma z K_a - 2c\sqrt{K_a}$$

(c)

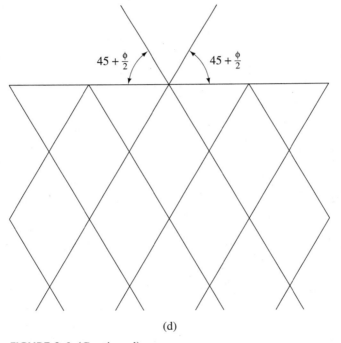

$$45+\frac{\phi}{2}$$ $$45+\frac{\phi}{2}$$

(d)

FIGURE 9.4 (Continued)

But

σ_o' = vertical effective overburden pressure = γz

$$\frac{1 - \sin \phi}{1 + \sin \phi} = \tan^2 \left(45 - \frac{\phi}{2} \right)$$

and

$$\frac{\cos \phi}{1 + \sin \phi} = \tan \left(45 - \frac{\phi}{2} \right)$$

Substituting the above into Eq. (9.14), we get

$$\sigma_a' = \gamma z \tan^2 \left(45 - \frac{\phi}{2} \right) - 2c \tan \left(45 - \frac{\phi}{2} \right) \qquad (9.15)$$

The variation of σ_a' with depth is shown in Figure 9.4c. For cohesionless soils, $c = 0$ and

$$\sigma_a' = \sigma_o' \tan^2 \left(45 - \frac{\phi}{2} \right) \qquad (9.16)$$

The ratio of σ_a' to σ_o' is called the *coefficient of Rankine's active earth pressure, K_a*, or

$$K_a = \frac{\sigma_a'}{\sigma_o'} = \tan^2 \left(45 - \frac{\phi}{2} \right) \qquad (9.17)$$

Again, from Figure 9.4b, we can see that the failure planes in the soil make $\pm(45 + \phi/2)$-degree angles with the direction of the major principal plane—that is, the horizontal. These failure planes are called *slip planes*. The slip planes are shown in Figure 9.4d.

Rankine's Passive State

Rankine's passive state is illustrated in Figure 9.5. *AB* is a frictionless wall that extends to an infinite depth. The initial stress condition on a soil element is represented by the Mohr's circle *a* in Figure 9.5b. If the wall is gradually pushed into the soil mass, the effective principal stress σ_h' will increase. Ultimately the wall will reach a state at which the stress condition in the soil element can be represented by the Mohr's circle *b*. At this time, failure of the soil will occur. This is referred to as *Rankine's passive state*. The effective lateral earth pressure σ_p', which is the

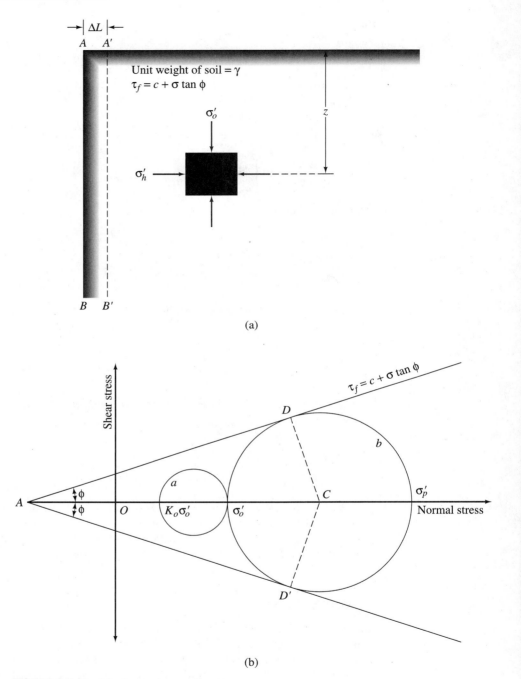

(a)

(b)

FIGURE 9.5 Rankine's passive earth pressure

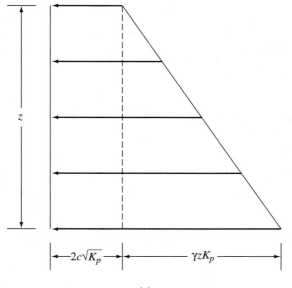

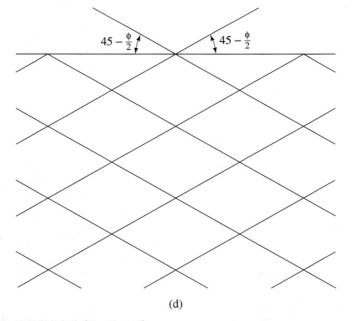

(c)

(d)

FIGURE 9.5 (Continued)

major principal stress, is called *Rankine's passive earth pressure*. From Figure 9.5b, it can be shown that

$$\sigma_p' = \sigma_o' \tan^2\left(45 + \frac{\phi}{2}\right) + 2c\tan\left(45 + \frac{\phi}{2}\right)$$

$$= \gamma z \tan^2\left(45 + \frac{\phi}{2}\right) + 2c\tan\left(45 + \frac{\phi}{2}\right) \tag{9.18}$$

The derivation is similar to that for Rankine's active state.

Figure 9.5c shows the variation of passive pressure with depth. For cohensionless soils ($c = 0$), we have

$$\sigma_p' = \sigma_o' \tan^2\left(45 + \frac{\phi}{2}\right)$$

or

$$\frac{\sigma_p'}{\sigma_o'} = K_p = \tan^2\left(45 + \frac{\phi}{2}\right) \tag{9.19}$$

K_p in the preceding equation is referred to as the *coefficient of Rankine's passive earth pressure*.

The points D and D' on the failure circle (Figure 9.5b) correspond to the slip planes in the soil. For Rankine's passive state, the slip planes make $\pm(45 - \phi/2)$-degree angles with the direction of the minor principal plane—that is, in the horizontal direction. Figure 9.5d shows the distribution of slip planes in the soil mass.

Effect of Wall Yielding

From the preceding discussion we know that sufficient movement of the wall is necessary to achieve a state of plastic equilibrium. However, the distribution of lateral earth pressure against a wall is very much influenced by the manner in which the wall actually yields. In most simple retaining walls (see Figure 9.6), movement may occur by simple translation or, more frequently, by rotation about the bottom.

For preliminary theoretical analysis, let us consider a frictionless retaining wall represented by a plane AB, as shown in Figure 9.7a. If the wall AB rotates sufficiently about its bottom to a position $A'B$, then a triangular soil mass ABC' adjacent to the wall will reach Rankine's active state. Since the slip planes in Rankine's active state make angles of $\pm(45 + \phi/2)$ degrees with the major principal plane, the soil mass in the state of plastic equilibrium is bounded by the plane BC', which makes an angle of $(45 + \phi/2)$ degrees with the horizontal. The soil inside the zone ABC' undergoes the same unit deformation in the horizontal direction

FIGURE 9.6 Cantilever retaining wall

everywhere, which is equal to $\Delta L_a/L_a$. The lateral earth pressure on the wall at any depth z from the ground surface can be calculated by Eq. (9.15).

In a similar manner, if the frictionless wall AB (Figure 9.7b) rotates sufficiently into the soil mass to a position $A''B$, then the triangular mass of soil ABC'' will reach Rankine's passive state. The slip plane BC'' bounding the soil wedge that is at a state of plastic equilibrium makes an angle of $(45 - \phi/2)$ degrees with the horizontal. Every point of the soil in the triangular zone ABC'' undergoes the same unit deformation in the horizontal direction, which is equal to $\Delta L_p/L_p$. The passive pressure on the wall at any depth z can be evaluated by using Eq. (9.18).

Typical values of the minimum wall tilt (ΔL_a and ΔL_p) required for achieving Rankine's state are given in Table 9.1. Figure 9.8 shows the variation of lateral earth pressure with wall tilt.

Table 9.1 Typical values of $\Delta L_a/H$ and $\Delta L_p/H$ for Rankine's state

Soil type	$\Delta L_a/H$	$\Delta L_p/H$
Loose sand	0.001–0.002	0.01
Dense sand	0.0005–0.001	0.005
Soft clay	0.02	0.04
Stiff clay	0.01	0.02

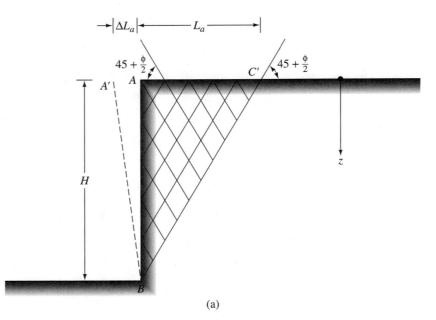

(a)

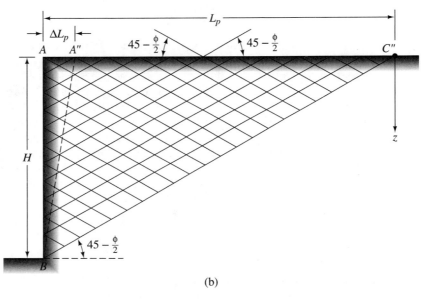

(b)

FIGURE 9.7 Rotation of frictionless wall about the bottom

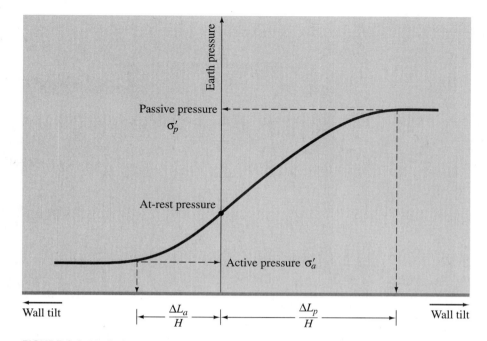

FIGURE 9.8 Variation of the magnitude of lateral earth pressure with wall tilt

9.3 Diagrams for Lateral Earth Pressure Distribution Against Retaining Walls

Backfill—Cohesionless Soil with Horizontal Ground Surface

Active Case Figure 9.9a shows a retaining wall with cohensionless soil backfill that has a horizontal ground surface. The unit weight and the angle of friction of the soil are γ and ϕ, respectively. For Rankine's active state, the earth pressure at any depth against the retaining wall can be given by Eq. (9.15):

$$\sigma_a = \sigma'_a = K_a \gamma z \qquad (Note: c = 0)$$

σ_a increases linearly with depth, and at the bottom of the wall, it will be

$$\sigma_a = K_a \gamma H \qquad (9.20)$$

The total force, P_a, per unit length of the wall is equal to the area of the pressure diagram, so

$$P_a = \frac{1}{2} K_a \gamma H^2 \qquad (9.21)$$

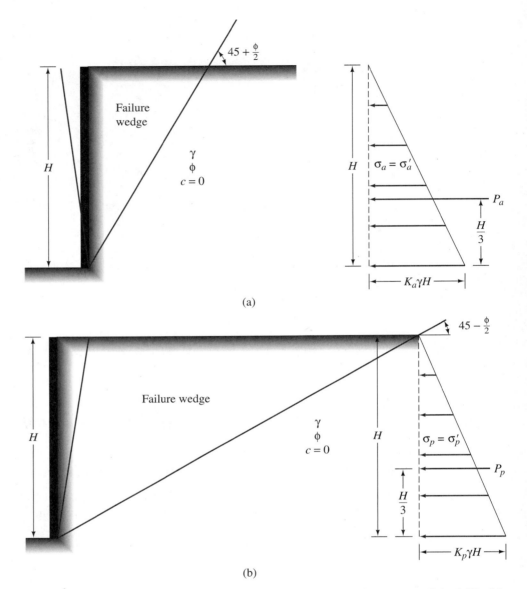

FIGURE 9.9 Pressure distribution against retaining wall for cohensionless soil backfill with horizontal ground surface: (a) Rankine's active state; (b) Rankine's passive state

Passive Case   The lateral pressure distribution against a retaining wall of height *H* for Rankine's passive state is shown in Figure 9.9b. The lateral earth pressure at any depth *z* [Eq. (9.19), *c* = 0] is

$$\sigma_p = \sigma'_p = K_p \gamma H \tag{9.22}$$

The total force, P_p, per unit length of the wall is

$$P_p = \frac{1}{2} K_p \gamma H^2 \tag{9.23}$$

Backfill—Partially Submerged Cohesionless Soil Supporting Surcharge

Active Case Figure 9.10a shows a frictionless retaining wall of height H and a backfill of cohesionless soil. The groundwater table is located at a depth of H_1 below the ground surface, and the backfill is supporting a surcharge pressure of q per unit area. From Eq. (9.17), we know that the effective active earth pressure at any depth can be given by

$$\sigma_a' = K_a \sigma_o' \tag{9.24}$$

where σ_o' and σ_a' are the effective vertical pressure and lateral pressure, respectively. At $z = 0$,

$$\sigma_o = \sigma_o' = q \tag{9.25}$$

and

$$\sigma_a = \sigma_a' = K_a q \tag{9.26}$$

At depth $z = H_1$,

$$\sigma_o = \sigma_o' = (q + \gamma H_1) \tag{9.27}$$

and

$$\sigma_a = \sigma_a' = K_a(q + \gamma H_1) \tag{9.28}$$

At depth $z = H$,

$$\sigma_o' = (q + \gamma H_1 + \gamma' H_2) \tag{9.29}$$

and

$$\sigma_a' = K_a(q + \gamma H_1 + \gamma' H_2) \tag{9.30}$$

where $\gamma' = \gamma_{\text{sat}} - \gamma_w$. The variation of σ_a' with depth is shown in Figure 9.10b.

The lateral pressure on the wall from the pore water between $z = 0$ and H_1 is 0, and for $z > H_1$, it increases linearly with depth (Figure 9.10c). At $z = H$,

$$u = \gamma_w H_2 \tag{9.31}$$

The total lateral pressure, σ_a, diagram (Figure 9.10d) is the sum of the pressure diagrams shown in Figures 9.10b and c. The total active force per unit length of the wall is the area of the total pressure diagram. Thus,

$$P_a = K_a q H + \frac{1}{2} K_a \gamma H_1^2 + K_a \gamma H_1 H_2 + \frac{1}{2}(K_a \gamma' + \gamma_w) H_2^2 \tag{9.32}$$

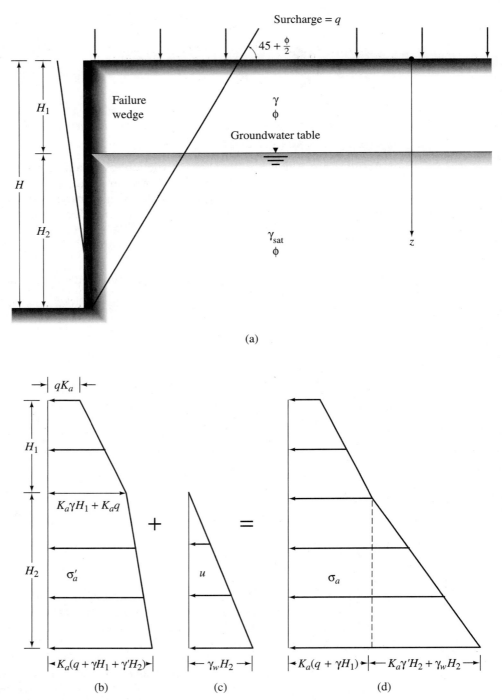

FIGURE 9.10 Rankine's active earth pressure distribution against retaining wall with partially submerged cohesionless soil backfill supporting surcharge

Passive Case   Figure 9.11a shows the same retaining wall as in Figure 9.10a. Rankine's passive pressure (effective) at any depth against the wall can be given by Eq. (9.19):

$$\sigma_p' = K_p \sigma_o'$$

Using the preceding equation, we can determine the variation of σ_p' with depth, as shown in Figure 9.11b. The variation of the pressure on the wall from water with depth is shown in Figure 9.11c. Figure 9.11d shows the distribution of the total pressure, σ_p, with depth. The total lateral passive force per unit length of the wall is the area of the diagram given in Figure 9.11d, or

$$P_p = K_p qH + \frac{1}{2} K_p \gamma H_1^2 + K_p \gamma H_1 H_2 + \frac{1}{2} (K_p \gamma' + \gamma_w) H_2^2 \qquad (9.33)$$

Backfill—Cohesive Soil with Horizontal Backfill

Active Case   Figure 9.12a shows a frictionless retaining wall with a cohesive soil backfill. The active pressure against the wall at any depth below the ground surface can be expressed as [Eq. (9.15)]

$$\sigma_a' = K_a \gamma z - 2c \sqrt{K_a}$$

The variation of $K_a \gamma z$ with depth is shown in Figure 9.12b, and the variation of $2c\sqrt{K_a}$ with depth is shown in Figure 9.12c. Note that $2c\sqrt{K_a}$ is not a function of z, and hence Figure 9.12c is a rectangle. The variation of the net value of σ_a with depth is plotted in Figure 9.12d. Also note that, because of the effect of cohesion, σ_a is negative in the upper part of the retaining wall. The depth z_o at which the active pressure becomes equal to 0 can be found from Eq. (9.15) as

$$K_a \gamma z_o - 2c \sqrt{K_a} = 0$$

or

$$z_o = \frac{2c}{\gamma \sqrt{K_a}} \qquad (9.34)$$

For the undrained condition—that is, $\phi = 0$, $K_a = \tan^2 45 = 1$, and $c = c_u$ (undrained cohesion)—we have

$$z_o = \frac{2c_u}{\gamma} \qquad (9.35)$$

So, with time, tensile cracks at the soil–wall interface will develop up to a depth of z_o.

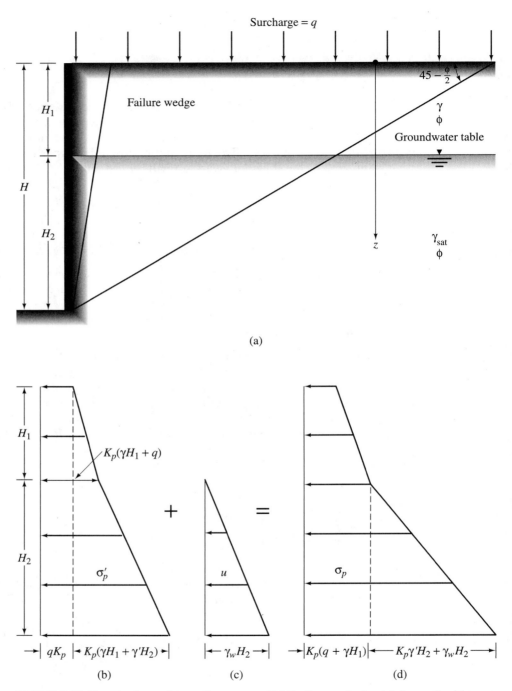

FIGURE 9.11 Rankine's passive earth pressure distribution against retaining wall with partially submerged cohesionless soil backfill supporting surcharge

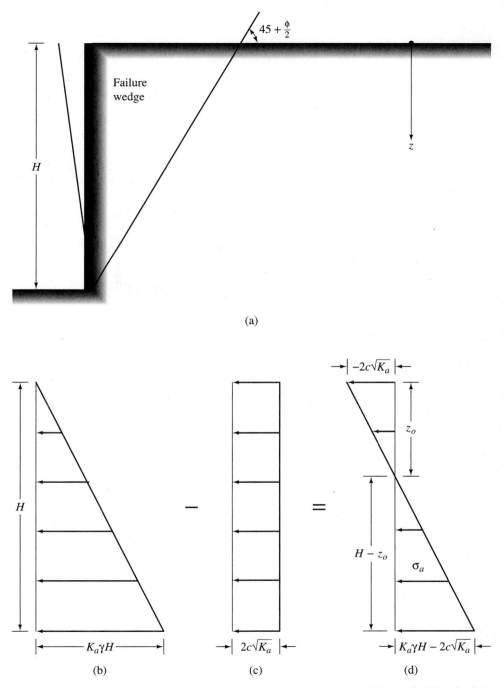

FIGURE 9.12 Rankine's active earth pressure distribution against retaining wall with cohesive soil backfill

The total active force per unit length of the wall can be found from the area of the total pressure diagram (Figure 9.12d), or

$$P_a = \frac{1}{2} K_a \gamma H^2 - 2\sqrt{K_a} c H \tag{9.36}$$

For $\phi = 0$ condition,

$$P_a = \frac{1}{2} \gamma H^2 - 2 c_u H \tag{9.37}$$

For calculation of the total active force, it is common practice to take the tensile cracks into account. Since there is no contact between the soil and the wall up to a depth of z_o after the development of tensile cracks, the active pressure distribution against the wall between $z = 2c/(\gamma\sqrt{K_a})$ and H (Figure 9.12d) only is considered. In that case,

$$
\begin{aligned}
P_a &= \frac{1}{2}(K_a \gamma H - 2\sqrt{K_a}c)\left(H - \frac{2c}{\gamma\sqrt{K_a}}\right) \\
&= \frac{1}{2} K_a \gamma H^2 - 2\sqrt{K_a}cH + 2\frac{c^2}{\gamma}
\end{aligned}
\tag{9.38}
$$

For the $\phi = 0$ condition,

$$P_a = \frac{1}{2} \gamma H^2 - 2 c_u H + 2\frac{c_u^2}{\gamma} \tag{9.39}$$

Note that, in Eq. (9.39), γ is the saturated unit weight of the soil.

Passive Case   Figure 9.13a shows the same retaining wall with backfill similar to that considered in Figure 9.12a. Rankine's passive pressure against the wall at depth z can be given by [Eq. (9.18)]

$$\sigma_p' = K_p \gamma z + 2\sqrt{K_p}c$$

At $z = 0$,

$$\sigma_p = 2\sqrt{K_p}c \tag{9.40}$$

and at $z = H$,

$$\sigma_p = K_p \gamma H + 2\sqrt{K_p}c \tag{9.41}$$

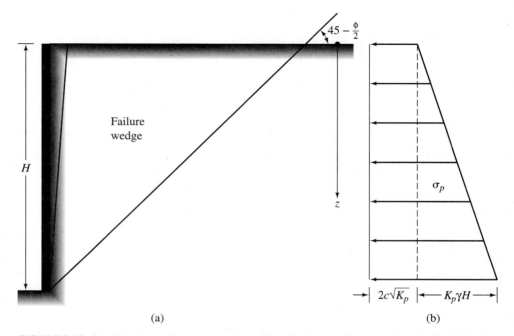

(a)   (b)

FIGURE 9.13 Rankine's passive earth pressure distribution against retaining wall with cohesive soil backfill

The variation of σ_p with depth is shown in Figure 9.13b. The passive force per unit length of the wall can be found from the area of the pressure diagrams as

$$P_p = \frac{1}{2} K_p \gamma H^2 + 2\sqrt{K_p}\, cH \qquad (9.42)$$

For the $\phi = 0$ condition, $K_p = 1$ and

$$P_p = \frac{1}{2} \gamma H^2 + 2c_u H \qquad (9.43)$$

In Eq. (9.43), γ is the saturated unit weight of the soil.

EXAMPLE 9.1

Calculate the Rankine active and passive forces per unit length of the wall shown in Figure 9.14a, and also determine the location of the resultant.

Solution   To determine the active force, since $c = 0$, we have

$$\sigma_a' = K_a \sigma_o' = K_a \gamma z$$

$$K_a = \frac{1 - \sin \phi}{1 + \sin \phi} = \frac{1 - \sin 30°}{1 + \sin 30°} = \frac{1}{3}$$

At $z = 0$, $\sigma_a' = 0$; at $z = 5$ m, $\sigma_a' = (1/3)(15.7)(5) = 26.2$ kN/m².

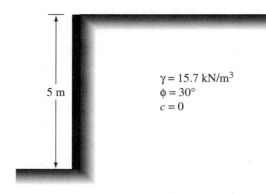

$\gamma = 15.7 \text{ kN/m}^3$
$\phi = 30°$
$c = 0$

5 m

(a)

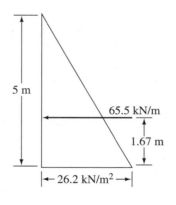

5 m

65.5 kN/m

1.67 m

26.2 kN/m^2

(b)

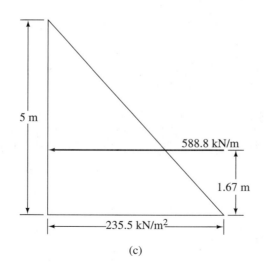

5 m

588.8 kN/m

1.67 m

235.5 kN/m^2

(c)

FIGURE 9.14

The active pressure distribution diagram is shown in Figure 9.14b:

$$\text{active force, } P_a = \frac{1}{2}(5)(26.2)$$

$$= \textbf{65.5 kN/m}$$

The total pressure distribution is triangular, and so P_a will act at a distance of $5/3 = 1.67$ m above the bottom of the wall.

To determine the passive force, we are given $c = 0$, so

$$\sigma'_p = \sigma_p = K_p\sigma'_o = K_p\gamma z$$

$$K_p = \frac{1 + \sin\phi}{1 - \sin\phi} = \frac{1 + 0.5}{1 - 0.5} = 3$$

At $z = 0$, $\sigma'_p = 0$; at $z = 5$ m, $\sigma'_p = 3(15.7)(5) = 235.5$ kN/m^2.

The total passive pressure distribution against the wall is shown in Figure 9.14c. Now,

$$P_p = \frac{1}{2}(5)(235.5) = \textbf{588.8 kN/m}$$

The resultant will act at a distance of $5/3 = 1.67$ m above the bottom of the wall.

∎

EXAMPLE 9.2

If the retaining wall shown in Figure 9.14a is restrained from moving, what will be the lateral force per unit length of the wall?

Solution If the wall is restrained from moving, the backfill will exert at-rest earth pressure. Thus,

$$\sigma'_h = \sigma_h = K_o\sigma'_o = K_o(\gamma z) \qquad \text{[Eq. (9.2)]}$$

$$K_o = 1 - \sin\phi \qquad \text{[Eq. (9.3)]}$$

or

$$K_o = 1 - \sin 30° = 0.5$$

and at $z = 0$, $\sigma'_h = 0$; at 5 m, $\sigma'_h = (0.5)(5)(15.7) = 39.3$ kN/m^2.

The total pressure distribution diagram is shown in Figure 9.15.

$$P_o = \frac{1}{2}(5)(39.3) = \textbf{98.3 kN/m}$$

∎

EXAMPLE 9.3

A retaining wall that has a soft, saturated clay backfill is shown in Figure 9.16. For the undrained condition ($\phi = 0$) of the backfill, determine the following values:

a. The maximum depth of the tensile crack
b. P_a before the tensile crack occurs
c. P_a after the tensile crack occurs

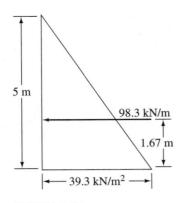

5 m

98.3 kN/m

1.67 m

|← 39.3 kN/m² →|

FIGURE 9.15

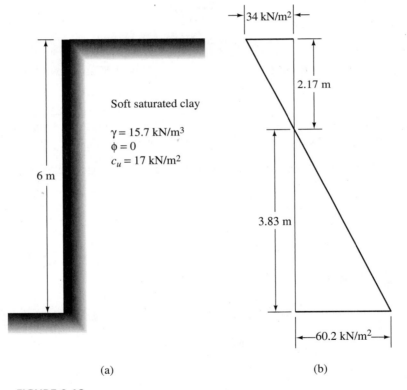

Soft saturated clay

$\gamma = 15.7$ kN/m³
$\phi = 0$
$c_u = 17$ kN/m²

6 m

→|34 kN/m²|←

2.17 m

3.83 m

|← 60.2 kN/m² →|

(a)   (b)

FIGURE 9.16

Solution For $\phi = 0$, $K_a = \tan^2 45 = 1$ and $c = c_u$. From Eq. (9.15), for the undrained condition, we have

$$\sigma_a = \gamma z - 2c_u$$

At $z = 0$,

$$\sigma_a = -2c_u = -(2)(17) = 34 \text{ kN/m}^2$$

At $z = 6$ m,

$$\sigma_a = (15.7)(6) - (2)(17) = 60.2 \text{ kN/m}^2$$

The variation of σ_a with depth is shown in Figure 9.16b.

a. From Eq. (9.35), the depth of the tensile crack equals

$$z_o = \frac{2c_u}{\gamma} = \frac{(2)(17)}{15.7} = \mathbf{2.17 \ m}$$

b. Before the tensile crack occurs [Eq.(9.37)],

$$P_a = \frac{1}{2}\gamma H^2 - 2c_u H$$

or

$$P_a = \frac{1}{2}(15.7)(6)^2 - 2(17)(6) = \mathbf{78.6 \ kN/m}$$

c. After the tensile crack occurs,

$$P_a = \frac{1}{2}(6 - 2.17)(60.2) = \mathbf{115.3 \ kN/m}$$

Note: The preceding P_a can also be obtained by substituting the proper values into Eq. (9.39). ∎

EXAMPLE 9.4

A frictionless retaining wall is shown in Figure 9.17a.

a. Determine the active force, P_a, after the tensile crack occurs.
b. What is the passive force, P_p?

Solution

a. Given $\phi = 26°$, we have

$$K_a = \frac{1 - \sin \phi}{1 + \sin \phi} = \frac{1 - \sin 26°}{1 + \sin 26°} = 0.39$$

From Eq. (9.15),

$$\sigma_a' = \sigma_a = K_a \sigma_o' - 2c\sqrt{K_a}$$

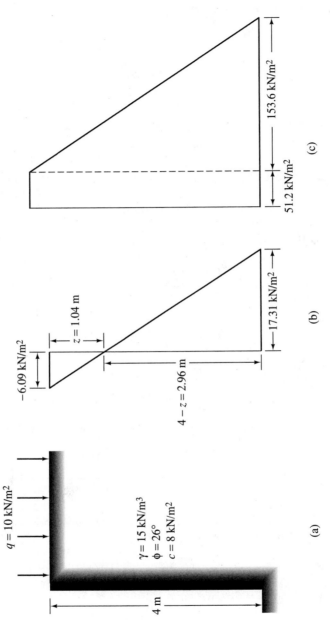

$q = 10$ kN/m^2

$\gamma = 15$ kN/m^3
$\phi = 26°$
$c = 8$ kN/m^2

4 m

(a)

-6.09 kN/m^2

$z = 1.04$ m

$4 - z = 2.96$ m

17.31 kN/m^2

(b)

153.6 kN/m^2

51.2 kN/m^2

(c)

FIGURE 9.17

At $z = 0$,

$$\sigma_a' = \sigma_a = (0.39)(10) - (2)(8)\sqrt{0.39} = 3.9 - 9.99 = -6.09 \text{ kN/m}^2$$

At $z = 4$ m,

$$\sigma_a' = \sigma_a = (0.39)[10 + (4)(15)] - (2)(8)\sqrt{0.39} = 27.3 - 9.99$$

$$= 17.31 \text{ kN/m}^2$$

The pressure distribution is shown in Figure 9.17b. From this diagram, we see that

$$\frac{6.09}{z} = \frac{17.31}{4 - z}$$

or

$$z = 1.04 \text{ m}$$

After the tensile crack occurs,

$$P_a = \frac{1}{2}(4 - z)(17.31) = \left(\frac{1}{2}\right)(2.96)(17.31) = \textbf{25.62 kN/m}$$

b. Given $\phi = 26°$, we have

$$K_p = \frac{1 + \sin\phi}{1 - \sin\phi} = \frac{1 + \sin 26°}{1 - \sin 26°} = \frac{1.4384}{0.5616} = 2.56$$

From Eq. (9.18),

$$\sigma_p' = \sigma_p = K_p\sigma_o' + 2\sqrt{K_p}c$$

At $z = 0$, $\sigma_o' = 10 \text{ kN/m}^2$ and

$$\sigma_p' = \sigma_p = (2.56)(10) + 2\sqrt{2.56}(8) = 25.6 + 25.6 = 51.2 \text{ kN/m}^2$$

Again, at $z = 4$ m, $\sigma_o = (10 + 4 \times 15) = 70 \text{ kN/m}^2$ and

$$\sigma_p' = \sigma_p = (2.56)(70) + 2\sqrt{2.56}(8) = 204.8 \text{ kN/m}^2$$

The distribution of σ_p ($=\sigma_p'$) is shown in Figure 9.17c. The lateral force per unit length of the wall is

$$P_p = (51.2)(4) + \frac{1}{2}(4)(153.6) = 204.8 + 307.2 = \textbf{512 kN/m}$$ ∎

EXAMPLE 9.5

A retaining wall is shown in Figure 9.18a. Determine Rankine's active force, P_a, per unit length of the wall. Also determine the location of the resultant.

Solution Given $c = 0$, we know that $\sigma_a' = K_a\sigma_o'$. For the upper layer of the soil, Rankine's active earth pressure coefficient is

$$K_a = K_{a(1)} = \frac{1 - \sin 30°}{1 + \sin 30°} = \frac{1}{3}$$

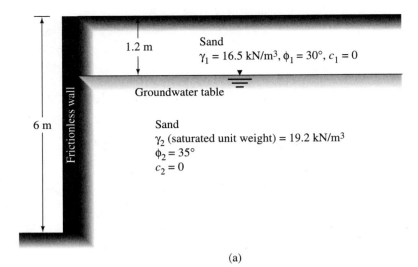

(a)

FIGURE 9.18

For the lower layer,

$$K_a = K_{a(2)} = \frac{1 - \sin 35°}{1 + \sin 35°} = \frac{0.4264}{1.5736} = 0.271$$

At $z = 0$, $\sigma_o = \sigma'_o = 0$. At $z = 1.2$ m (just inside the bottom of the upper layer), $\sigma_o = \sigma'_o = (1.2)(16.5) = 19.8$ kN/m². So

$$\sigma_a = \sigma'_a = K_{a(1)}\sigma'_o = \left(\frac{1}{3}\right)(19.8) = 6.6 \text{ kN/m}^2$$

Again, at $z = 1.2$ m (in the lower layer), $\sigma_o = \sigma'_o = (1.2)(16.5) = 19.8$ kN/m², and

$$\sigma_a = \sigma'_a = K_{a(2)}\sigma'_o = (0.271)(19.8) = 5.37 \text{ kN/m}^2$$

At $z = 6$ m,

$$\sigma'_o = (1.2)(16.5) + (4.8)(19.2 - 9.81) = 64.87 \text{ kN/m}^2$$
$$\uparrow$$
$$\gamma_w$$

and

$$\sigma'_a = K_{a(2)}\sigma'_o = (0.271)(64.87) = 17.58 \text{ kN/m}^2$$

The variation of σ'_a with depth is shown in Figure 9.18b.
    The lateral pressures from the pore water are as follows:

    At $z = 0, u = 0$

    At $z = 1.2$ m, $u = 0$

    At $z = 6$ m, $u = (4.8)(\gamma_w) = (4.8)(9.81) = 47.1$ kN/m²

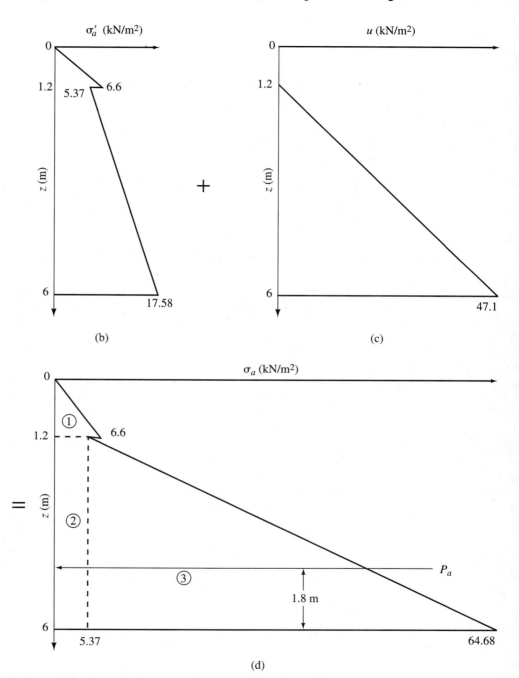

FIGURE 9.18 (Continued)

The variation of u with depth is shown in Figure 9.18c, and the variation for σ_a (total active pressure) is shown in Figure 9.18d. Thus,

$$P_a = \left(\frac{1}{2}\right)(6.6)(1.2) + (4.8)(5.37) + \left(\frac{1}{2}\right)(4.8)(64.68 - 5.37)$$

$$= 3.96 + 25.78 + 142.34 = \textbf{172.08 kN/m}$$

The location of the resultant can be found by taking the moment about the bottom of the wall. Thus,

$$\bar{z} = \frac{3.96\left(4.8 + \dfrac{1.2}{3}\right) + (25.78)(2.4) + (142.34)\left(\dfrac{4.8}{3}\right)}{172.08} = \textbf{1.8 m} \qquad \blacksquare$$

9.4   *Retaining Walls with Friction*

So far in our study of active and passive earth pressures, we have considered the case of frictionless walls. In reality, retaining walls are rough, and shear forces develop between the face of the wall and the backfill. To understand the effect of wall friction on the failure surface, let us consider a rough retaining wall AB with a horizontal granular backfill, as shown in Figure 9.19.

In the active case (Figure 9.19a), when the wall AB moves to a position $A'B$, the soil mass in the active zone will be stretched outward. This will cause a downward motion of the soil relative to the wall. This motion causes a downward shear on the wall (Figure 9.19b), and it is called *positive wall friction in the active case.* If δ is the angle of friction between the wall and the backfill, then the resultant active force, P_a, will be inclined at an angle δ to the normal drawn to the back face of the retaining wall. Advanced studies show that the failure surface in the backfill can be represented by BCD, as shown in Figure 9.19a. The portion BC is curved, and the portion CD of the failure surface is a straight line. Rankine's active state exists in the zone ACD.

Under certain conditions, if the wall shown in Figure 9.19a is forced downward relative to the backfill, then the direction of the active force, P_a, will change as shown in Figure 9.19c. This is a situation of *negative wall friction in the active case* $(-\delta)$. Figure 9.19c also shows the nature of the failure surface in the backfill.

The effect of wall friction for the passive state is shown in Figures 9.19d and e. When the wall AB is pushed to a position $A'B$ (Figure 9.19d), the soil in the passive zone will be compressed. The result is an upward motion relative to the wall. The upward motion of the soil will cause an upward shear on the retaining wall (Figure 9.19e). This is referred to as *positive wall friction in the passive case.* The resultant passive force, P_p, will be inclined at an angle δ to the normal drawn to the back face of the wall. The failure surface in the soil has a curved lower portion BC and a straight upper portion CD. Rankine's passive state exists in the zone ACD.

If the wall shown in Figure 9.19d is forced upward relative to the backfill, then the direction of the passive force, P_p, will change as shown in Figure 9.19f.

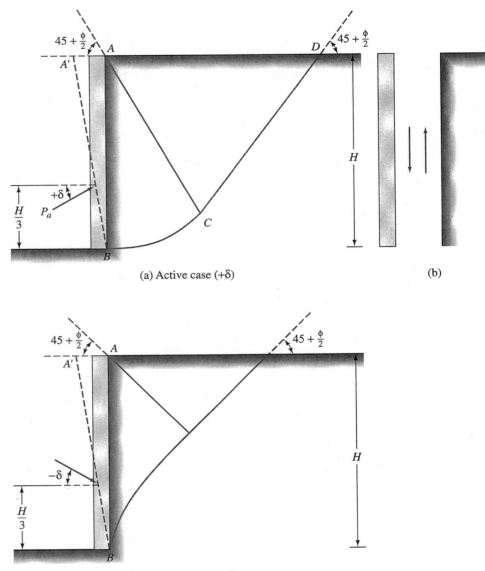

(a) Active case (+δ)

(b)

(c) Active case (−δ)

FIGURE 9.19 Effect of wall friction on failure surface

This is *negative wall friction in the passive case* $(-\delta)$. Figure 9.19f also shows the nature of the failure surface in the backfill under such a condition.

For practical considerations, in the case of loose granular backfill, the angle of wall friction δ is taken to be equal to the angle of friction of the soil, ϕ. For dense granular backfills, δ is smaller than ϕ and is in the range $\phi/2 \leq \delta \leq (2/3)\phi$.

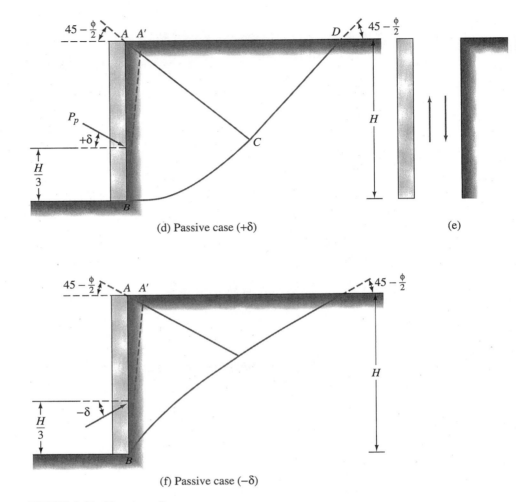

(d) Passive case (+δ)   (e)

(f) Passive case (−δ)

FIGURE 9.19 (Continued)

9.5   *Coulomb's Earth Pressure Theory*

More than 200 years ago, Coulomb (1776) presented a theory for active and passive earth pressures against retaining walls. In this theory, Coulomb assumed that the failure surface is a plane. The wall friction was taken into consideration. The general principles of the derivation of Coulomb's earth pressure theory for a cohesionless backfill (shear strength defined by the equation $\tau_f = \sigma' \tan \phi$) are given in this section.

Active Case

Let AB (Figure 9.20a) be the back face of a retaining wall supporting a granular soil, the surface of which is constantly sloping at an angle α with the horizontal.

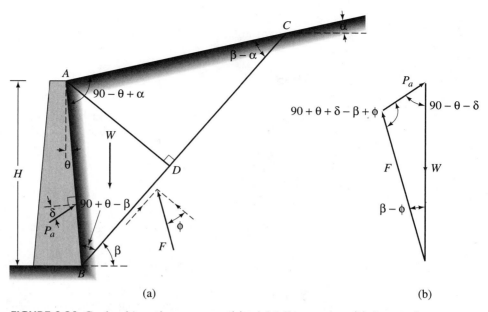

FIGURE 9.20 Coulomb's active pressure: (a) trial failure wedge; (b) force polygon

BC is a trial failure surface. In the stability consideration of the probable failure wedge ABC, the following forces are involved (per unit length of the wall):

1. W, the effective weight of the soil wedge.
2. F, the resultant of the shear and normal forces on the surface of failure, BC. This is inclined at an angle ϕ to the normal drawn to the plane BC.
3. P_a, the active force per unit length of the wall. The direction of P_a is inclined at an angle δ to the normal drawn to the face of the wall that supports the soil. δ is the angle of friction between the soil and the wall.

The force triangle for the wedge is shown in Figure 9.20b. From the law of sines, we have

$$\frac{W}{\sin(90 + \theta + \delta - \beta + \phi)} = \frac{P_a}{\sin(\beta - \phi)} \tag{9.44}$$

or

$$P_a = \frac{\sin(\beta - \phi)}{\sin(90 + \theta + \delta - \beta + \phi)} W \tag{9.45}$$

The preceding equation can be written in the form

$$P_a = \frac{1}{2}\gamma H^2 \left[\frac{\cos(\theta - \beta)\cos(\theta - \alpha)\sin(\beta - \phi)}{\cos^2\theta \sin(\beta - \alpha)\sin(90 + \theta + \delta - \beta + \phi)} \right] \tag{9.46}$$

where γ = unit weight of the backfill. The values of γ, H, θ, α, ϕ, and δ are constants, and β is the only variable. To determine the critical value of β for maximum P_a, we have

$$\frac{dP_a}{d\beta} = 0 \tag{9.47}$$

After solving Eq. (9.47), when the relationship of β is substituted into Eq. (9.46), we obtain Coulomb's active earth pressure as

$$P_a = \frac{1}{2} K_a \gamma H^2 \tag{9.48}$$

where K_a is Coulomb's active earth pressure coefficient, given by

$$K_a = \frac{\cos^2(\phi - \theta)}{\cos^2 \theta \cos(\delta + \theta) \left[1 + \sqrt{\dfrac{\sin(\delta + \phi) \sin(\phi - \alpha)}{\cos(\delta + \theta) \cos(\theta - \alpha)}} \right]^2} \tag{9.49}$$

Note that when $\alpha = 0°$, $\theta = 0°$, and $\delta = 0°$, Coulomb's active earth pressure coefficient becomes equal to $(1 - \sin \phi)/(1 + \sin \phi)$, which is the same as Rankine's earth pressure coefficient given earlier in this chapter.

The variation of the values of K_a for retaining walls with a vertical back ($\theta = 0$) and horizontal backfill ($\alpha = 0$) is given in Table 9.2. From this table note that for a given value of ϕ, the effect of wall friction is to reduce somewhat the active earth pressure coefficient.

Tables 9.3 and 9.4 give the values of K_a [Eq. (9.49)] for $\delta = \frac{2}{3}\phi$ and $\delta = \phi/2$. These tables may be useful in retaining wall design (see Chapter 11).

Table 9.2 Values of K_a [Eq. (9.49)] for $\theta = 0°$, $\alpha = 0°$

	δ (deg) $\rightarrow$					
$\downarrow \phi$ (deg)	0	5	10	15	20	25
28	0.3610	0.3448	0.3330	0.3251	0.3203	0.3186
30	0.3333	0.3189	0.3085	0.3014	0.2973	0.2956
32	0.3073	0.2945	0.2853	0.2791	0.2755	0.2745
34	0.2827	0.2714	0.2633	0.2579	0.2549	0.2542
36	0.2596	0.2497	0.2426	0.2379	0.2354	0.2350
38	0.2379	0.2292	0.2230	0.2190	0.2169	0.2167
40	0.2174	0.2089	0.2045	0.2011	0.1994	0.1995
42	0.1982	0.1916	0.1870	0.1841	0.1828	0.1831

Table 9.3 Values of K_a [Eq. (9.49)]. *Note:* $\delta = \frac{2}{3}\phi$

α (deg)	ϕ (deg)	θ (deg)					
		0	5	10	15	20	25
0	28	0.3213	0.3588	0.4007	0.4481	0.5026	0.5662
	29	0.3091	0.3467	0.3886	0.4362	0.4908	0.5547
	30	0.2973	0.3349	0.3769	0.4245	0.4794	0.5435
	31	0.2860	0.3235	0.3655	0.4133	0.4682	0.5326
	32	0.2750	0.3125	0.3545	0.4023	0.4574	0.5220
	33	0.2645	0.3019	0.3439	0.3917	0.4469	0.5117
	34	0.2543	0.2916	0.3335	0.3813	0.4367	0.5017
	35	0.2444	0.2816	0.3235	0.3713	0.4267	0.4919
	36	0.2349	0.2719	0.3137	0.3615	0.4170	0.4824
	37	0.2257	0.2626	0.3042	0.3520	0.4075	0.4732
	38	0.2168	0.2535	0.2950	0.3427	0.3983	0.4641
	39	0.2082	0.2447	0.2861	0.3337	0.3894	0.4553
	40	0.1998	0.2361	0.2774	0.3249	0.3806	0.4468
	41	0.1918	0.2278	0.2689	0.3164	0.3721	0.4384
	42	0.1840	0.2197	0.2606	0.3080	0.3637	0.4302
5	28	0.3431	0.3845	0.4311	0.4843	0.5461	0.6190
	29	0.3295	0.3709	0.4175	0.4707	0.5325	0.6056
	30	0.3165	0.3578	0.4043	0.4575	0.5194	0.5926
	31	0.3039	0.3451	0.3916	0.4447	0.5067	0.5800
	32	0.2919	0.3329	0.3792	0.4324	0.4943	0.5677
	33	0.2803	0.3211	0.3673	0.4204	0.4823	0.5558
	34	0.2691	0.3097	0.3558	0.4088	0.4707	0.5443
	35	0.2583	0.2987	0.3446	0.3975	0.4594	0.5330
	36	0.2479	0.2881	0.3338	0.3866	0.4484	0.5221
	37	0.2379	0.2778	0.3233	0.3759	0.4377	0.5115
	38	0.2282	0.2679	0.3131	0.3656	0.4273	0.5012
	39	0.2188	0.2582	0.3033	0.3556	0.4172	0.4911
	40	0.2098	0.2489	0.2937	0.3458	0.4074	0.4813
	41	0.2011	0.2398	0.2844	0.3363	0.3978	0.4718
	42	0.1927	0.2311	0.2753	0.3271	0.3884	0.4625
10	28	0.3702	0.4164	0.4686	0.5287	0.5992	0.6834
	29	0.3548	0.4007	0.4528	0.5128	0.5831	0.6672
	30	0.3400	0.3857	0.4376	0.4974	0.5676	0.6516
	31	0.3259	0.3713	0.4230	0.4826	0.5526	0.6365
	32	0.3123	0.3575	0.4089	0.4683	0.5382	0.6219
	33	0.2993	0.3442	0.3953	0.4545	0.5242	0.6078
	34	0.2868	0.3314	0.3822	0.4412	0.5107	0.5942
	35	0.2748	0.3190	0.3696	0.4283	0.4976	0.5810
	36	0.2633	0.3072	0.3574	0.4158	0.4849	0.5682
	37	0.2522	0.2957	0.3456	0.4037	0.4726	0.5558
	38	0.2415	0.2846	0.3342	0.3920	0.4607	0.5437
	39	0.2313	0.2740	0.3231	0.3807	0.4491	0.5321
	40	0.2214	0.2636	0.3125	0.3697	0.4379	0.5207
	41	0.2119	0.2537	0.3021	0.3590	0.4270	0.5097
	42	0.2027	0.2441	0.2921	0.3487	0.4164	0.4990

Table 9.3 (Continued)

α (deg)	ϕ (deg)	θ (deg)					
		0	5	10	15	20	25
15	28	0.4065	0.4585	0.5179	0.5868	0.6685	0.7670
	29	0.3881	0.4397	0.4987	0.5672	0.6483	0.7463
	30	0.3707	0.4219	0.4804	0.5484	0.6291	0.7265
	31	0.3541	0.4049	0.4629	0.5305	0.6106	0.7076
	32	0.3384	0.3887	0.4462	0.5133	0.5930	0.6895
	33	0.3234	0.3732	0.4303	0.4969	0.5761	0.6721
	34	0.3091	0.3583	0.4150	0.4811	0.5598	0.6554
	35	0.2954	0.3442	0.4003	0.4659	0.5442	0.6393
	36	0.2823	0.3306	0.3862	0.4513	0.5291	0.6238
	37	0.2698	0.3175	0.3726	0.4373	0.5146	0.6089
	38	0.2578	0.3050	0.3595	0.4237	0.5006	0.5945
	39	0.2463	0.2929	0.3470	0.4106	0.4871	0.5805
	40	0.2353	0.2813	0.3348	0.3980	0.4740	0.5671
	41	0.2247	0.2702	0.3231	0.3858	0.4613	0.5541
	42	0.2146	0.2594	0.3118	0.3740	0.4491	0.5415
20	28	0.4602	0.5205	0.5900	0.6714	0.7689	0.8880
	29	0.4364	0.4958	0.5642	0.6445	0.7406	0.8581
	30	0.4142	0.4728	0.5403	0.6195	0.7144	0.8303
	31	0.3935	0.4513	0.5179	0.5961	0.6898	0.8043
	32	0.3742	0.4311	0.4968	0.5741	0.6666	0.7799
	33	0.3559	0.4121	0.4769	0.5532	0.6448	0.7569
	34	0.3388	0.3941	0.4581	0.5335	0.6241	0.7351
	35	0.3225	0.3771	0.4402	0.5148	0.6044	0.7144
	36	0.3071	0.3609	0.4233	0.4969	0.5856	0.6947
	37	0.2925	0.3455	0.4071	0.4799	0.5677	0.6759
	38	0.2787	0.3308	0.3916	0.4636	0.5506	0.6579
	39	0.2654	0.3168	0.3768	0.4480	0.5342	0.6407
	40	0.2529	0.3034	0.3626	0.4331	0.5185	0.6242
	41	0.2408	0.2906	0.3490	0.4187	0.5033	0.6083
	42	0.2294	0.2784	0.3360	0.4049	0.4888	0.5930

Passive Case

Figure 9.21a shows a retaining wall with a sloping cohensionless backfill similar to that considered in Figure 9.20a. The force polygon for equilibrium of the wedge ABC for the passive state is shown in Figure 9.21b. P_p is the notation for the passive force. Other notations used are the same as those for the active case considered in this section. In a procedure similar to the one we followed in the active case, we get

$$P_p = \frac{1}{2} K_p \gamma H^2 \tag{9.50}$$

Table 9.4 Values of K_a [Eq. (9.49)]. *Note:* $\delta = \phi/2$

α (deg)	ϕ (deg)	θ (deg) 0	5	10	15	20	25
0	28	0.3264	0.3629	0.4034	0.4490	0.5011	0.5616
	29	0.3137	0.3502	0.3907	0.4363	0.4886	0.5492
	30	0.3014	0.3379	0.3784	0.4241	0.4764	0.5371
	31	0.2896	0.3260	0.3665	0.4121	0.4645	0.5253
	32	0.2782	0.3145	0.3549	0.4005	0.4529	0.5137
	33	0.2671	0.3033	0.3436	0.3892	0.4415	0.5025
	34	0.2564	0.2925	0.3327	0.3782	0.4305	0.4915
	35	0.2461	0.2820	0.3221	0.3675	0.4197	0.4807
	36	0.2362	0.2718	0.3118	0.3571	0.4092	0.4702
	37	0.2265	0.2620	0.3017	0.3469	0.3990	0.4599
	38	0.2172	0.2524	0.2920	0.3370	0.3890	0.4498
	39	0.2081	0.2431	0.2825	0.3273	0.3792	0.4400
	40	0.1994	0.2341	0.2732	0.3179	0.3696	0.4304
	41	0.1909	0.2253	0.2642	0.3087	0.3602	0.4209
	42	0.1828	0.2168	0.2554	0.2997	0.3511	0.4117
5	28	0.3477	0.3879	0.4327	0.4837	0.5425	0.6115
	29	0.3337	0.3737	0.4185	0.4694	0.5282	0.5972
	30	0.3202	0.3601	0.4048	0.4556	0.5144	0.5833
	31	0.3072	0.3470	0.3915	0.4422	0.5009	0.5698
	32	0.2946	0.3342	0.3787	0.4292	0.4878	0.5566
	33	0.2825	0.3219	0.3662	0.4166	0.4750	0.5437
	34	0.2709	0.3101	0.3541	0.4043	0.4626	0.5312
	35	0.2596	0.2986	0.3424	0.3924	0.4505	0.5190
	36	0.2488	0.2874	0.3310	0.3808	0.4387	0.5070
	37	0.2383	0.2767	0.3199	0.3695	0.4272	0.4954
	38	0.2282	0.2662	0.3092	0.3585	0.4160	0.4840
	39	0.2185	0.2561	0.2988	0.3478	0.4050	0.4729
	40	0.2090	0.2463	0.2887	0.3374	0.3944	0.4620
	41	0.1999	0.2368	0.2788	0.3273	0.3840	0.4514
	42	0.1911	0.2276	0.2693	0.3174	0.3738	0.4410
10	28	0.3743	0.4187	0.4688	0.5261	0.5928	0.6719
	29	0.3584	0.4026	0.4525	0.5096	0.5761	0.6549
	30	0.3432	0.3872	0.4368	0.4936	0.5599	0.6385
	31	0.3286	0.3723	0.4217	0.4782	0.5442	0.6225
	32	0.3145	0.3580	0.4071	0.4633	0.5290	0.6071
	33	0.3011	0.3442	0.3930	0.4489	0.5143	0.5920
	34	0.2881	0.3309	0.3793	0.4350	0.5000	0.5775
	35	0.2757	0.3181	0.3662	0.4215	0.4862	0.5633
	36	0.2637	0.3058	0.3534	0.4084	0.4727	0.5495
	37	0.2522	0.2938	0.3411	0.3957	0.4597	0.5361
	38	0.2412	0.2823	0.3292	0.3833	0.4470	0.5230
	39	0.2305	0.2712	0.3176	0.3714	0.4346	0.5103
	40	0.2202	0.2604	0.3064	0.3597	0.4226	0.4979
	41	0.2103	0.2500	0.2956	0.3484	0.4109	0.4858
	42	0.2007	0.2400	0.2850	0.3375	0.3995	0.4740

Table 9.4 (Continued)

α (deg)	ϕ (deg)	θ (deg)					
		0	5	10	15	20	25
15	28	0.4095	0.4594	0.5159	0.5812	0.6579	0.7498
	29	0.3908	0.4402	0.4964	0.5611	0.6373	0.7284
	30	0.3730	0.4220	0.4777	0.5419	0.6175	0.7080
	31	0.3560	0.4046	0.4598	0.5235	0.5985	0.6884
	32	0.3398	0.3880	0.4427	0.5059	0.5803	0.6695
	33	0.3244	0.3721	0.4262	0.4889	0.5627	0.6513
	34	0.3097	0.3568	0.4105	0.4726	0.5458	0.6338
	35	0.2956	0.3422	0.3953	0.4569	0.5295	0.6168
	36	0.2821	0.3282	0.3807	0.4417	0.5138	0.6004
	37	0.2692	0.3147	0.3667	0.4271	0.4985	0.5846
	38	0.2569	0.3017	0.3531	0.4130	0.4838	0.5692
	39	0.2450	0.2893	0.3401	0.3993	0.4695	0.5543
	40	0.2336	0.2773	0.3275	0.3861	0.4557	0.5399
	41	0.2227	0.2657	0.3153	0.3733	0.4423	0.5258
	42	0.2122	0.2546	0.3035	0.3609	0.4293	0.5122
20	28	0.4614	0.5188	0.5844	0.6608	0.7514	0.8613
	29	0.4374	0.4940	0.5586	0.6339	0.7232	0.8313
	30	0.4150	0.4708	0.5345	0.6087	0.6968	0.8034
	31	0.3941	0.4491	0.5119	0.5851	0.6720	0.7772
	32	0.3744	0.4286	0.4906	0.5628	0.6486	0.7524
	33	0.3559	0.4093	0.4704	0.5417	0.6264	0.7289
	34	0.3384	0.3910	0.4513	0.5216	0.6052	0.7066
	35	0.3218	0.3736	0.4331	0.5025	0.5851	0.6853
	36	0.3061	0.3571	0.4157	0.4842	0.5658	0.6649
	37	0.2911	0.3413	0.3991	0.4668	0.5474	0.6453
	38	0.2769	0.3263	0.3833	0.4500	0.5297	0.6266
	39	0.2633	0.3120	0.3681	0.4340	0.5127	0.6085
	40	0.2504	0.2982	0.3535	0.4185	0.4963	0.5912
	41	0.2381	0.2851	0.3395	0.4037	0.4805	0.5744
	42	0.2263	0.2725	0.3261	0.3894	0.4653	0.5582

where K_p = coefficient of passive earth pressure for Coulomb's case, or

$$K_p = \frac{\cos^2(\phi + \theta)}{\cos^2 \theta \cos(\delta - \theta)\left[1 - \sqrt{\dfrac{\sin(\phi - \delta)\sin(\phi + \alpha)}{\cos(\delta - \theta)\cos(\alpha - \theta)}}\right]^2} \tag{9.51}$$

For a frictionless wall with the vertical back face supporting granular soil backfill with a horizontal surface (that is, $\theta = 0°$, $\alpha = 0°$, and $\delta = 0°$), Eq. (9.51) yields

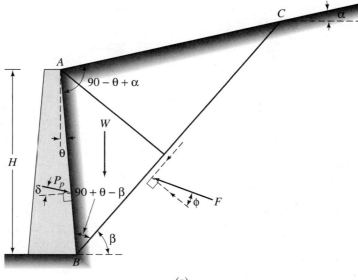

(a)

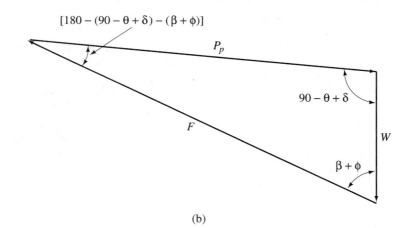

(b)

FIGURE 9.21 Coulomb's passive pressure: (a) trial failure wedge;
(b) force polygon

$$K_p = \frac{1 + \sin \phi}{1 - \sin \phi} = \tan^2 \left(45 + \frac{\phi}{2} \right)$$

This is the same relationship that was obtained for the passive earth pressure coefficient in Rankine's case given by Eq. (9.19).

Table 9.5 Values of K_p [Eq. (9.51)] for $\theta = 0°$ and $\alpha = 0°$

$\downarrow \phi$ (deg)	δ (deg) $\rightarrow$				
	0	5	10	15	20
15	1.698	1.900	2.130	2.405	2.735
20	2.040	2.313	2.636	3.030	3.525
25	2.464	2.830	3.286	3.855	4.597
30	3.000	3.506	4.143	4.977	6.105
35	3.690	4.390	5.310	6.854	8.324
40	4.600	5.590	6.946	8.870	11.772

The variation of K_p with ϕ and δ (for $\theta = 0$ and $\alpha = 0$) is given in Table 9.5. It can be observed from this table that, for given values of α and ϕ, the value of K_p increases with the wall friction. *Note that making the assumption that the failure surface is a plane in Coulomb's theory grossly overestimates the passive resistance of walls, particularly for $\delta > \phi/2$.* This error is somewhat unsafe for all design purposes.

9.6   *Approximate Analysis of Active Force on Retaining Walls*

In practical design considerations, the active force on a retaining wall can be calculated by using either Coulomb's or Rankine's method. The procedure for such calculations for a gravity retaining wall with granular backfill is shown in Figure 9.22.

Figure 9.22a shows a gravity retaining wall with its backfill that has a horizontal ground surface. If Coulomb's method is used, the active thrust per unit length of the wall, P_a, can be determined by Eq. (9.48). This force will act at an angle δ to the normal drawn to the back face of the wall. If Rankine's method is used, the active thrust is calculated on a vertical plane drawn through the heel of the wall [Eq. (9.21)]:

$$P_a = \frac{1}{2} K_a \gamma H^2$$

where $K_a = \dfrac{1 - \sin \phi}{1 + \sin \phi} = \tan^2 \left(45 - \dfrac{\phi}{2} \right)$.

In such a case, $P_{a(\text{Rankine})}$ is added vectorally to the weight of the wedge of soil, W_s, for the stability analysis.

Figure 9.22b shows a similar retaining wall with a granular backfill that has an inclined ground surface. Equation (9.48), or Rankine's solution, may be used to determine the active force on a vertical plane through the heel of the wall, which can then be added vectorally to the weight of the wedge of soil ABC_2 for stability analysis. Note, however, that the direction of Rankine's active force in this case is

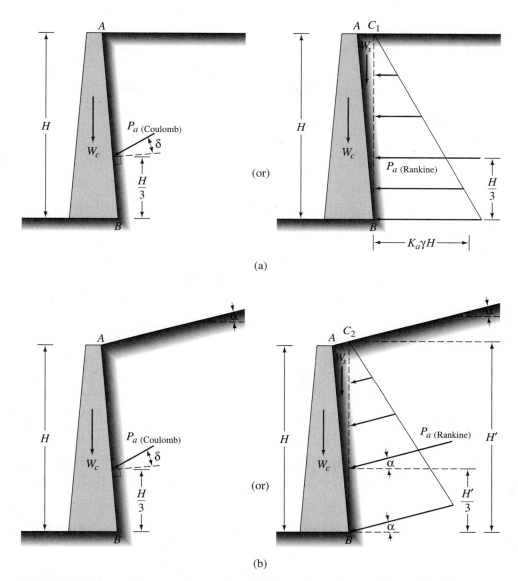

FIGURE 9.22 Approximate analyses of active force on gravity retaining walls with granular backfill

not horizontal anymore, and the vertical plane BC_2 is not the minor principal plane. The value of $P_{a(\text{Rankine})}$ can be given by the relationship

$$P_a = \frac{1}{2} K_a \gamma H'^2 \tag{9.52}$$

Table 9.6 Values of K_a [Eq. (9.53)]

↓ α (deg)	ϕ (deg) →						
	28	30	32	34	36	38	40
0	0.361	0.333	0.307	0.283	0.260	0.238	0.217
5	0.366	0.337	0.311	0.286	0.262	0.240	0.219
10	0.380	0.350	0.321	0.294	0.270	0.246	0.225
15	0.409	0.373	0.341	0.311	0.283	0.258	0.235
20	0.461	0.414	0.374	0.338	0.306	0.277	0.250
25	0.573	0.494	0.434	0.385	0.343	0.307	0.275

where $H' = \overline{BC_2}$ and

$$K_a = \text{Rankine's active pressure coefficient}$$
$$= \cos \alpha \, \frac{\cos \alpha - \sqrt{\cos^2 \alpha - \cos^2 \phi}}{\cos \alpha + \sqrt{\cos^2 \alpha - \cos^2 \phi}} \tag{9.53}$$

where α = slope of the ground surface.

P_a obtained from Eq. (9.52) acts at a distance of $H'/3$ measured vertically from B and inclined at an angle α with the horizontal. The values of K_a defined by Eq. (9.53) for various slope angles and soil friction angles are given in Table 9.6. For a horizontal ground surface (that is, $\alpha = 0$), Eq. (9.53) translates to

$$K_a = \frac{1 - \sin \phi}{1 + \sin \phi} = \tan^2 \left(45 - \frac{\phi}{2} \right)$$

Problems

9.1 Assuming that the wall shown in Figure 9.23 is restrained from yielding, find the magnitude and location of the resultant lateral force per unit length of the wall for the following cases:
  a. $H = 5$ m, $\gamma = 14.4$ kN/m³, $\phi = 31°$
  b. $H = 4$ m, $\gamma = 13.4$ kN/m³, $\phi = 28°$

9.2 Figure 9.23 shows a retaining wall with cohesionless soil backfill. For the following cases, determine the total active force per unit length of the wall for Rankine's state and the location of the resultant.
  a. $H = 4.5$ m, $\gamma = 17.6$ kN/m³, $\phi = 36°$
  b. $H = 5$ m, $\gamma = 17.0$ kN/m³, $\phi = 38°$
  c. $H = 4$ m, $\gamma = 19.95$ kN/m³, $\phi = 42°$

FIGURE 9.23

9.3 From Figure 9.23, determine the passive force, P_p, per unit length of the wall for Rankine's case. Also state Rankine's passive pressure at the bottom of the wall. Consider the following cases:
   a. $H = 2.45$ m, $\gamma = 16.67$ kN/m³, $\phi = 33°$
   b. $H = 4$ m, $\rho = 1800$ kg/m³, $\phi = 38°$

9.4 A retaining wall is shown in Figure 9.24. Determine Rankine's active force,

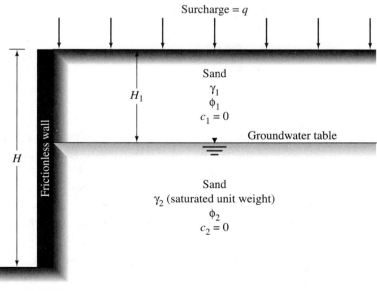

FIGURE 9.24

P_a, per unit length of the wall and the location of the resultant for each of the following cases:

a. $H = 6$ m, $H_1 = 2$ m, $\gamma_1 = 16$ kN/m³, $\gamma_2 = 19$ kN/m³, $\phi_1 = 32°$, $\phi_2 = 36°$, $q = 15$ kN/m²

b. $H = 5$ m, $H_1 = 1.5$ m, $\gamma_1 = 17.2$ kN/m³, $\gamma_2 = 20.4$ kN/m³, $\phi_1 = 30°$, $\phi_2 = 34°$, $q = 19.15$ kN/m²

9.5 Refer to Figure 9.24. Determine Rankine's passive force, P_p, per unit length of the wall for the following cases. Also find the location of the resultant for each case.

a. $H = 5$ m, $H_1 = 1.5$ m, $\gamma_1 = 16.5$ kN/m³, $\gamma_2 = 19$ kN/m³, $\phi_1 = 30°$, $\phi_2 = 36°$, $q = 0$

b. $H = 6$ m, $H_1 = 2$ m, $\gamma_1 = 17$ kN/m³, $\gamma_2 = 19.8$ kN/m³, $\phi_1 = 34°$, $\phi_2 = 34°$, $q = 14$ kN/m²

9.6 A retaining wall 6 m high with a vertical back face retains a homogeneous saturated soft clay. The saturated unit weight of the clay is 19 kN/m³. Laboratory tests showed that the undrained shear strength, c_u, of the clay is 16.8 kN/m².

a. Do the necessary calculations and draw the variation of Rankine's active pressure on the wall with depth.

b. Find the depth up to which a tensile crack can occur.

c. Determine the total active force per unit length of the wall before the tensile crack occurs.

d. Determine the total active force per unit length of the wall after the tensile crack occurs. Also find the location of the resultant.

9.7 Redo Problem 9.6, assuming that the backfill is supporting a surcharge of 9.6 kN/m².

9.8 Repeat Problem 9.6 with the following values:

$$\text{height of wall} = 6 \text{ m}$$

$$\gamma_{sat} = 19.8 \text{ kN/m}^3$$

$$c_u = 14.7 \text{ kN/m}^2$$

9.9 A retaining wall 6 m high with a vertical back face has a c-ϕ soil for backfill. For the backfill, $\gamma = 18.1$ kN/m³, $c = 29$ kN/m², and $\phi = 18°$. Taking the existence of the tensile crack into consideration, determine the active force, P_a, per unit length of the wall for Rankine's active state.

9.10 For the wall described in Problem 9.9, determine the passive force, P_p, per unit length for Rankine's passive state.

9.11 For the retaining wall shown in Figure 9.25, determine the active force, P_a, for Rankine's state. Also find the location of the resultant. Assume that the tensile crack exists.

a. $\rho = 2300$ kg/m³, $\phi = 0°$, $c = c_u = 32$ kN/m²

b. $\rho = 1850$ kg/m³, $\phi = 16°$, $c = 15$ kN/m²

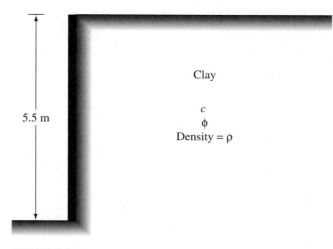

Clay

c

ϕ

Density = ρ

5.5 m

FIGURE 9.25

9.12 A retaining wall is shown in Figure 9.26. The height of the wall is 6 m, and the unit weight of the sand backfill is 18.9 kN/m³. Calculate the active force, P_a, on the wall using Coulomb's equation for the following values of the angle of wall friction:

a. $\delta = 0°$
b. $\delta = 10°$
c. $\delta = 20°$

Comment on the direction and location of the resultant.

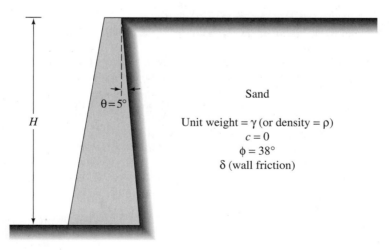

$\theta = 5°$

Sand

Unit weight = γ (or density = ρ)
$c = 0$
$\phi = 38°$
δ (wall friction)

H

FIGURE 9.26

9.13   For the retaining wall described in Problem 9.12, determine the passive force, P_p, per unit length of the wall using Colomb's equation for the following values of the angle of wall friction:
  a. $\delta = 0°$
  b. $\delta = 10°$
  c. $\delta = 20°$

References

Coulomb, C. A. (1776). "Essai sur une Application des Règles de Maximis et Minimis à quelques Problèmes de Statique, relatifs a l'Architecture," *Mem. Roy. des Sciences,* Paris, Vol. 3, 38.

Jaky, J. (1944). "The Coefficient of Earth Pressure at Rest," *Journal of the Society of Hungarian Architects and Engineers,* Vol. 7, 355–358.

Massarsch, K. R. (1979). "Lateral Earth Pressure in Normally Consolidated Clay," *Proceedings of the Seventh European Conference on Soil Mechanics and Foundation Engineering,* Brighton, England, Vol. 2, 245–250.

Rankine, W. M. J. (1857). "On Stability on Loose Earth," *Philosophic Transactions of Royal Society,* London, Part I, 9–27.

Supplementary References for Further Study

Brooker, E. W., and Ireland, H. O. (1965). "Earth Pressure at Rest Related to Stress History," *Canadian Geotechnical Journal,* Vol. 2, No. 1, 1–15.

Caquot, A., and Kerisel, J. (1948). *Tables for the Calculation of Passive Pressure, Active Pressure, and Bearing Capacity of Foundations,* Gauthier-Villars, Paris.

Dubrova, G. A. (1963). "Interaction of Soil and Structures," Izd. *Rechnoy Transport,* Moscow.

Matsuzawa, H., and Hazarika, H. (1996). "Analysis of Active Earth Pressure Against Rigid Retaining Wall Subjected to Different Modes of Movement," *Soils and Foundations,* Tokyo, Japan, Vol. 36, No. 3, 51–66.

Mayne, P. W., and Kulhawy, F. H. (1982). "K_o–OCR Relationships in Soil," *Journal of the Geotechnical Engineering Division,* ASCE, Vol. 108, No. GT6, 851–872.

Mazindrani, Z. H., and Ganjali, M. H. (1997). "Lateral Earth Pressure Problem of Cohesive Backfill with Inclined Surface," *Journal of Geotechnical and Geoenvironmental Engineering,* ASCE, Vol. 123, No. 2. 110–112.

Sherif, M. A., and Fang, Y. S. (1984). "Dynamic Earth Pressure on Walls Rotating About the Top," *Soils and Foundations,* Vol. 24, No. 4, 109–117.

Sherif, M. A., Fang, Y. S., and Sherif, R. I. (1984). "K_A and K_O Behind Rotating and Non-Yielding Walls," *Journal of Geotechnical Engineering,* ASCE, Vol. 110, No. GT1, 41–56.

Spangler, M. G. (1938). "Horizontal Pressures on Retaining Walls Due to Concentrated Surface Loads," Iowa State University Engineering Experiment Station, *Bulletin,* No. 140.

Terzaghi, K. (1941). "General Wedge Theory of Earth Pressure," *Transactions,* ASCE, Vol. 106, 68–97.

Terzaghi, K., and Peck, R. B. (1967). *Soil Mechanics in Engineering Practice,* 2nd ed., Wiley, New York.

10

Slope Stability

An exposed ground surface that stands at an angle with the horizontal is called an *unrestrained slope*. The slope can be natural or constructed. If the ground surface is not horizontal, a component of gravity will cause the soil to move downward, as shown in Figure 10.1. If the component of gravity is large enough, slope failure can occur; that is, the soil mass in zone *abcdea* can slide downward. The driving force overcomes the resistance from the shear strength of the soil along the rupture surface.

In many cases, civil engineers are expected to make calculations to check the safety of natural slopes, slopes of excavations, and compacted embankments. This process, called *slope stability analysis,* involves determining and comparing the shear stress developed along the most likely rupture surface with the shear strength of the soil.

The stability analysis of a slope is not an easy task. Evaluating variables such as the soil stratification and its in-place shear strength parameters may prove to be a formidable task. Seepage through the slope and the choice of a potential slip surface add to the complexity of the problem. This chapter explains the basic principles involved in slope stability analysis.

10.1 Factor of Safety

The task of the engineer charged with analyzing slope stability is to determine the factor of safety. Generally, the factor of safety is defined as

$$FS_s = \frac{\tau_f}{\tau_d} \tag{10.1}$$

where FS_s = factor of safety with respect to strength

τ_f = average shear strength of the soil

τ_d = average shear stress developed along the potential failure surface

339

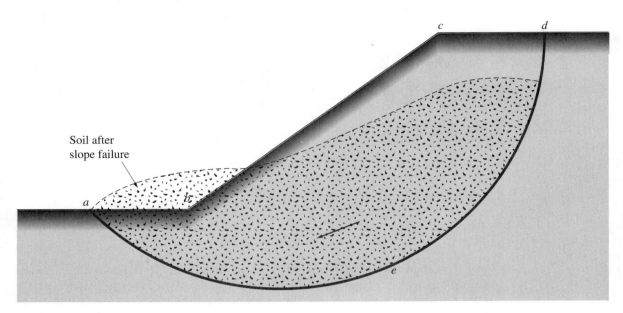

FIGURE 10.1 Slope failure

The shear strength of a soil consists of two components, cohesion and friction, and may be expressed as

$$\tau_f = c + \sigma' \tan \phi \qquad (10.2)$$

where c = cohesion
ϕ = drained angle of friction
σ' = effective normal stress on the potential failure surface

In a similar manner, we can also write

$$\tau_d = c_d + \sigma' \tan \phi_d \qquad (10.3)$$

where c_d and ϕ_d are, respectively, the effective cohesion and the angle of friction that develop along the potential failure surface. Substituting Eqs. (10.2) and (10.3) into Eq. (10.1), we get

$$FS_s = \frac{c + \sigma' \tan \phi}{c_d + \sigma' \tan \phi_d} \qquad (10.4)$$

Now we can introduce some other aspects of the factor of safety—that is, the factor of safety with respect to cohesion, FS_c, and the factor of safety with respect to friction, FS_ϕ. They are defined as follows:

$$FS_c = \frac{c}{c_d} \tag{10.5}$$

and

$$FS_\phi = \frac{\tan \phi}{\tan \phi_d} \tag{10.6}$$

When Eqs. (10.4), (10.5), and (10.6) are compared, we see that when FS_c becomes equal to FS_ϕ, that is the factor of safety with respect to strength. Or, if

$$\frac{c}{c_d} = \frac{\tan \phi}{\tan \phi_d}$$

we can write

$$FS_s = FS_c = FS_\phi \tag{10.7}$$

When F_s is equal to 1, the slope is in a state of impending failure. Generally, a value of 1.5 for the factor of safety with respect to strength is acceptable for the design of a stable slope.

10.2 *Stability of Infinite Slopes without Seepage*

In considering the problem of slope stability, we may start with the case of an infinite slope, as shown in Figure 10.2. An infinite slope is one in which H is much greater than the slope height. The shear strength of the soil may be given by [Eq. (10.2)]

$$\tau_f = c + \sigma' \tan \phi$$

We will evaluate the factor of safety against a possible slope failure along a plane AB located at a depth H below the ground surface. The slope failure can occur by the movement of soil above the plane AB from right to left.

Let us consider a slope element, *abcd*, that has a unit length perpendicular to the plane of the section shown. The forces, F, that act on the faces *ab* and *cd* are equal and opposite and may be ignored. The effective weight of the soil element is (with pore water pressure equal to 0)

$$W = \text{(volume of the soil element)} \times \text{(unit weight of soil)} = \gamma LH \tag{10.8}$$

The weight, W, can be resolved into two components:

1. Force perpendicular to the plane $AB = N_a = W \cos \beta = \gamma LH \cos \beta$.
2. Force parallel to the plane $AB = T_a = W \sin \beta = \gamma LH \sin \beta$. Note that this is the force that tends to cause the slip along the plane.

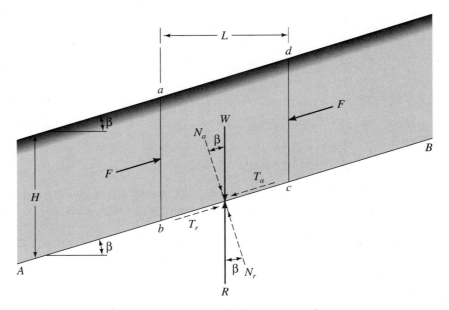

FIGURE 10.2 Analysis of infinite slope (without seepage)

Thus, the effective normal stress σ' and the shear stress τ at the base of the slope element can be given as

$$\sigma' = \frac{N_a}{\text{area of the base}} = \frac{\gamma L H \cos \beta}{\left(\dfrac{L}{\cos \beta}\right)} = \gamma H \cos^2 \beta \tag{10.9}$$

and

$$\tau = \frac{T_a}{\text{area of the base}} = \frac{\gamma L H \sin \beta}{\left(\dfrac{L}{\cos \beta}\right)} = \gamma H \cos \beta \sin \beta \tag{10.10}$$

The reaction to the weight W is an equal and opposite force R. The normal and tangential components of R with respect to the plane AB are N_r and T_r:

$$N_r = R \cos \beta = W \cos \beta \tag{10.11}$$

$$T_r = R \sin \beta = W \sin \beta \tag{10.12}$$

For equilibrium, the resistive shear stress that develops at the base of the element is equal to $(T_r)/(\text{area of the base}) = \gamma H \sin \beta \cos \beta$. This may also be written in the form [Eq. (10.3)]

$$\tau_d = c_d + \sigma' \tan \phi_d$$

The value of the effective normal stress is given by Eq. (10.9). Substitution of Eq. (10.9) into Eq. (10.3) yields

$$\tau_d = c_d + \gamma H \cos^2 \beta \tan \phi_d \tag{10.13}$$

Thus,

$$\gamma H \sin \beta \cos \beta = c_d + \gamma H \cos^2 \beta \tan \phi_d$$

or

$$\frac{c_d}{\gamma H} = \sin \beta \cos \beta - \cos^2 \beta \tan \phi_d$$

$$= \cos^2 \beta (\tan \beta - \tan \phi_d) \tag{10.14}$$

The factor of safety with respect to strength was defined in Eq. (10.7), from which

$$\tan \phi_d = \frac{\tan \phi}{FS_s} \quad \text{and} \quad c_d = \frac{c}{FS_s} \tag{10.15}$$

Substituting the preceding relationships into Eq. (10.14), we obtain

$$FS_s = \frac{c}{\gamma H \cos^2 \beta \tan \beta} + \frac{\tan \phi}{\tan \beta} \tag{10.16}$$

For granular soils, $c = 0$, and the factor of safety, FS_s, becomes equal to $(\tan \phi)/(\tan \beta)$. This indicates that, in an infinite slope in sand, the value of FS_s is independent of the height H, and the slope is stable as long as $\beta < \phi$. The angle ϕ for cohesionless soils is called the *angle of repose*.

If a soil possesses cohesion and friction, the depth of the plane along which critical equilibrium occurs may be determined by substituting $FS_s = 1$ and $H = H_{cr}$ into Eq. (10.16). Thus,

$$H_{cr} = \frac{c}{\gamma} \frac{1}{\cos^2 \beta (\tan \beta - \tan \phi)} \tag{10.17}$$

**EXAMPLE
10.1**

Consider the infinite slope shown in Figure 10.3.

    a. Determine the factor of safety against sliding along the soil–rock interface given $H = 2.4$ m.

    b. What height, H, will give a factor of safety, FS_s, of 2 against sliding along the soil–rock interface?

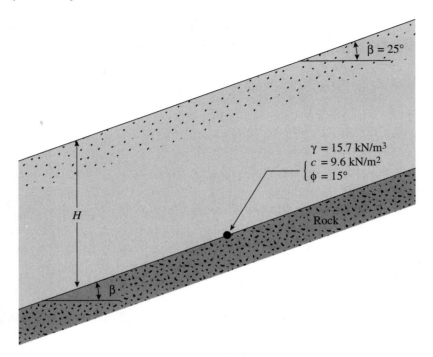

FIGURE 10.3

Solution

a. Equation (10.16) is

$$FS_s = \frac{c}{\gamma H \cos^2 \beta \tan \beta} + \frac{\tan \phi}{\tan \beta}$$

Given $c = 9.6$ kN/m², $\gamma = 15.7$ kN/m³, $\phi = 15°$, $\beta = 25°$, and $H = 2.4$ m, we have

$$FS_s = \frac{9.6}{(15.7)(2.4)(\cos^2 25)(\tan 25)} + \frac{\tan 15}{\tan 25} = \mathbf{1.24}$$

b.
$$FS_s = \frac{c}{\gamma H \cos^2 \beta \tan \beta} + \frac{\tan \phi}{\tan \beta}$$

$$2 = \frac{9.6}{(15.7)(H)(\cos^2 25)(\tan 25)} + \frac{\tan 15}{\tan 25}$$

$$H = \mathbf{1.12\ m} \qquad\blacksquare$$

10.3 *Stability of Infinite Slopes with Seepage*

Figure 10.4a shows an infinite slope. It is assumed that there is seepage through the soil and that the groundwater level coincides with the ground surface. The shear

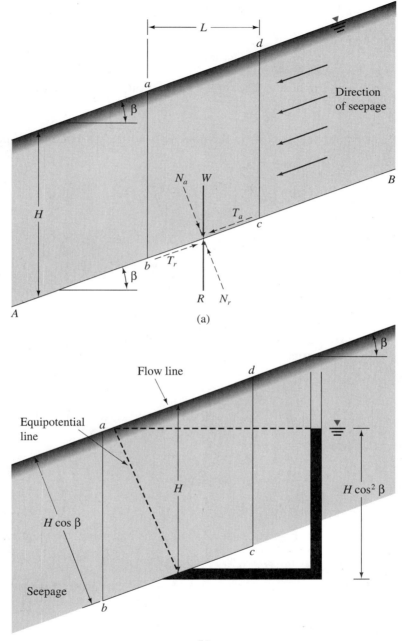

FIGURE 10.4 Analysis of infinite slope (with seepage)

strength of the soil is given by

$$\tau_f = c + \sigma' \tan \phi$$

To determine the factor of safety against failure along the plane AB, consider the slope element $abcd$. The forces that act on the vertical faces ab and cd are equal and opposite. The total weight of the slope element of unit length is

$$W = \gamma_{\text{sat}} LH \qquad (10.18)$$

The components of W in the directions normal and parallel to plane AB are

$$N_a = W \cos \beta = \gamma_{\text{sat}} LH \cos \beta \qquad (10.19)$$

and

$$T_a = W \sin \beta = \gamma_{\text{sat}} LH \sin \beta \qquad (10.20)$$

The reaction to weight W is equal to R. Thus,

$$N_r = R \cos \beta = W \cos \beta = \gamma_{\text{sat}} LH \cos \beta \qquad (10.21)$$

and

$$T_r = R \sin \beta = W \sin \beta = \gamma_{\text{sat}} LH \sin \beta \qquad (10.22)$$

We give the total normal stress and the shear stress at the base of the element. The total normal stress is

$$\sigma = \frac{N_r}{\left(\dfrac{L}{\cos \beta}\right)} = \gamma_{\text{sat}} H \cos^2 \beta \qquad (10.23)$$

and the shear stress is

$$\tau = \frac{T_r}{\left(\dfrac{L}{\cos \beta}\right)} = \gamma_{\text{sat}} H \cos \beta \sin \beta \qquad (10.24)$$

The resistive shear stress developed at the base of the element can also be given by

$$\tau_d = c_d + \sigma' \tan \phi_d = c_d + (\sigma - u) \tan \phi_d \qquad (10.25)$$

where $u =$ the pore water pressure $= \gamma_w H \cos^2 \beta$ (see Figure 10.4b). Substituting the values of σ [Eq. (10.23)] and u into Eq. (10.25), we get

$$\tau_d = c_d + (\gamma_{\text{sat}} H \cos^2 \beta - \gamma_w H \cos^2 \beta) \tan \phi_d$$

$$= c_d + \gamma' H \cos^2 \beta \tan \phi_d \qquad (10.26)$$

Now, setting the right-hand sides of Eqs. (10.24) and (10.26) equal to each other gives

$$\gamma_{\text{sat}} H \cos \beta \sin \beta = c_d + \gamma' H \cos^2 \beta \tan \phi_d$$

where γ' = effective unit weight of soil, or

$$\frac{c_d}{\gamma_{sat}H} = \cos^2 \beta \left(\tan \beta - \frac{\gamma'}{\gamma_{sat}} \tan \phi_d \right)$$ (10.27)

The factor of safety with respect to strength can be found by substituting $\tan \phi_d = (\tan \phi)/FS_s$ and $c_d = c/FS_s$ into Eq. (10.27), or

$$FS_s = \frac{c}{\gamma_{sat}H \cos^2 \beta \tan \beta} + \frac{\gamma'}{\gamma_{sat}} \frac{\tan \phi}{\tan \beta}$$ (10.28)

EXAMPLE 10.2

Refer to Figure 10.3. If there is seepage through the soil and the groundwater table coincides with the ground surface, what is the factor of safety, FS_s, given $H = 1.2$ m and $\gamma_{sat} = 18.5$ kN/m³?

Solution Equation (10.28) is

$$FS_s = \frac{c}{\gamma_{sat}H \cos^2 \beta \tan \beta} + \frac{\gamma'}{\gamma_{sat}} \frac{\tan \phi}{\tan \beta}$$

so we have

$$FS_s = \frac{9.6}{(18.5)(1.2)(\cos^2 25)(\tan 25)} + \frac{(18.5 - 9.81)}{18.5} \left(\frac{\tan 15}{\tan 25} \right) = \mathbf{1.4}$$ ∎

10.4 Finite Slopes

When the value of H_{cr} approaches the height of the slope, the slope is generally considered finite. When analyzing the stability of a finite slope in a homogeneous soil, for simplicity, we need to make an assumption about the general shape of the surface of potential failure. Although there is considerable evidence that slope failures usually occur on curved failure surfaces, Culmann (1875) approximated the surface of potential failure as a plane. The factor of safety, FS_s, calculated using Culmann's approximation gives fairly good results for near-vertical slopes only. After extensive investigation of slope failures in the 1920s, a Swedish geotechnical commission recommended that the actual surface of sliding may be approximated to be circularly cylindrical.

Since that time, most conventional stability analyses of slopes have been made by assuming that the curve of potential sliding is an arc of a circle. However, in many circumstances (for example, zoned dams and foundations on weak strata), stability analysis using plane failure of sliding is more appropriate and yields excellent results.

Analysis of Finite Slope with Plane Failure Surface (Culmann's Method)

This analysis is based on the assumption that the failure of a slope occurs along a plane when the average shearing stress that tends to cause the slip is greater than the shear strength of the soil. Also, the most critical plane is the one that has a minimum ratio of the average shearing stress that tends to cause failure to the shear strength of soil.

Figure 10.5 shows a slope of height H. The slope rises at an angle β with the horizontal. AC is a trial failure plane. If we consider a unit length perpendicular to the section of the slope, the weight of the wedge $ABC = W$:

$$W = \frac{1}{2}(H)(\overline{BC})(1)(\gamma)$$

$$= \frac{1}{2}H(H\cot\theta - H\cot\beta)\gamma$$

$$= \frac{1}{2}\gamma H^2\left[\frac{\sin(\beta - \theta)}{\sin\beta\sin\theta}\right] \tag{10.29}$$

The normal and tangential components of W with respect to the plane AC are as follows:

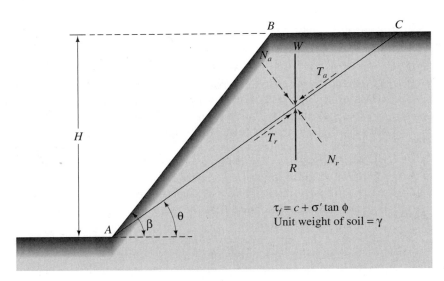

FIGURE 10.5 Finite slope analysis—Culmann's method

$$N_a = \text{normal component} = W \cos \theta$$

$$= \frac{1}{2} \gamma H^2 \left[\frac{\sin(\beta - \theta)}{\sin \beta \sin \theta} \right] \cos \theta \tag{10.30}$$

$$T_a = \text{tangential component} = W \sin \theta$$

$$= \frac{1}{2} \gamma H^2 \left[\frac{\sin(\beta - \theta)}{\sin \beta \sin \theta} \right] \sin \theta \tag{10.31}$$

The average effective normal stress and shear stress on the plane *AC* may be given by

$$\sigma' = \text{average effective normal stress}$$

$$= \frac{N_a}{(\overline{AC})(1)} = \frac{N_a}{\left(\dfrac{H}{\sin \theta} \right)}$$

$$= \frac{1}{2} \gamma H \left[\frac{\sin(\beta - \theta)}{\sin \beta \sin \theta} \right] \cos \theta \sin \theta \tag{10.32}$$

and

$$\tau = \text{average shear stress}$$

$$= \frac{T_a}{(\overline{AC})(1)} = \frac{T_a}{\left(\dfrac{H}{\sin \theta} \right)}$$

$$= \frac{1}{2} \gamma H \left[\frac{\sin(\beta - \theta)}{\sin \beta \sin \theta} \right] \sin^2 \theta \tag{10.33}$$

The average resistive shearing stress developed along the plane *AC* may also be expressed as

$$\tau_d = c_d + \sigma' \tan \phi_d$$

$$= c_d + \frac{1}{2} \gamma H \left[\frac{\sin(\beta - \theta)}{\sin \beta \sin \theta} \right] \cos \theta \sin \theta \tan \phi_d \tag{10.34}$$

Now, from Eqs. (10.33) and (10.34), we have

$$\frac{1}{2} \gamma H \left[\frac{\sin(\beta - \theta)}{\sin \beta \sin \theta} \right] \sin^2 \theta = c_d + \frac{1}{2} \gamma H \left[\frac{\sin(\beta - \theta)}{\sin \beta \sin \theta} \right] \cos \theta \sin \theta \tan \phi_d \tag{10.35}$$

or

$$c_d = \frac{1}{2}\gamma H \left[\frac{\sin(\beta - \theta)(\sin\theta - \cos\theta\tan\phi_d)}{\sin\beta} \right] \tag{10.36}$$

The expression in Eq. (10.36) is derived for the trial failure plane AC. In an effort to determine the critical failure plane, we use the principle of maxima and minima (for a given value of ϕ_d) to find the angle θ at which the developed cohesion would be maximum. Thus, the first derivative of c_d with respect to θ is set equal to 0, or

$$\frac{\partial c_d}{\partial\theta} = 0 \tag{10.37}$$

Since γ, H, and β are constants in Eq. (10.36), we have

$$\frac{\partial}{\partial\theta}[\sin(\beta - \theta)(\sin\theta - \cos\theta\tan\phi_d)] = 0 \tag{10.38}$$

Solving Eq. (10.38) gives the critical value of θ, or

$$\theta_{cr} = \frac{\beta + \phi_d}{2} \tag{10.39}$$

Substitution of the value of $\theta = \theta_{cr}$ into Eq. (10.36) yields

$$c_d = \frac{\gamma H}{4}\left[\frac{1 - \cos(\beta - \phi_d)}{\sin\beta\cos\phi_d} \right] \tag{10.40}$$

The maximum height of the slope for which critical equilibrium occurs can be obtained by substituting $c_d = c$ and $\phi_d = \phi$ into Eq. (10.40). Thus,

$$H_{cr} = \frac{4c}{\gamma}\left[\frac{\sin\beta\cos\phi}{1 - \cos(\beta - \phi)} \right] \tag{10.41}$$

EXAMPLE 10.3

A cut is to be made in a soil that has $\gamma = 16.5$ kN/m^3, $c = 29$ kN/m^2, and $\phi = 15°$. The side of the cut slope will make an angle of $45°$ with the horizontal. What depth of the cut slope will have a factor of safety, FS_s, of 3?

Solution We are given $\phi = 15°$ and $c = 29 \text{ kN/m}^2$. If $FS_s = 3$, then FS_c and FS_ϕ should both be equal to 3. We have

$$FS_c = \frac{c}{c_d}$$

or

$$c_d = \frac{c}{FS_c} = \frac{c}{FS_s} = \frac{29}{3} = 9.67 \text{ kN/m}^2$$

Similarly,

$$FS_\phi = \frac{\tan \phi}{\tan \phi_d}$$

$$\tan \phi_d = \frac{\tan \phi}{FS_\phi} = \frac{\tan \phi}{FS_s} = \frac{\tan 15}{3}$$

or

$$\phi_d = \tan^{-1}\left[\frac{\tan 15}{3}\right] = 5.1°$$

Substituting the preceding values of c_d and ϕ_d into Eq. (10.40) gives

$$H = \frac{4c_d}{\gamma}\left[\frac{\sin \beta \cos \phi_d}{1 - \cos(\beta - \phi_d)}\right] = \frac{4 \times 9.67}{16.5}\left[\frac{\sin 45 \cos 5.1}{1 - \cos(45 - 5.1)}\right] \approx \textbf{7.1 m} \quad ■$$

10.5 *Analysis of Finite Slope with Circularly Cylindrical Failure Surface—General*

In general, slope failure occurs in one of the following modes (Figure 10.6):

1. When the failure occurs in such a way that the surface of sliding intersects the slope at or above its toe, it is called a *slope failure* (Figure 10.6a). The failure circle is referred to as a *toe circle* if it passes through the toe of the slope and as a *slope circle* if it passes above the toe of the slope. Under certain circumstances, it is possible to have a shallow slope failure, as shown in Figure 10.6b.
2. When the failure occurs in such a way that the surface of sliding passes at some distance below the toe of the slope, it is called a *base failure* (Figure 10.6c). The failure circle in the case of base failure is called a *midpoint circle*.

Various procedures of stability analysis may, in general, be divided into two major classes:

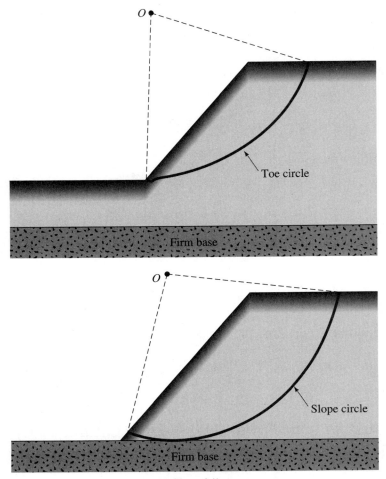

(a) Slope failure

FIGURE 10.6 Modes of failure of finite slope

1. *Mass procedure.* In this case, the mass of the soil above the surface of sliding is taken as a unit. This procedure is useful when the soil that forms the slope is assumed to be homogeneous, although this is hardly the case in most natural slopes.

2. *Method of slices.* In this procedure, the soil above the surface of sliding is divided into a number of vertical parallel slices. The stability of each of the slices is calculated separately. This is a versatile technique in which the nonhomogeneity of the soils and pore water pressure can be taken into consideration. It also accounts for the variation of the normal stress along the potential failure surface.

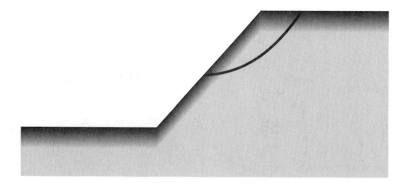

(b) Shallow slope failure

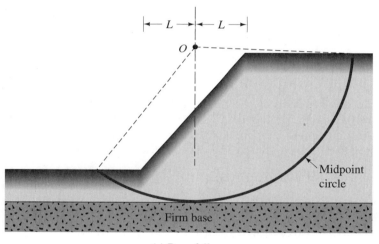

(c) Base failure

FIGURE 10.6 (Continued)

The fundamentals of the analysis of slope stability by mass procedure and method of slices are presented in the following sections.

10.6 *Mass Procedure of Stability Analysis (Circularly Cylindrical Failure Surface)*

Slopes in Homogeneous Clay Soil with $\phi = 0$ (Undrained Condition)

Figure 10.7 shows a slope in a homogeneous soil. The undrained shear strength of the soil is assumed to be constant with depth and may be given by $\tau_f = c_u$. To make

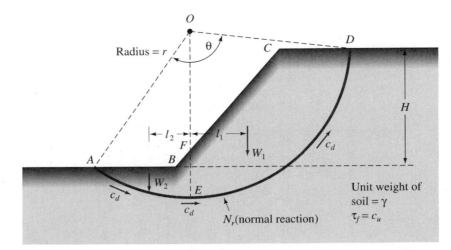

FIGURE 10.7 Stability analysis of slope in homogeneous clay soil ($\phi = 0$)

the stability analysis, we choose a trial potential curve of sliding AED, which is an arc of a circle that has a radius r. The center of the circle is located at O. Considering the unit length perpendicular to the section of the slope, we can give the total weight of the soil above the curve AED as $W = W_1 + W_2$, where

$$W_1 = (\text{area of } FCDEF)(\gamma)$$

and

$$W_2 = (\text{area of } ABFEA)(\gamma)$$

Note that γ = saturated unit weight of the soil.

Failure of the slope may occur by the sliding of the soil mass. The moment of the driving force about O to cause slope instability is

$$M_d = W_1 l_1 - W_2 l_2 \tag{10.42}$$

where l_1 and l_2 are the moment arms.

The resistance to sliding is derived from the cohesion that acts along the potential surface of sliding. If c_d is the cohesion that needs to be developed, then the moment of the resisting forces about O is

$$M_R = c_d(\widehat{AED})(1)(r) = c_d r^2 \theta \tag{10.43}$$

For equilibrium, $M_R = M_d$; thus,

$$c_d r^2 \theta = W_1 l_1 - W_2 l_2$$

or

$$c_d = \frac{W_1 l_1 - W_2 l_2}{r^2 \theta} \tag{10.44}$$

The factor of safety against sliding may now be found:

$$FS_s = \frac{\tau_f}{c_d} = \frac{c_u}{c_d} \tag{10.45}$$

Note that the potential curve of sliding, AED, was chosen arbitrarily. The critical surface is the one for which the ratio of c_u to c_d is a minimum. In other words, c_d is maximum. To find the critical surface for sliding, a number of trials are made for different trial circles. The minimum value of the factor of safety thus obtained is the factor of safety against sliding for the slope, and the corresponding circle is the critical circle.

Stability problems of this type were solved analytically by Fellenius (1927) and Taylor (1937). For the case of *critical circles*, the developed cohesion can be expressed by the relationship

$$c_d = \gamma H m$$

or

$$\frac{c_d}{\gamma H} = m \tag{10.46}$$

Note that the term m on the right-hand side of the preceding equation is nondimensional and is referred to as the *stability number*. The critical height (that is, $FS_s = 1$) of the slope can be evaluated by substituting $H = H_{cr}$ and $c_d = c_u$ (full mobilization of the undrained shear strength) into Eq. (10.46). Thus,

$$H_{cr} = \frac{c_u}{\gamma m} \tag{10.47}$$

Values of the stability number m for various slope angles β are given in Figure 10.8. Terzaghi and Peck (1967) used the term $\gamma H / c_d$, the reciprocal of m, and called it the *stability factor*. Figure 10.8 should be used carefully. Note that it is valid for slopes of saturated clay and is applicable to only undrained conditions ($\phi = 0$).

In reference to Figure 10.8, consider these issues:

1. For slope angle β greater than 53°, the critical circle is always a toe circle. The location of the center of the critical toe circle may be found with the aid of Figure 10.9.

For β > 53°:
All circles are toe circles.

For β < 53°:

Toe circle   ───────

Midpoint circle   ▬ ▪ ▬ ▪ ▬

Slope circle   ▬ ▬ ▬ ▬

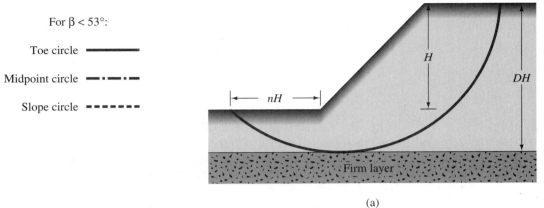

(a)

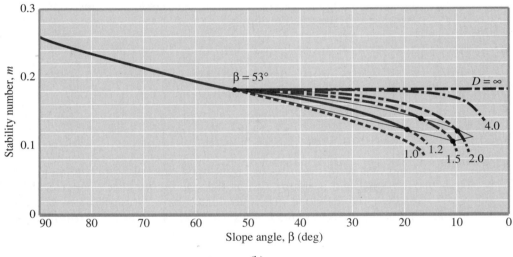

(b)

FIGURE 10.8 (a) Definition of parameters for midpoint circle-type failure; (b) plot of stability number against slope angle (Redrawn from Terzaghi and Peck, 1967)

2. For $β < 53°$, the critical circle may be a toe, slope, or midpoint circle, depending on the location of the firm base under the slope. This is called the *depth function,* which is defined as

$$D = \frac{\text{vertical distance from the top of the slope to the firm base}}{\text{height of the slope}} \quad (10.48)$$

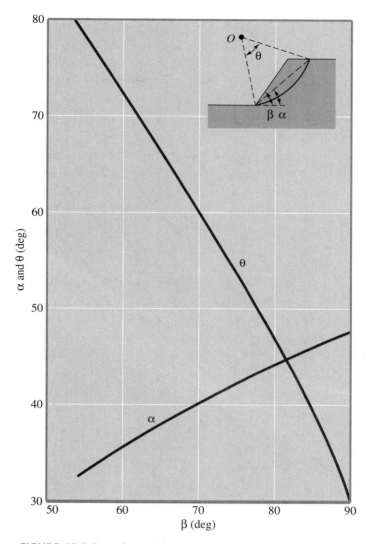

FIGURE 10.9 Location of the center of critical circles for $\beta > 53°$

3. When the critical circle is a midpoint circle (that is, the failure surface is tangent to the firm base), its position can be determined with the aid of Figure 10.10.
4. The maximum possible value of the stability number for failure at the midpoint circle is 0.181.

Fellenius (1927) also investigated the case of critical toe circles for slopes with $\beta < 53°$. The location of these can be determined using Figure 10.11 and Table

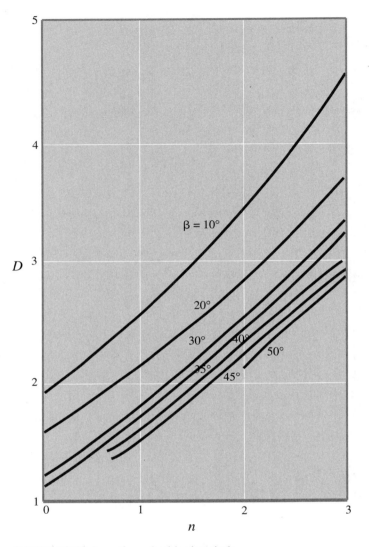

FIGURE 10.10 Location of midpoint circle

10.1. Note that these critical toe circles are not necessarily the most critical circles that exist.

**EXAMPLE
10.4**

A cut slope in saturated clay (Figure 10.12) makes an angle of 56° with the horizontal.

    a. Determine the maximum depth up to which the cut could be made. Assume that the critical surface for sliding is circularly cylindrical. What will be the nature of the critical circle (that is, toe, slope, or midpoint)?

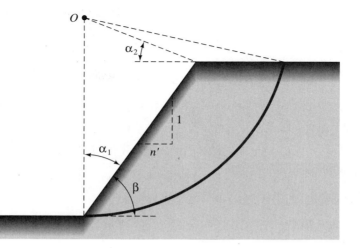

FIGURE 10.11 Location of the center of critical toe circles for $\beta < 53°$

b. Referring to part a, determine the distance of the point of intersection of the critical failure circle from the top edge of the slope.

c. How deep should the cut be made if a factor of safety of 2 against sliding is required?

Solution

a. Since the slope angle $\beta = 56° > 53°$, the critical circle is a **toe circle**. From Figure 10.8, for $\beta = 56°$, $m = 0.185$. Using Eq. (10.47), we have

$$H_{cr} = \frac{c_u}{\gamma m} = \frac{24}{(15.7)(0.185)} = 8.26 \text{ m} \approx \textbf{8.25 m}$$

Table 10.1 Location of the center of critical toe circles ($\beta < 53°$)

n'	β (deg)	α_1 (deg)	α_2 (deg)
1.0	45	28	37
1.5	33.68	26	35
2.0	26.57	25	35
3.0	18.43	25	35
5.0	11.32	25	37

Note: For notations of n', β, α_1, and α_2, see Figure 10.11.

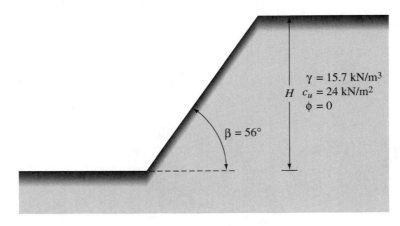

$\gamma = 15.7$ kN/m³
H $c_u = 24$ kN/m²
$\phi = 0$

$\beta = 56°$

FIGURE 10.12

b. Refer to Figure 10.13. For the critical circle, we have

$$\overline{BC} = \overline{EF} = \overline{AF} - \overline{AE} = H_{cr}(\cot\alpha - \cot 56°)$$

From Figure 10.9, for $\beta = 56°$, the magnitude of α is 33°, so

$$\overline{BC} = 8.25(\cot 33 - \cot 56) = 7.14 \text{ m} \approx \mathbf{7.15 \text{ m}}$$

c. Developed cohesion is

$$c_d = \frac{c_u}{FS_s} = \frac{24}{2} = 12 \text{ kN/m}^2$$

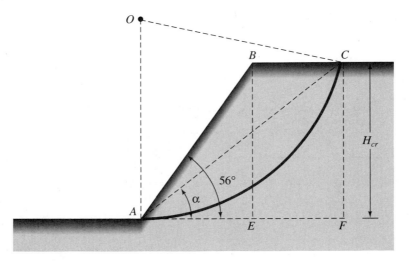

FIGURE 10.13

From Figure 10.8, for $\beta = 56°$, $m = 0.185$. Thus, we have

$$H = \frac{c_d}{\gamma m} = \frac{12}{(15.7)(0.185)} = \textbf{4.13 m}$$ ∎

EXAMPLE 10.5

A cut slope was excavated in a saturated clay. The slope made an angle of 40° with the horizontal. Slope failure occurred when the cut reached a depth of 6.1 m. Previous soil explorations showed that a rock layer was located at a depth of 9.15 m below the ground surface. Assume an undrained condition and $\gamma_{sat} = 17.29$ kN/m³.

    a. Determine the undrained cohesion of the clay (use Figure 10.8).
    b. What was the nature of the critical circle?
    c. With reference to the toe of the slope, at what distance did the surface of sliding intersect the bottom of the excavation?

Solution

    a. Referring to Figure 10.8, we find

$$D = \frac{9.15}{6.1} = 1.5$$

$$\gamma_{sat} = 17.29 \text{ kN/m}^3$$

$$H_{cr} = \frac{c_u}{\gamma m}$$

From Figure 10.8, for $\beta = 40°$ and $D = 1.5$, $m = 0.175$, so

$$c_u = (H_{cr})(\gamma)(m) = (6.1)(17.29)(0.175) = \textbf{18.5 kN/m}^2$$

    b. Midpoint circle
    c. From Figure 10.10, for $D = 1.5$ and $\beta = 40°$, $n = 0.9$, so

$$\text{distance} = (n)(H_{cr}) = (0.9)(6.1) = \textbf{5.49 m}$$ ∎

Slopes in Homogeneous Soil with $\phi > 0$

A slope in a homogeneous soil is shown in Figure 10.14a. The shear strength of the soil is given by

$$\tau_f = c + \sigma' \tan \phi$$

The pore water pressure is assumed to be 0. $\widehat{AC}$ is a trial circular arc that passes through the toe of the slope, and O is the center of the circle. Considering unit length

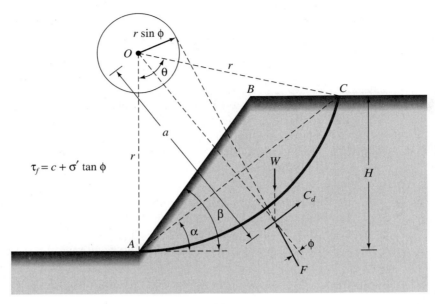

$$\tau_f = c + \sigma' \tan \phi$$

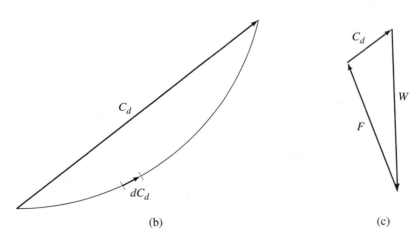

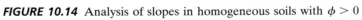

(b)

(c)

FIGURE 10.14 Analysis of slopes in homogeneous soils with $\phi > 0$

perpendicular to the section of the slope, we find

weight of the soil wedge $ABC = W = (\text{area of } ABC)(\gamma)$

For equilibrium, the following other forces are acting on the wedge:

1. C_d—the resultant of the cohesive force that is equal to the unit cohesion developed times the length of the cord $\overline{AC}$. The magnitude of C_d is given by (Figure 10.14b).

$$C_d = c_d(\overline{AC}) \tag{10.49}$$

C_d acts in a direction parallel to the cord AC (Figure 10.14b) and at a distance a from the center of the circle O such that

$$C_d(a) = c_d(\widehat{AC})r$$

or

$$a = \frac{c_d(\widehat{AC})r}{C_d} = \frac{\widehat{AC}}{\overline{AC}}r \tag{10.50}$$

2. F—the resultant of the normal and frictional forces along the surface of sliding. For equilibrium, the line of action of F will pass through the point of intersection of the line of action of W and C_d.

Now, if we assume the full friction is mobilized ($\phi_d = \phi$ or $FS_\phi = 1$), then the line of action of F will make an angle ϕ with a normal to the arc, and thus it will be a tangent to a circle with its center at O and having a radius of $r \sin \phi$. This circle is called the *friction circle*. Actually, the radius of the friction circle is a little larger than $r \sin \phi$.

Since the directions of W, C_d, and F are known and the magnitude of W is known, we can plot a force polygon, as shown in Figure 10.14c. The magnitude of C_d can be determined from the force polygon. So the unit cohesion developed can be found:

$$c_d = \frac{C_d}{AC}$$

Determining the magnitude of c_d described previously is based on a trial surface of sliding. Several trials must be made to obtain the most critical sliding surface along which the developed cohesion is a maximum. So it is possible to express the maximum cohesion developed along the critical surface as

$$c_d = \gamma H[f(\alpha, \beta, \theta, \phi)] \tag{10.51}$$

For critical equilibrium—that is, $FS_c = FS_\phi = FS_s = 1$—we can substitute $H = H_{cr}$ and $c_d = c$ into Eq. (10.51):

$$c = \gamma H_{cr}[f(\alpha, \beta, \theta, \phi)]$$

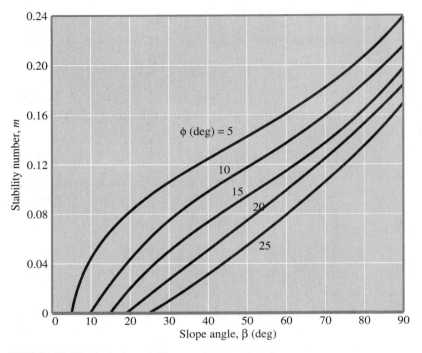

FIGURE 10.15 Taylor's stability number for $\phi > 0$

or

$$\frac{c}{\gamma H_{cr}} = f(\alpha, \beta, \theta, \phi) = m \qquad (10.52)$$

where m = stability number. The values of m for various values of ϕ and β (Taylor, 1937) are given in Figure 10.15. Example 10.6 illustrates the use of this chart.

Calculations have shown that, for ϕ greater than about 3°, the critical circles are all *toe circles*. Using Taylor's method of slope stability (as shown in Example 10.6), Singh (1970) provided graphs of equal factors of safety, FS_s, for various slopes, and these are given in Figure 10.16. In these charts, the pore water pressure was assumed to be 0.

EXAMPLE 10.6

A slope with $\beta = 45°$ is to be constructed with a soil that has $\phi = 20°$ and $c = 24$ kN/m². The unit weight of the compacted soil will be 18.9 kN/m³.

a. Find the critical height of the slope.
b. If the height of the slope is 10 m, determine the factor of safety with respect to strength.

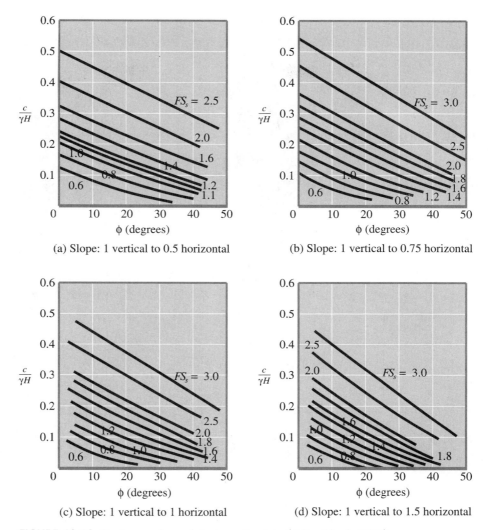

FIGURE 10.16 Contours of equal factors of safety (After Singh, 1970)

Solution

a. We have

$$m = \frac{c}{\gamma H_{cr}}$$

From Figure 10.15, for $\beta = 45°$ and $\phi = 20°$, $m = 0.06$. So

$$H_{cr} = \frac{c}{\gamma m} = \frac{24}{(18.9)(0.06)} = \textbf{21.1 m}$$

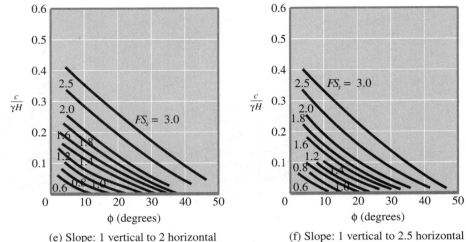

(e) Slope: 1 vertical to 2 horizontal   (f) Slope: 1 vertical to 2.5 horizontal

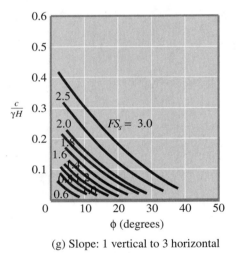

(g) Slope: 1 vertical to 3 horizontal

FIGURE 10.16 (Continued)

b. If we assume that full friction is mobilized, then, referring to Figure 10.15 (for $\beta = 45°$ and $\phi_d = \phi = 20°$), we have

$$m = 0.06 = \frac{c_d}{\gamma H}$$

or

$$c_d = (0.06)(18.9)(10) = 11.34 \text{ kN/m}^2$$

Thus,

$$FS_\phi = \frac{\tan \phi}{\tan \phi_d} = \frac{\tan 20}{\tan 20} = 1$$

and

$$FS_c = \frac{c}{c_d} = \frac{24}{11.34} = 2.12$$

Since $FS_c \neq FS_\phi$, this is not the factor of safety with respect to strength.

Now we can make another trial. Let the developed angle of friction, ϕ_d, be equal to 15°. For $\beta = 45°$ and the friction angle equal to 15°, we find from Figure 10.15

$$m = 0.085 = \frac{c_d}{\gamma H}$$

or

$$c_d = (0.085)(18.9)(10) = 16.07 \text{ kN/m}^2$$

For this trial,

$$FS_\phi = \frac{\tan \phi}{\tan \phi_d} = \frac{\tan 20}{\tan 15} = 1.36$$

and

$$FS_c = \frac{c}{c_d} = \frac{24}{16.07} = 1.49$$

Similar calculations of FS_ϕ and FS_c for various assumed values of ϕ_d are given in the table.

ϕ_d	tan ϕ_d	FS_ϕ	m	c_d (kN/m²)	FS_c
20	0.364	1.0	0.06	11.34	2.12
15	0.268	1.36	0.085	16.07	1.49
10	0.176	2.07	0.11	20.79	1.15
5	0.0875	4.16	0.136	25.70	0.93

The values of FS_ϕ are plotted against their corresponding values of FS_c in Figure 10.17, from which we find

$$FS_c = FS_\phi = FS_s = \textbf{1.45}$$

■

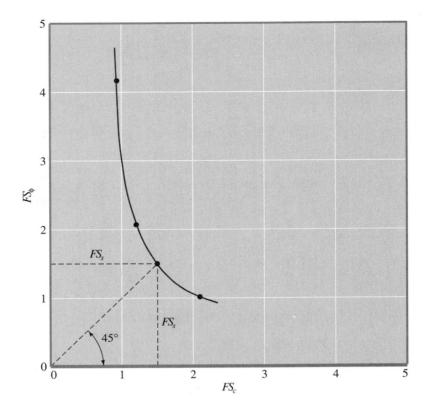

FIGURE 10.17

10.7   *Method of Slices*

Stability analysis using the method of slices can be explained by referring to Figure 10.18a, in which AC is an arc of a circle representing the trial failure surface. The soil above the trial failure surface is divided into several vertical slices. The width of each slice need not be the same. Considering unit length perpendicular to the cross-section shown, the forces that act on a typical slice (nth slice) are shown in Figure 10.18b. W_n is the effective weight of the slice. The forces N_r and T_r are the normal and tangential components of the reaction R, respectively. P_n and P_{n+1} are the normal forces that act on the sides of the slice. Similarly, the shearing forces that act on the sides of the slice are T_n and T_{n+1}. For simplicity, the pore water pressure is assumed to be 0. The forces P_n, P_{n+1}, T_n, and T_{n+1} are difficult to determine. However, we can make an approximate assumption that the resultants of P_n and T_n are equal in magnitude to the resultants of P_{n+1} and T_{n+1} and also that their lines of action coincide.

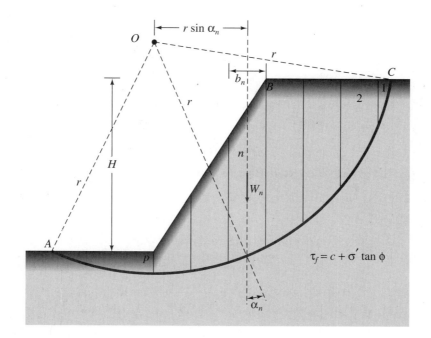

(a)

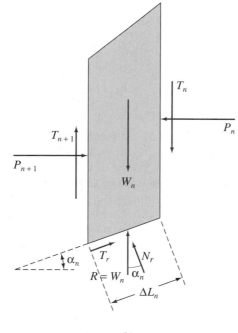

(b)

FIGURE 10.18 Stability analysis by ordinary method of slices: (a) trial failure surface; (b) forces acting on nth slice

For equilibrium consideration, we have

$$N_r = W_n \cos \alpha_n$$

The resisting shear force can be expressed as

$$T_r = \tau_d (\Delta L_n) = \frac{\tau_f (\Delta L_n)}{FS_s} = \frac{1}{FS_s} [c + \sigma' \tan \phi] \Delta L_n \qquad (10.53)$$

The effective normal stress, σ', in Eq. (10.53) is equal to

$$\frac{N_r}{\Delta L_n} = \frac{W_n \cos \alpha_n}{\Delta L_n}$$

For equilibrium of the trial wedge ABC, the moment of the driving force about O equals the moment of the resisting force about O, or

$$\sum_{n=1}^{n=p} W_n r \sin \alpha_n = \sum_{n=1}^{n=p} \frac{1}{FS_s} \left(c + \frac{W_n \cos \alpha_n}{\Delta L_n} \tan \phi \right) (\Delta L_n)(r)$$

or

$$FS_s = \frac{\displaystyle\sum_{n=1}^{n=p} (c \, \Delta L_n + W_n \cos \alpha_n \tan \phi)}{\displaystyle\sum_{n=1}^{n=p} W_n \sin \alpha_n} \qquad (10.54)$$

Note: ΔL_n in Eq. (10.54) is approximately equal to $(b_n)/(\cos \alpha_n)$, where b_n = width of the nth slice.

Note that the value of α_n may be either positive or negative. The value of α_n is positive when the slope of the arc is in the same quadrant as the ground slope. To find the minimum factor of safety—that is, the factor of safety for the critical circle—several trials are made by changing the center of the trial circle. This method is generally referred to as the *ordinary method of slices*.

For convenience, a slope in a homogeneous soil is shown in Figure 10.18. However, the method of slices can be extended to slopes with layered soil, as shown in Figure 10.19. The general procedure of stability analysis is the same. There are some minor points that should be kept in mind. When Eq. (10.54) is used for the factor of safety calculation, the values of ϕ and c will not be the same for all slices. For example, for slice no. 3 (Figure 10.19), we have to use a friction angle of $\phi = \phi_3$ and cohesion $c = c_3$; similarly, for slice no. 2, $\phi = \phi_2$ and $c = c_2$.

Bishop's Simplified Method of Slices

In 1955, Bishop proposed a more refined solution to the ordinary method of slices. In this method, the effect of forces on the sides of each slice is accounted for to

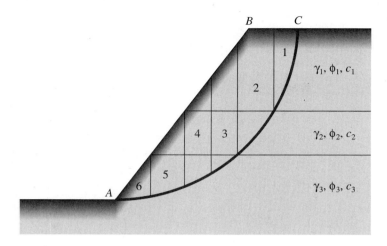

FIGURE 10.19 Stability analysis, by ordinary method of slices, for slopes in layered soils

some degree. We can study this method by referring to the slope analysis presented in Figure 10.18. The forces that act on the *n*th slice shown in Figure 10.18b have been redrawn in Figure 10.20a. Now, let $P_n - P_{n+1} = \Delta P$ and $T_n - T_{n+1} = \Delta T$. Also, we can write

$$T_r = N_r(\tan \phi_d) + c_d \Delta L_n = N_r \left(\frac{\tan \phi}{FS_s} \right) + \frac{c \Delta L_n}{FS_s} \tag{10.55}$$

Figure 10.20b shows the force polygon for equilibrium of the *n*th slice. Summing the forces in the vertical direction gives

$$W_n + \Delta T = N_r \cos \alpha_n + \left[\frac{N_r \tan \phi}{FS_s} + \frac{c \Delta L_n}{FS_s} \right] \sin \alpha_n$$

or

$$N_r = \frac{W_n + \Delta T - \dfrac{c \Delta L_n}{FS_s} \sin \alpha_n}{\cos \alpha_n + \dfrac{\tan \phi \sin \alpha_n}{FS_s}} \tag{10.56}$$

For equilibrium of the wedge *ABC* (Figure 10.18a), taking the moment about *O* gives

$$\sum_{n=1}^{n=p} W_n r \sin \alpha_n = \sum_{n=1}^{n=p} T_r \gamma \tag{10.57}$$

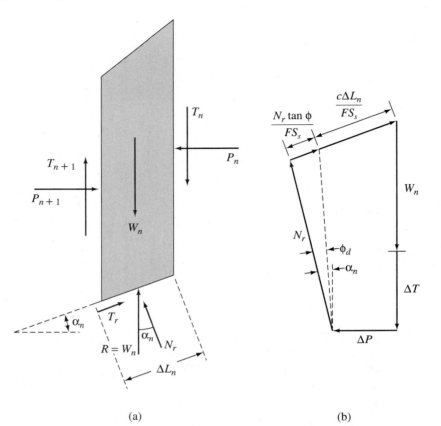

(a)   (b)

FIGURE 10.20 Bishop's simplified method of slices: (a) forces acting on the *n*th slice; (b) force polygon for equilibrium

where $T_r = \dfrac{1}{FS_s}(c + \sigma' \tan \phi)\, \Delta L_n$

$$= \frac{1}{FS_s}(c\,\Delta L_n + N_r \tan \phi) \qquad (10.58)$$

Substitution of Eqs. (10.56) and (10.58) into Eq. (10.57) gives

$$FS_s = \frac{\displaystyle\sum_{n=1}^{n=p}(cb_n + W_n \tan \phi + \Delta T \tan \phi)\,\dfrac{1}{m_{\alpha(n)}}}{\displaystyle\sum_{n=1}^{n=p}W_n \sin \alpha_n} \qquad (10.59)$$

where

$$m_{\alpha(n)} = \cos \alpha_n + \frac{\tan \phi \sin \alpha_n}{FS_s} \qquad (10.60)$$

For simplicity, if we let $\Delta T = 0$, then Eq. (10.59) becomes

$$FS_s = \frac{\sum\limits_{n=1}^{n=p}(cb_n + W_n \tan\phi)\dfrac{1}{m_{\alpha(n)}}}{\sum\limits_{n=1}^{n=p}W_n \sin\alpha_n} \qquad (10.61)$$

Note that the term FS_s is present on both sides of Eq. (10.61). Hence, a trial-and-error procedure needs to be adopted to find the value of FS_s. As in the method of ordinary slices, a number of failure surfaces must be investigated to find the critical surface that provides the minimum factor of safety.

Bishop's simplified method is probably the most widely used method. When incorporated into computer programs, it yields satisfactory results in most cases. The ordinary method of slices is presented in this chapter as a learning tool. It is rarely used now because it is too conservative.

EXAMPLE 10.7

For the slope shown in Figure 10.21, find the factor of safety against sliding for the trial slip surface AC. Use the ordinary method of slices.

Solution   The sliding wedge is divided into seven slices. Other calculations are shown in the table.

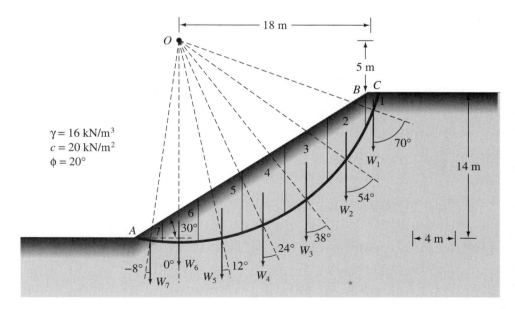

FIGURE 10.21

Slice no. (1)	W (kN/m) (2)	α_n (deg) (3)	$\sin \alpha_n$ (4)	$\cos \alpha_n$ (5)	ΔL_n (m) (6)	$W_n \sin \alpha_n$ (kN/m) (7)	$W_n \cos \alpha_n$ (kN/m) (8)
1	22.4	70	0.94	0.342	2.924	21.1	6.7
2	294.4	54	0.81	0.588	6.803	238.5	173.1
3	435.2	38	0.616	0.788	5.076	268.1	342.94
4	435.2	24	0.407	0.914	4.376	177.1	397.8
5	390.4	12	0.208	0.978	4.09	81.2	381.8
6	268.8	0	0	1	4	0	268.8
7	66.58	−8	−0.139	0.990	3.232	−9.25	65.9
					Σ col. 6 = 30.501 m	Σ col. 7 = 776.75 kN/m	Σ col. 8 = 1638.04 kN/m

$$FS_s = \frac{(\Sigma \text{ col. } 6)(c) + (\Sigma \text{ col. } 8) \tan \phi}{\Sigma \text{ col. } 7}$$

$$= \frac{(30.501)(20) + (1638.04)(\tan 20)}{776.75} = \mathbf{1.55} \qquad \blacksquare$$

10.8   *Stability Analysis by Method of Slices for Steady-State Seepage*

The fundamentals of the ordinary method of slices and Bishop's simplified method of slices were presented in Section 10.7, and we assumed the pore water pressure to be 0. However, for steady-state seepage through slopes, as is the situation in many practical cases, the pore water pressure has to be taken into consideration when effective shear strength parameters are used. So we need to modify Eqs. (10.54) and (10.61) slightly.

Figure 10.22 shows a slope through which there is steady-state seepage. For the nth slice, the average pore water pressure at the bottom of the slice is equal to $u_n = h_n \gamma_w$. The total force caused by the pore water pressure at the bottom of the nth slice is equal to $u_n \Delta L_n$. Thus, Eq. (10.54) for the ordinary method of slices will be modified to read

$$FS_s = \frac{\displaystyle\sum_{n=1}^{n=p} [c \Delta L_n + (W_n \cos \alpha_n - u_n \Delta L_n)] \tan \phi}{\displaystyle\sum_{n=1}^{n=p} W_n \sin \alpha_n} \qquad (10.62)$$

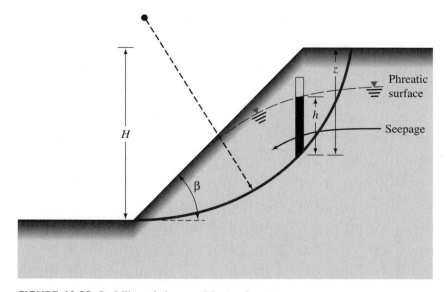

FIGURE 10.22 Stability of slopes with steady-state seepage

Similarly, Eq. (10.61) for Bishop's simplified method of slices will be modified to the form

$$FS_s = \frac{\displaystyle\sum_{n=1}^{n=p} [cb_n + (W_n - u_n b_n) \tan \phi] \frac{1}{m_{(\alpha)n}}}{\displaystyle\sum_{n=1}^{n=p} W_n \sin \alpha_n} \tag{10.63}$$

Note that W_n in Eqs. (10.62) and (10.63) is the total weight of the slice.

Using the method of slices, Bishop and Morgenstern (1960) provided charts to determine the factor of safety of simple slopes that takes into account the effects of pore water pressure. These solutions are given in the next section.

10.9 *Bishop and Morgenstern's Solution for Stability of Simple Slopes with Seepage*

Using Eq. (10.63), Bishop and Morgenstern developed tables for the calculation of FS_s for simple slopes. The principles of these developments can be explained as follows: In Eq. (10.63), we have

$$W_n = \text{total weight of the } n\text{th slice} = \gamma b_n z_n \tag{10.64}$$

where z_n = average height of the nth slice
  $u_n = h_n \gamma_w$

So we can let

$$r_{u(n)} = \frac{u_n}{\gamma z_n} = \frac{h_n \gamma_w}{\gamma z_n} \qquad (10.65)$$

Note that $r_{u(n)}$ is a nondimensional quantity. Substituting Eqs. (10.64) and (10.65) into Eq. (10.63) and simplifying, we obtain

$$FS_s = \left[\frac{1}{\displaystyle\sum_{n=1}^{n=p} \frac{b_n}{H} \frac{z_n}{H} \sin \alpha_n} \right] \times \sum_{n=1}^{n=p} \left\{ \frac{\dfrac{c}{\gamma H} \dfrac{b_n}{H} + \dfrac{b_n}{H} \dfrac{z_n}{H} [1 - r_{u(n)}] \tan \phi}{m_{\alpha(n)}} \right\} \qquad (10.66)$$

For a steady-state seepage condition, a weighted average value of $r_{u(n)}$ can be taken, which is a constant. Let the weighted average value of $r_{u(n)}$ be r_u. For most practical cases, the value of r_u may range up to 0.5. So

$$FS_s = \left[\frac{1}{\displaystyle\sum_{n=1}^{n=p} \frac{b_n}{H} \frac{z_n}{H} \sin \alpha_n} \right] \times \sum_{n=1}^{n=p} \left\{ \frac{\dfrac{c}{\gamma H} \dfrac{b_n}{H} + \dfrac{b_n}{H} \dfrac{z_n}{H} (1 - r_u) \tan \phi}{m_{\alpha(n)}} \right\} \qquad (10.67)$$

The factor of safety based on the preceding equation can be solved and expressed in the form

$$FS_s = m' - n' r_u \qquad (10.68)$$

where m' and n' are stability coefficients. Table 10.2 gives the values of m' and n' for various combinations of $c/\gamma H$, D, ϕ, and β.

To determine FS_s from Table 10.2, use the following step-by-step procedure:

1. Obtain ϕ, β, and $c/\gamma H$.
2. Obtain r_u (weighted average value).
3. From Table 10.2, obtain the values of m' and n' for $D = 1$, 1.25, and 1.50 (for the required parameters ϕ, β, r_u, and $c/\gamma H$).
4. Determine FS_s using the values of m' and n' for each value of D.
5. The required value of FS_s is the smallest one obtained in Step 4.

Table 10.2 Bishop and Morgenstern's values of m' and n'

a. *Stability coefficients m' and n' for $c/\gamma H = 0$*

Stability coefficients for earth slopes

ϕ	Slope 2:1		Slope 3:1		Slope 4:1		Slope 5:1	
	m'	n'	m'	n'	m'	n'	m'	n'
10.0	0.353	0.441	0.529	0.588	0.705	0.749	0.882	0.917
12.5	0.443	0.554	0.665	0.739	0.887	0.943	1.109	1.153
15.0	0.536	0.670	0.804	0.893	1.072	1.139	1.340	1.393
17.5	0.631	0.789	0.946	1.051	1.261	1.340	1.577	1.639
20.0	0.728	0.910	1.092	1.213	1.456	1.547	1.820	1.892
22.5	0.828	1.035	1.243	1.381	1.657	1.761	2.071	2.153
25.0	0.933	1.166	1.399	1.554	1.865	1.982	2.332	2.424
27.5	1.041	1.301	1.562	1.736	2.082	2.213	2.603	2.706
30.0	1.155	1.444	1.732	1.924	2.309	2.454	2.887	3.001
32.5	1.274	1.593	1.911	2.123	2.548	2.708	3.185	3.311
35.0	1.400	1.750	2.101	2.334	2.801	2.977	3.501	3.639
37.5	1.535	1.919	2.302	2.558	3.069	3.261	3.837	3.989
40.0	1.678	2.098	2.517	2.797	3.356	3.566	4.196	4.362

b. *Stability coefficients m' and n' for $c/\gamma H = 0.025$ and $D = 1.00$*

Stability coefficients for earth slopes

ϕ	Slope 2:1		Slope 3:1		Slope 4:1		Slope 5:1	
	m'	n'	m'	n'	m'	n'	m'	n'
10.0	0.678	0.534	0.906	0.683	1.130	0.846	1.365	1.031
12.5	0.790	0.655	1.066	0.849	1.337	1.061	1.620	1.282
15.0	0.901	0.776	1.224	1.014	1.544	1.273	1.868	1.534
17.5	1.012	0.898	1.380	1.179	1.751	1.485	2.121	1.789
20.0	1.124	1.022	1.542	1.347	1.962	1.698	2.380	2.050
22.5	1.239	1.150	1.705	1.518	2.177	1.916	2.646	2.317
25.0	1.356	1.282	1.875	1.696	2.400	2.141	2.921	2.596
27.5	1.478	1.421	2.050	1.882	2.631	2.375	3.207	2.886
30.0	1.606	1.567	2.235	2.078	2.873	2.622	3.508	3.191
32.5	1.739	1.721	2.431	2.285	3.127	2.883	3.823	3.511
35.0	1.880	1.885	2.635	2.505	3.396	3.160	4.156	3.849
37.5	2.030	2.060	2.855	2.741	3.681	3.458	4.510	4.209
40.0	2.190	2.247	3.090	2.993	3.984	3.778	4.885	4.592

Table 10.2 (Continued)

c.   *Stability coefficients m' and n' for c/γH = 0.025 and D = 1.25*

Stability coefficients for earth slopes

	Slope 2:1		Slope 3:1		Slope 4:1		Slope 5:1	
ϕ	*m'*	*n'*	*m'*	*n'*	*m'*	*n'*	*m'*	*n'*
10.0	0.737	0.614	0.901	0.726	1.085	0.867	1.285	1.014
12.5	0.878	0.759	1.076	0.908	1.299	1.098	1.543	1.278
15.0	1.019	0.907	1.253	1.093	1.515	1.311	1.803	1.545
17.5	1.162	1.059	1.433	1.282	1.736	1.541	2.065	1.814
20.0	1.309	1.216	1.618	1.478	1.961	1.775	2.334	2.090
22.5	1.461	1.379	1.808	1.680	2.194	2.017	2.610	2.373
25.0	1.619	1.547	2.007	1.891	2.437	2.269	2.879	2.669
27.5	1.783	1.728	2.213	2.111	2.689	2.531	3.196	2.976
30.0	1.956	1.915	2.431	2.342	2.953	2.806	3.511	3.299
32.5	2.139	2.112	2.659	2.686	3.231	3.095	3.841	3.638
35.0	2.331	2.321	2.901	2.841	3.524	3.400	4.191	3.998
37.5	2.536	2.541	3.158	3.112	3.835	3.723	4.563	4.379
40.0	2.753	2.775	3.431	3.399	4.164	4.064	4.958	4.784

d.   *Stability coefficients m' and n' for c/γH = 0.05 and D = 1.00*

Stability coefficients for earth slopes

	Slope 2:1		Slope 3:1		Slope 4:1		Slope 5:1	
ϕ	*m'*	*n'*	*m'*	*n'*	*m'*	*n'*	*m'*	*n'*
10.0	0.913	0.563	1.181	0.717	1.469	0.910	1.733	1.069
12.5	1.030	0.690	1.343	0.878	1.688	1.136	1.995	1.316
15.0	1.145	0.816	1.506	1.043	1.904	1.353	2.256	1.567
17.5	1.262	0.942	1.671	1.212	2.117	1.565	2.517	1.825
20.0	1.380	1.071	1.840	1.387	2.333	1.776	2.783	2.091
22.5	1.500	1.202	2.014	1.568	2.551	1.989	3.055	2.365
25.0	1.624	1.338	2.193	1.757	2.778	2.211	3.336	2.651
27.5	1.753	1.480	1.380	1.952	3.013	2.444	3.628	2.948
30.0	1.888	1.630	2.574	2.157	3.261	2.693	3.934	3.259
32.5	2.029	1.789	2.777	2.370	3.523	2.961	4.256	3.585
35.0	2.178	1.958	2.990	2.592	3.803	3.253	4.597	3.927
37.5	2.336	2.138	3.215	2.826	4.103	3.574	4.959	4.288
40.0	2.505	2.332	3.451	3.071	4.425	3.926	5.344	4.668

Table 10.2 (Continued)

e. *Stability coefficients m' and n' for c/γH = 0.05 and D = 1.25*

Stability coefficients for earth slopes

	Slope 2:1		Slope 3:1		Slope 4:1		Slope 5:1	
ϕ	m'	n'	m'	n'	m'	n'	m'	n'
10.0	0.919	0.633	1.119	0.766	1.344	0.886	1.594	1.042
12.5	1.065	0.792	1.294	0.941	1.563	1.112	1.850	1.300
15.0	1.211	0.950	1.471	1.119	1.782	1.338	2.109	1.562
17.5	1.359	1.108	1.650	1.303	2.004	1.567	2.373	1.831
20.0	1.509	1.266	1.834	1.493	2.230	1.799	2.643	2.107
22.5	1.663	1.428	2.024	1.690	2.463	2.038	2.921	2.392
25.0	1.822	1.595	2.222	1.897	2.705	2.287	3.211	2.690
27.5	1.988	1.769	2.428	2.113	2.957	2.546	3.513	2.999
30.0	2.161	1.950	2.645	2.342	3.221	2.819	3.829	3.324
32.5	2.343	2.141	2.873	2.583	3.500	3.107	4.161	3.665
35.0	2.535	2.344	3.114	2.839	3.795	3.413	4.511	4.025
37.5	2.738	2.560	3.370	3.111	4.109	3.740	4.881	4.405
40.0	2.953	2.791	3.642	3.400	4.442	4.090	5.273	4.806

f. *Stability coefficients m' and n' for c/γH = 0.05 and D = 1.50*

Stability coefficients for earth slopes

	Slope 2:1		Slope 3:1		Slope 4:1		Slope 5:1	
ϕ	m'	n'	m'	n'	m'	n'	m'	n'
10.0	1.022	0.751	1.170	0.828	1.343	0.974	1.547	1.108
12.5	1.202	0.936	1.376	1.043	1.589	1.227	1.829	1.399
15.0	1.383	1.122	1.583	1.260	1.835	1.480	2.112	1.690
17.5	1.565	1.309	1.795	1.480	2.084	1.734	2.398	1.983
20.0	1.752	1.501	2.011	1.705	2.337	1.993	2.690	2.280
22.5	1.943	1.698	2.234	1.937	2.597	2.258	2.990	2.585
25.0	2.143	1.903	2.467	2.179	2.867	2.534	3.302	2.902
27.5	2.350	2.117	2.709	2.431	3.148	2.820	3.626	3.231
30.0	2.568	2.342	2.964	2.696	3.443	3.120	3.967	3.577
32.5	2.798	2.580	3.232	2.975	3.753	3.436	4.326	3.940
35.0	3.041	2.832	3.515	3.269	4.082	3.771	4.707	4.325
37.5	3.299	3.102	3.817	3.583	4.431	4.128	5.112	4.735
40.0	3.574	3.389	4.136	3.915	4.803	4.507	5.543	5.171

**EXAMPLE
10.8**

Use the following values:

> slope: 3 horizontal : 1 vertical
>
> $H = 12.6$ m
>
> $\phi = 25°$
>
> $c = 12$ kN/m^2
>
> $\gamma = 19$ kN/m^3
>
> $r_u = 0.25$

Determine the minimum factor of safety using Bishop and Morgenstern's method.

Solution Given slope $= 3\text{H} : 1\text{V}$, $\phi = 25°$, and $r_u = 0.25$, we find

$$\frac{c}{\gamma H} = \frac{12}{(19)(12.6)} = 0.05$$

From Tables 10.2a, b, and c, we can prepare the following table:

D	m'	n'	FS$_s$ = m' – n'r$_u$
1	2.193	1.757	1.754
1.25	2.222	1.897	1.748
1.5	2.467	2.179	1.922

So the minimum factor of safety is $1.748 \approx$ **1.75.**   ■

Problems

10.1 For the slope shown in Figure 10.23, find the height, H, for critical equilibrium given $\beta = 25°$.

10.2 Refer to Figure 10.23.
  a. If $\beta = 25°$ and $H = 3$ m, what is the factor of safety of the slope against sliding along the soil–rock interface?
  b. For $\beta = 30°$, find the height, H, that will have a factor of safety of 1.5 against sliding along the soil–rock interface.

10.3 Refer to Figure 10.23. Plot a graph of H_{cr} versus the slope angle β (for β varying from 20° to 40°).

10.4 An infinite slope is shown in Figure 10.24. The shear strength parameters at the interface of soil and rock are $c = 18$ kN/m^2 and $\phi = 25°$.
  a. If $H = 8$ m and $\beta = 20°$, find the factor of safety against sliding along the rock surface.
  b. If $\beta = 30°$, find the height, H, for which $FS_s = 1$. (Assume the pore water pressure is 0.)

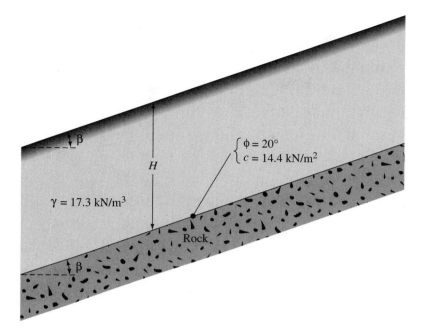

$$\phi = 20°$$
$$c = 14.4 \text{ kN/m}^2$$

H

$\gamma = 17.3 \text{ kN/m}^3$

β

Rock

β

FIGURE 10.23

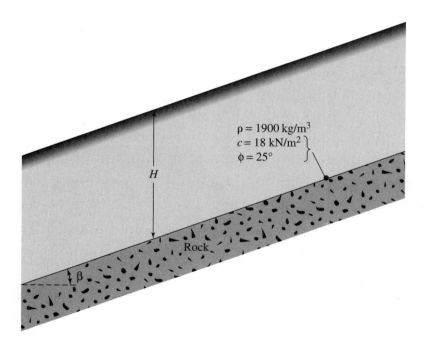

$$\rho = 1900 \text{ kg/m}^3$$
$$c = 18 \text{ kN/m}^2$$
$$\phi = 25°$$

H

β

Rock

FIGURE 10.24

10.5   Refer to Figure 10.24. If there were seepage through the soil and the ground-water table coincided with the ground surface, what would be the value of FS_s? Use $H = 8$ m, $\rho_{sat} = 1900$ kg/m³, and $\beta = 20°$.

10.6   For the infinite slope shown in Figure 10.25, find the factor of safety against sliding along the plane AB given $H = 3$m. Note that there is seepage through the soil, and the groundwater table coincides with the ground surface.

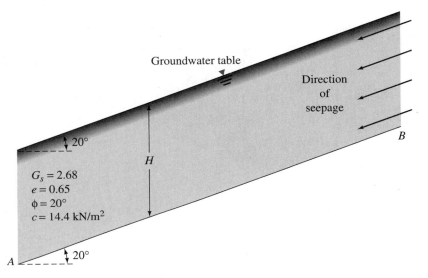

Groundwater table

Direction
of
seepage

B

$20°$

H

$G_s = 2.68$
$e = 0.65$
$\phi = 20°$
$c = 14.4$ kN/m²

$20°$

A

FIGURE 10.25

10.7   A slope is shown in Figure 10.26. AC represents a trial failure plane. For the wedge ABC, find the factor of safety against sliding.

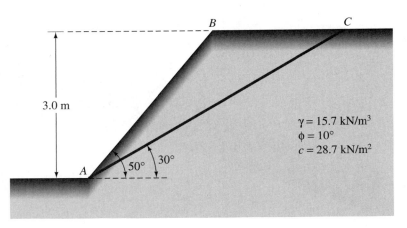

B C

3.0 m

$\gamma = 15.7$ kN/m³
$\phi = 10°$
$c = 28.7$ kN/m²

$50°$ $30°$
A

FIGURE 10.26

10.8 A finite slope is shown in Figure 10.27. Assuming that the slope failure would occur along a plane (Culmann's assumption), find the height of the slope for critical equilibrium given $\phi = 10°$, $c = 12$ kN/m², $\gamma = 17.3$ kN/m³, and $\beta = 50°$.

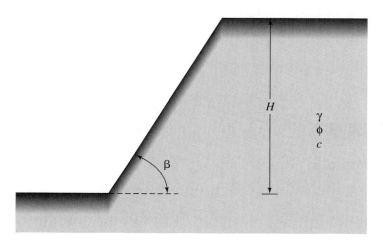

FIGURE 10.27

10.9   Repeat Problem 10.8 with $\phi = 20°$, $c = 25$ kN/m², $\gamma = 18$ kN/m³, and $\beta = 45°$.

10.10 Refer to Figure 10.27. Using the soil parameters given in Problem 10.8, find the height of the slope, H, that will have a factor of safety of 2.5 against sliding. Assume that the critical surface for sliding is a plane.

10.11 Refer to Figure 10.27. Given $\phi = 15°$, $c = 9.6$ kN/m², $\gamma = 18.0$ kN/m³, $\beta = 60°$, and $H = 2.7$ m, determine the factor of safety with respect to sliding. Assume that the critical surface for sliding is a plane.

10.12 Refer to Problem 10.11. Find the height of the slope, H, that will have $FS_s = 1.5$. Assume that the critical surface for sliding is a plane.

10.13 A cut slope is to be made in a soft clay with its sides rising at an angle of 75° to the horizontal (Figure 10.28). Assume that $c_u = 31.1$ kN/m² and $\gamma = 17.3$ kN/m³.

  a. Determine the maximum depth up to which the excavation can be carried out.

  b. Find the radius, r, of the critical circle when the factor of safety is equal to 1 (part a).

  c. Find the distance $\overline{BC}$.

10.14 If the cut described in Problem 10.13 is made to a depth of only 3.0 m, what will be the factor of safety of the slope against sliding?

10.15 Using the graph given in Figure 10.8, determine the height of a slope, 1 vertical to $\frac{1}{2}$ horizontal, in saturated clay that has an undrained shear strength

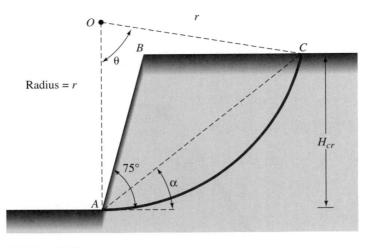

FIGURE 10.28

of 32.6 kN/m². The desired factor of safety against sliding is 2. Assume $\gamma = 18.9$ kN/m³.

10.16 Refer to Problem 10.15. What should be the critical height of the slope? What will be the nature of the critical circle? Also, find the radius of the critical circle.

10.17 For the slope shown in Figure 10.29, find the factor of safety against sliding for the trial surface $\overset{\frown}{AC}$.

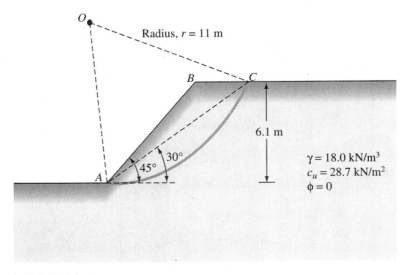

FIGURE 10.29

10.18 A cut slope was excavated in a saturated clay. The slope angle β is equal to 35° with respect to the horizontal. Slope failure occurred when the cut reached a depth of 8.2 m. Previous soil explorations showed that a rock layer was located at a depth of 11 m below the ground surface. Assume an undrained condition and $\gamma_{\text{sat}} = 19.2$ kN/m³.
    a. Determine the undrained cohesion of the clay (use Figure 10.8).
    b. What was the nature of the critical circle?
    c. With reference to the toe of the slope, at what distance did the surface of sliding intersect the bottom of the excavation?

10.19 If the cut slope described in Problem 10.18 is to be excavated in a manner such that $H_{cr} = 9$ m, what angle should the slope make with the horizontal? (Use Figure 10.8 and the results of Problem 10.18a.)

10.20 Refer to Figure 10.30. Use Taylor's chart for $\phi > 0$ (Figure 10.15) to find the critical height of the slope in each case:
    a. $n' = 2$, $\phi = 15°$, $c = 31.1$ kN/m², and $\gamma = 18.0$ kN/m³
    b. $n' = 1$, $\phi = 25°$, $c = 24$ kN/m², and $\gamma = 18.0$ kN/m³
    c. $n' = 2.5$, $\phi = 12°$, $c = 25$ kN/m², and $\gamma = 17$ kN/m³
    d. $n' = 1.5$, $\phi = 18°$, $c = 18$ kN/m², and $\gamma = 16.5$ kN/m³

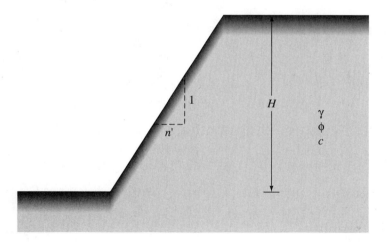

FIGURE 10.30

10.21 Referring to Figure 10.30 and using Figure 10.15, find the factor of safety with respect to sliding for the following cases:
    a. $n' = 2.5$, $\phi = 12°$, $c = 23.9$ kN/m², $\gamma = 16.5$ kN/m³, and $H = 10$ m
    b. $n' = 1.5$, $\phi = 15°$, $c = 18$ kN/m², $\gamma = 16.5$ kN/m³, and $H = 6$ m

10.22 Refer to Figure 10.30 and Figure 10.16.
    a. If $n' = 2$, $\phi = 10°$, $c = 33.5$ kN/m², and $\gamma = 17.3$ kN/m³, draw a graph of the height of the slope, H, against FS_s (varying from 1 to 3).
    b. If $n' = 1$, $\phi = 15°$, $c = 18$ kN/m², and $\gamma = 17.1$ kN/m³, draw a graph of the height of the slope, H, against FS_s (varying from 1 to 3).

10.23 Referring to Figure 10.31 and using the ordinary method of slices, find the factor of safety against sliding for the trial case $\beta = 45°$, $\phi = 15°$, $c = 18$ kN/m^2, $\gamma = 17.1$ kN/m^3, $H = 5$ m, $\alpha = 30°$, and $\theta = 80°$.

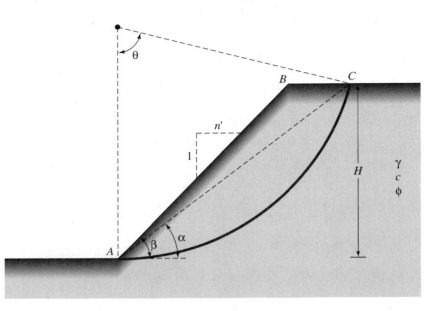

FIGURE 10.31

10.24 Determine the minimum factor of safety of a slope with the following parameters: $H = 6.1$ m, $\beta = 26.57°$, $\phi = 25°$, $c = 5.5$ kN/m^2, $\gamma = 18$ kN/m^3, and $r_u = 0.5$. Use Bishop and Morgenstern's method.

References

Bishop, A. W. (1955). "The Use of Slip Circle in the Stability Analysis of Earth Slopes," *Geotechnique*, Vol. 5, No. 1, 7–17.

Bishop, A. W., and Morgenstern, N. R. (1960). "Stability Coefficients for Earth Slopes," *Geotechnique*, Vol. 10, No. 4, 129–147.

Culmann, C. (1875). *Die Graphische Statik*, Meyer and Zeller, Zurich.

Fellenius, W. (1927). *Erdstatische Berechnungen*, revised edition, W. Ernst u. Sons, Berlin.

Singh, A. (1970). "Shear Strength and Stability of Man-Made Slopes," *Journal of the Soil Mechanics and Foundations Division*, ASCE, Vol. 96, No. SM6, 1879–1892.

Taylor, D. W. (1937). "Stability of Earth Slopes," *Journal of the Boston Society of Civil Engineers,* Vol. 24, 197–246.

Terzaghi, K., and Peck, R. B. (1967). *Soil Mechanics in Engineering Practice*, 2nd ed., Wiley, New York.

Supplementary References for Further Study

Ladd, C. C. (1972). "Test Embankment on Sensitive Clay," *Proceedings,* Conference on Performance of Earth and Earth-Supported Structures, ASCE, Vol. 1, Part 1, 101–128.

Morgenstern, N. R. (1963). "Stability Charts for Earth Slopes During Rapid Drawdown," *Geotechnique,* Vol. 13, No. 2, 121–133.

Morgenstern, N. R., and Price, V. E. (1965). "The Analysis of the Stability of General Slip Surfaces," *Geotechnique,* Vol. 15, No. 1, 79–93.

O'Connor, M. J., and Mitchell, R. J. (1977). "An Extension of the Bishop and Morgenstern Slope Stability Charts," *Canadian Geotechnical Journal,* Vol. 14, No. 1, 144–151.

Spencer, E. (1967). "A Method of Analysis of the Stability of Embankments Assuming Parallel Inter-Slice Forces," *Geotechnique,* Vol. 17, No. 1, 11–26.

11

Shallow Foundations—Bearing Capacity and Settlement

The lowest part of a structure is generally referred to as the *foundation.* Its function is to transfer the load of the structure to the soil on which it is resting. A properly designed foundation is one that transfers the load throughout the soil without overstressing the soil. Overstressing the soil can result in either excessive settlement or shear failure of the soil, both of which cause damage to the structure. Thus, geotechnical and structural engineers who design foundations must evaluate the bearing capacity of soils.

Depending on the structure and soil encountered, various types of foundations are used. Figure 11.1 shows the most common types of foundations. A *spread footing* is simply an enlargement of a load-bearing wall or column that makes it possible to spread the load of the structure over a larger area of the soil. In soil with low load-bearing capacity, the size of the spread footings required is impractically large. In that case, it is more economical to contruct the entire structure over a concrete pad. This is called a *mat foundation.*

Pile and *drilled shaft foundations* are used for heavier structures when great depth is required for supporting the load. Piles are structural members made of timber, concrete, or steel that transmit the load of the superstructure to the lower layers of the soil. According to how they transmit their load into the subsoil, piles can be divided into two categories: friction piles and end-bearing piles. In the case of friction piles, the superstructure load is resisted by the shear stresses generated along the surface of the pile. In the end-bearing pile, the load carried by the pile is transmitted at its tip to a firm stratum.

In the case of drilled shafts, a shaft is drilled into the subsoil and is then filled with concrete. A metal casing may be used while the shaft is being drilled. The casing may be left in place or withdrawn during the placing of concrete. Generally, the diameter of a drilled shaft is much larger than that of a pile. The distinction between piles and drilled shafts becomes hazy at an approximate diameter of 1 m, and then the definitions and nomenclature are inaccurate.

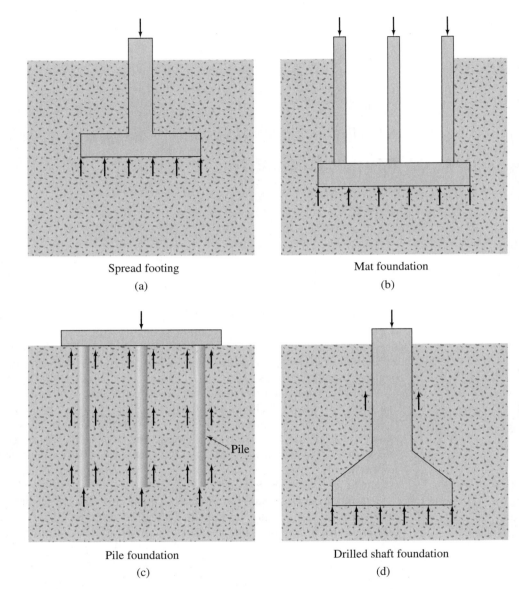

Spread footing
(a)

Mat foundation
(b)

Pile foundation
(c)

Drilled shaft foundation
(d)

FIGURE 11.1 Common types of foundations

Spread footings and mat foundations are generally referred to as shallow foundations, and pile and drilled shaft foundations are classified as deep foundations. In a more general sense, shallow foundations are those foundations that have a depth-of-embedment-to-width ratio of approximately less than four. When the depth-of-embedment-to-width ratio of a foundation is greater than four, it may be classified as a deep foundation.

In this chapter, we discuss the soil-bearing capacity for shallow foundations. As mentioned before, for a foundation to function properly, (1) the settlement of soil caused by the load must be within the tolerable limit, and (2) shear failure of the soil supporting the foundation must not occur. Compressibility of soil—consolidation and elasticity theory—was introduced in Chapter 6. This chapter introduces the load-carrying capacity of shallow foundations based on the criterion of shear failure in soil.

ULTIMATE BEARING CAPACITY OF SHALLOW FOUNDATIONS

11.1 *General Concepts*

Consider a strip (i.e., theoretically length is infinity) foundation resting on the surface of a dense sand or stiff cohesive soil, as shown in Figure 11.2a, with a width of *B*. Now, if load is gradually applied to the foundation, settlement will increase. The variation of the load per unit area on the foundation, q, with the foundation settlement is also shown in Figure 11.2a. At a certain point—when the load per unit area equals q_u—a sudden failure in the soil supporting the foundation will take place, and the failure surface in the soil will extend to the ground surface. This load per unit area, q_u, is usually referred to as the *ultimate bearing capacity of the foundation.* When this type of sudden failure in soil takes place, it is called *general shear failure.*

If the foundation under consideration rests on sand or clayey soil of medium compaction (Figure 11.2b), an increase of load on the foundation will also be accompanied by an increase of settlement. However, in this case the failure surface in the soil will gradually extend outward from the foundation, as shown by the solid lines in Figure 11.2b. When the load per unit area on the foundation equals $q_{u(1)}$, the foundation movement will be accompanied by sudden jerks. A considerable movement of the foundation is then required for the failure surface in soil to extend to the ground surface (as shown by the broken lines in Figure 11.2b). The load per unit area at which this happens is the *ultimate bearing capacity,* q_u. Beyond this point, an increase of load will be accompanied by a large increase of foundation settlement. The load per unit area of the foundation, $q_{u(1)}$, is referred to as the *first failure load* (Vesic, 1963). Note that a peak value of q is not realized in this type of failure, which is called *local shear failure* in soil.

If the foundation is supported by a fairly loose soil, the load–settlement plot will be like the one in Figure 11.2c. In this case, the failure surface in soil will not extend to the ground surface. Beyond the ultimate failure load, q_u, the load–settlement plot will be steep and practically linear. This type of failure in soil is called *punching shear failure.*

Based on experimental results, Vesic (1973) proposed a relationship for the mode of bearing capacity failure of foundations resting on sands. Figure 11.3 shows this relationship, which involves the following notation:

D_r = relative density of sand

D_f = depth of foundation measured from the ground surface

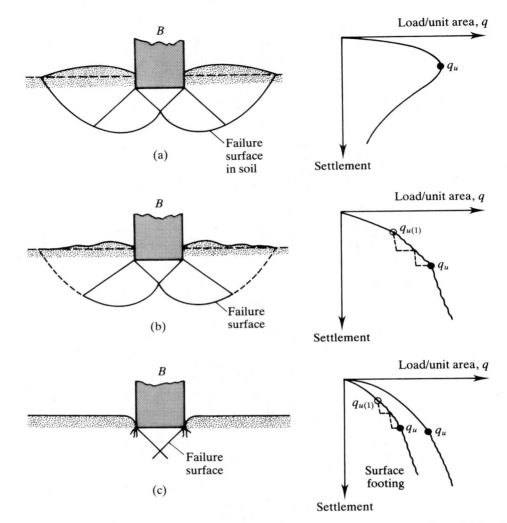

FIGURE 11.2 Nature of bearing capacity failure in soil: (a) general shear failure; (b) local shear failure; (c) punching shear failure

$$B^* = \frac{2BL}{B + L} \tag{11.1}$$

where B = width of foundation
L = length of foundation

(*Note:* L is always greater than B.)
   For square foundations, $B = L$; for circular foundations, $B = L$ = diameter. So

$$B^* = B \tag{11.2}$$

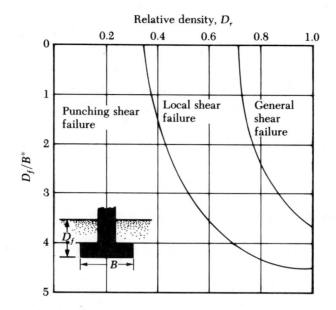

FIGURE 11.3 Modes of foundation failure in sand
(After Vesic, 1973)

For foundations at a shallow depth (that is, small D_f/B^*), the ultimate load may occur at a foundation settlement of 4% to 10% of B. This condition occurs with general shear failure in soil; however, with local or punching shear failure, the ultimate load may occur at settlements of 15% to 25% of the width of foundation (B).

11.2   *Ultimate Bearing Capacity Theory*

Terzaghi (1943) was the first to present a comprehensive theory for evaluating the ultimate bearing capacity of rough shallow foundations. According to this theory, a foundation is *shallow* if the depth, D_f (Figure 11.4), of the foundation is less than or equal to the width of the foundation. Later investigators, however, have suggested that foundations with D_f equal to 3 to 4 times the width of the foundation may be defined as *shallow foundations*.

Terzaghi suggested that for a *continuous*, or *strip, foundation* (that is, the width-to-length ratio of the foundation approaches 0), the failure surface in soil at ultimate load may be assumed to be similar to that shown in Figure 11.4. (Note that this is the case of general shear failure as defined in Figure 11.2a.) The effect of soil above the bottom of the foundation may also be assumed to be replaced by an equivalent surcharge, $q = \gamma D_f$ (where γ = unit weight of soil). The failure zone under the foundation can be separated into three parts (see Figure 11.4):

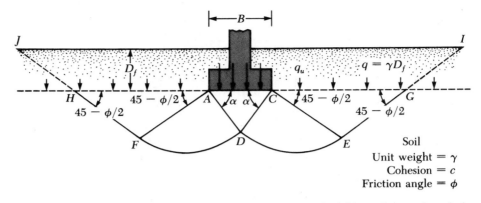

FIGURE 11.4 Bearing capacity failure in soil under a rough rigid continuous foundation

1. The *triangular zone ACD* immediately under the foundation
2. The *radial shear zones ADF* and *CDE*, with the curves *DE* and *DF* being arcs of a logarithmic spiral
3. Two triangular *Rankine passive zones AFH* and *CEG*

The angles *CAD* and *ACD* are assumed to be equal to the soil friction angle (that is, $\alpha = \phi$). Note that, with the replacement of the soil above the bottom of the foundation by an equivalent surcharge q, the shear resistance of the soil along the failure surfaces *GI* and *HJ* was neglected.

Using the equilibrium analysis, Terzaghi expressed the ultimate bearing capacity in the form

$$q_u = cN_c + qN_q + \frac{1}{2}\gamma BN_\gamma \quad \text{(strip foundation)} \qquad (11.3)$$

where
$$c = \text{cohesion of soil}$$
$$\gamma = \text{unit weight of soil}$$
$$q = \gamma D_f$$
$$N_c, N_q, N_\gamma = \text{bearing capacity factors that are nondimensional and are only functions of the soil friction angle, } \phi$$

Based on laboratory and field studies of bearing capacity, the basic nature of the failure surface in soil suggested by Terzaghi now appears to be correct (Vesic, 1973). However, the angle α shown in Figure 11.4 is closer to $45 + \phi/2$ than to ϕ, as was originally assumed by Terzaghi. With $\alpha = 45 + \phi/2$, the relations for N_c and N_q can be derived as

$$N_q = \tan^2\left(45 + \frac{\phi}{2}\right)e^{\pi \tan \phi} \qquad (11.4)$$

$$N_c = (N_q - 1) \cot \phi \qquad (11.5)$$

The equation for N_c given by Eq. (11.5) was originally derived by Prandtl (1921), and the relation for N_q [Eq. (11.4)] was presented by Reissner (1924). Caquot and Kerisel (1953) and Vesic (1973) gave the relation for N_γ as

$$N_\gamma = 2(N_q + 1) \tan \phi \qquad (11.6)$$

Table 11.1 shows the variation of the preceding bearing capacity factors with soil friction angles.

Table 11.1 Bearing capacity factors*

ϕ	N_c	N_q	N_γ	N_q/N_c	$\tan \phi$	ϕ	N_c	N_q	N_γ	N_q/N_c	$\tan \phi$
0	5.14	1.00	0.00	0.20	0.00	26	22.25	11.85	12.54	0.53	0.49
1	5.38	1.09	0.07	0.20	0.02	27	23.94	13.20	14.47	0.55	0.51
2	5.63	1.20	0.15	0.21	0.03	28	25.80	14.72	16.72	0.57	0.53
3	5.90	1.31	0.24	0.22	0.05	29	27.86	16.44	19.34	0.59	0.55
4	6.19	1.43	0.34	0.23	0.07	30	30.14	18.40	22.40	0.61	0.58
5	6.49	1.57	0.45	0.24	0.09	31	32.67	20.63	25.99	0.63	0.60
6	6.81	1.72	0.57	0.25	0.11	32	35.49	23.18	30.22	0.65	0.62
7	7.16	1.88	0.71	0.26	0.12	33	38.64	26.09	35.19	0.68	0.65
8	7.53	2.06	0.86	0.27	0.14	34	42.16	29.44	41.06	0.70	0.67
9	7.92	2.25	1.03	0.28	0.16	35	46.12	33.30	48.03	0.72	0.70
10	8.35	2.47	1.22	0.30	0.18	36	50.59	37.75	56.31	0.75	0.73
11	8.80	2.71	1.44	0.31	0.19	37	55.63	42.92	66.19	0.77	0.75
12	9.28	2.97	1.69	0.32	0.21	38	61.35	48.93	78.03	0.80	0.78
13	9.81	3.26	1.97	0.33	0.23	39	67.87	55.96	92.25	0.82	0.81
14	10.37	3.59	2.29	0.35	0.25	40	75.31	64.20	109.41	0.85	0.84
15	10.98	3.94	2.65	0.36	0.27	41	83.86	73.90	130.22	0.88	0.87
16	11.63	4.34	3.06	0.37	0.29	42	93.71	85.38	155.55	0.91	0.90
17	12.34	4.77	3.53	0.39	0.31	43	105.11	99.02	186.54	0.94	0.93
18	13.10	5.26	4.07	0.40	0.32	44	118.37	115.31	224.64	0.97	0.97
19	13.93	5.80	4.68	0.42	0.34	45	133.88	134.88	271.76	1.01	1.00
20	14.83	6.40	5.39	0.43	0.36	46	152.10	158.51	330.35	1.04	1.04
21	15.82	7.07	6.20	0.45	0.38	47	173.64	187.21	403.67	1.08	1.07
22	16.88	7.82	7.13	0.46	0.40	48	199.26	222.31	496.01	1.12	1.11
23	18.05	8.66	8.20	0.48	0.42	49	229.93	265.51	613.16	1.15	1.15
24	19.32	9.60	9.44	0.50	0.45	50	266.89	319.07	762.89	1.20	1.19
25	20.72	10.66	10.88	0.51	0.47						

*After Vesic (1973)

The ultimate bearing capacity expression presented in Eq. (11.3) is for a continuous foundation only. It does not apply to the case of rectangular foundations. Also, the equation does not take into account the shearing resistance along the failure surface in soil above the bottom of the foundation (portion of the failure surface marked as *GI* and *HJ* in Figure 11.4). In addition, the load on the foundation may be inclined. To account for all these shortcomings, Meyerhof (1963) suggested the following form of the general bearing capacity equation:

$$q_u = cN_cF_{cs}F_{cd}F_{ci} + qN_qF_{qs}F_{qd}F_{qi} + \frac{1}{2}\gamma BN_\gamma F_{\gamma s}F_{\gamma d}F_{\gamma i}$$

(11.7)

where

c = cohesion
q = effective stress at the level of the bottom of the foundation
γ = unit weight of soil
B = width of foundation (= diameter for a circular foundation)
$F_{cs}, F_{qs}, F_{\gamma s}$ = shape factors
$F_{cd}, F_{qd}, F_{\gamma d}$ = depth factors
$F_{ci}, F_{qi}, F_{\gamma i}$ = load inclination factors
N_c, N_q, N_γ = bearing capacity factors

The relationships for the shape factors, depth factors, and inclination factors *recommended for use* are given in Table 11.2.

Net Ultimate Bearing Capacity

The net ultimate bearing capacity is defined as the ultimate pressure per unit area of the foundation that can be supported by the soil in excess of the pressure caused by the surrounding soil at the foundation level. If the difference between the unit weight of concrete used in the foundation and the unit weight of soil surrounding the foundation is assumed to be negligible, then

$$q_{\text{net}(u)} = q_u - q$$

(11.8)

where $q_{\text{net}(u)}$ = net ultimate bearing capacity.

11.3   Modification of Bearing Capacity Equations for Water Table

Equation (11.7) was developed for determining the ultimate bearing capacity based on the assumption that the water table is located well below the foundation. However, if the water table is close to the foundation, some modifications of the bearing capacity equation are necessary, depending on the location of the water table (see Figure 11.5).

Table 11.2 Shape, depth, and inclination factors recommended for use

Factor	Relationship	Source
Shape*	$F_{cs} = 1 + \dfrac{B}{L}\dfrac{N_q}{N_c}$	De Beer (1970)
	$F_{qs} = 1 + \dfrac{B}{L}\tan\phi$	
	$F_{\gamma s} = 1 - 0.4\dfrac{B}{L}$	
	where L = length of the foundation ($L > B$)	
Depth[†]	*Condition (a):* $D_f/B \le 1$	Hansen (1970)
	$F_{cd} = 1 + 0.4\dfrac{D_f}{B}$	
	$F_{qd} = 1 + 2\tan\phi(1 - \sin\phi)^2\dfrac{D_f}{B}$	
	$F_{\gamma d} = 1$	
	Condition (b): $D_f/B > 1$	
	$F_{cd} = 1 + (0.4)\tan^{-1}\left(\dfrac{D_f}{B}\right)$	
	$F_{qd} = 1 + 2\tan\phi(1 - \sin\phi)^2\tan^{-1}\left(\dfrac{D_f}{B}\right)$	
	$F_{\gamma d} = 1$	
Inclination	$F_{ci} = F_{qi} = \left(1 - \dfrac{\beta^\circ}{90^\circ}\right)^2$	Meyerhof (1963); Hanna and Meyerhof (1981)
	$F_{\gamma i} = \left(1 - \dfrac{\beta}{\phi}\right)^2$	
	where β = inclination of the load on the foundation with respect to the vertical	

*These shape factors are empirical relations based on extensive laboratory tests.
[†]The factor $\tan^{-1}(D_f/B)$ is in radians.

Case I: If the water table is located so that $0 \le D_1 \le D_f$, the factor q in the bearing capacity equations takes the form

$$q = \text{effective surcharge} = D_1\gamma + D_2(\gamma_{\text{sat}} - \gamma_w) \qquad (11.9)$$

where γ_{sat} = saturated unit weight of soil
γ_w = unit weight of water

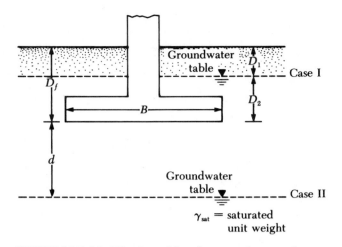

FIGURE 11.5 Modification of bearing capacity equations for water table

Also, the value of γ in the last term of the equations has to be replaced by $\gamma' = \gamma_{sat} - \gamma_w$.

Case II: For a water table located so that $0 \le d \le B$,

$$q = \gamma D_f \tag{11.10}$$

The factor γ in the last term of the bearing capacity equations must be replaced by the factor

$$\bar{\gamma} = \gamma' + \frac{d}{B}(\gamma - \gamma') \tag{11.11}$$

The preceding modifications are based on the assumption that there is no seepage force in the soil.

Case III: When the water table is located so that $d \ge B$, the water will have no effect on the ultimate bearing capacity.

11.4   *The Factor of Safety*

Calculating the gross allowable load-bearing capacity of shallow foundations requires the application of a factor of safety (*FS*) to the gross ultimate bearing capacity, or

$$q_{all} = \frac{q_u}{FS} \tag{11.12}$$

However, some practicing engineers prefer to use a factor of safety of

$$\text{net stress increase on soil} = \frac{\text{net ultimate bearing capacity}}{FS} \qquad (11.13)$$

The net ultimate bearing capacity was defined in Eq. (11.8) as

$$q_{net(u)} = q_u - q$$

Substituting this equation into Eq. (11.13) yields

net stress increase on soil

= load from the superstructure per unit area of the foundation

$$= q_{all(net)} = \frac{q_u - q}{FS} \qquad (11.14)$$

The factor of safety defined by Eq. (11.14) may be at least 3 in all cases.

**EXAMPLE
11.1**

A square column foundation to be constructed on a sandy soil has to carry a gross allowable total load of 150 kN. The depth of the foundation will be 0.7 m. The load will be inclined at an angle of 20° to the vertical (Figure 11.6). The standard penetration resistances, N_F, obtained from field exploration are listed in the table.

Depth (m)	N_F
1.5	3
3.0	6
4.5	9
6	10
7.5	10
9	8

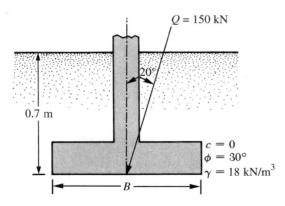

FIGURE 11.6

Assume that the unit weight of the soil is 18 kN/m³. Determine the width of the foundation, B. Use a factor of safety of 3.

Solution The standard penetration resistances can be corrected by using Eq. (8.9) and the Liao and Whitman equation given in Table 8.4. See the following table.

Depth (m)	Effective overburden pressure, σ'_o (kN/m²)	C_N	N_F	$N_{cor} = N_F C_N$
1.5	27	1.88	3	≈6
3.0	54	1.33	6	≈8
4.5	81	1.09	9	≈10
6	108	0.94	10	≈9
7.5	135	0.84	10	≈8
9	162	0.77	8	≈6

The average corrected N_{cor} value obtained is about 8. Now, referring to Eq. (8.8), we can conservatively assume the soil friction angle ϕ to be about 30°. With $c = 0$, the ultimate bearing capacity [Eq. (11.7)] becomes

$$q_u = qN_q F_{qs}F_{qd}F_{qi} + \frac{1}{2}\gamma BN_\gamma F_{\gamma s}F_{\gamma d}F_{\gamma i}$$

$$q = (0.7)(18) = 12.6 \text{ kN/m}^2$$

$$\gamma = 18 \text{ kN/m}^3$$

From Table 11.1, for $\phi = 30°$, we find

$$N_q = 18.4$$

$$N_\gamma = 22.4$$

From Table 11.2,

$$F_{qs} = 1 + \left(\frac{B}{L}\right)\tan\phi = 1 + 0.577 = 1.577$$

$$F_{\gamma s} = 1 - 0.4\left(\frac{B}{L}\right) = 0.6$$

$$F_{qd} = 1 + 2\tan\phi(1 - \sin\phi)^2\frac{D_f}{B} = 1 + \frac{(0.289)(0.7)}{B} = 1 + \frac{0.202}{B}$$

$$F_{\gamma d} = 1$$

$$F_{qi} = \left(1 - \frac{\beta°}{90°}\right)^2 = \left(1 - \frac{20}{90}\right)^2 = 0.605$$

$$F_{\gamma i} = \left(1 - \frac{\beta^\circ}{\phi}\right)^2 = \left(1 - \frac{20}{30}\right)^2 = 0.11$$

Hence,

$$q_u = (12.6)(18.4)(1.577)\left(1 + \frac{0.202}{B}\right)(0.605) + (0.5)(18)(B)(22.4)(0.6)(1)(0.11)$$

$$= 212.2 + \frac{44.68}{B} + 13.3B \tag{a}$$

Thus,

$$q_{all} = \frac{q_u}{3} = 73.73 + \frac{14.89}{B} + 4.43B \tag{b}$$

For Q = total allowable load = $q_{all} \times B^2$ or

$$q_{all} = \frac{150}{B^2} \tag{c}$$

Equating the right-hand sides of Eqs. (b) and (c) gives

$$\frac{150}{B^2} = 73.73 + \frac{14.89}{B} + 4.43B$$

By trial and error, we find $B \approx$ **1.3 m.** ■

11.5 *Eccentrically Loaded Foundations*

As with the base of a retaining wall, there are several instances in which foundations are subjected to moments in addition to the vertical load, as shown in Figure 11.7a. In such cases, the distribution of pressure by the foundation on the soil is not uniform. The distribution of nominal pressure is

$$q_{max} = \frac{Q}{BL} + \frac{6M}{B^2L} \tag{11.15}$$

and

$$q_{min} = \frac{Q}{BL} - \frac{6M}{B^2L} \tag{11.16}$$

where Q = total vertical load
       M = moment on the foundation

The exact distribution of pressure is difficult to estimate.

The factor of safety for such types of loading against bearing capacity failure can be evaluated using the procedure suggested by Meyerhof (1953), which is generally referred to as the *effective area* method. The following is Meyerhof's step-

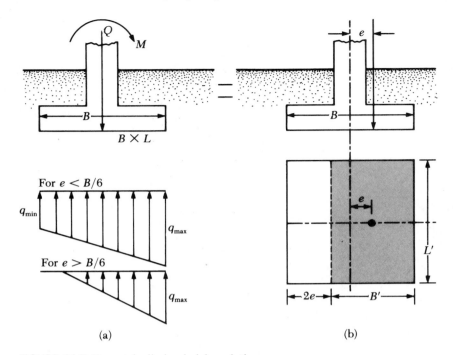

FIGURE 11.7 Eccentrically loaded foundations

by-step procedure for determining the ultimate load that the soil can support and the factor of safety against bearing capacity failure.

1. Figure 11.7b shows a force system equivalent to that shown in Figure 11.7a. The distance e is the eccentricity, or

$$e = \frac{M}{Q} \tag{11.17}$$

Substituting Eq. (11.17) in Eqs. (11.15) and (11.16) gives

$$q_{max} = \frac{Q}{BL}\left(1 + \frac{6e}{B}\right) \tag{11.18}$$

and

$$q_{min} = \frac{Q}{BL}\left(1 - \frac{6e}{B}\right) \tag{11.19}$$

Note that, in these equations, when the eccentricity e becomes $B/6$, q_{min} is 0. For $e > B/6$, q_{min} will be negative, which means that tension will develop. Because soil cannot take any tension, there will be a separation between the foundation and the soil underlying it. The nature of the pressure distribution on the soil will be as shown in Figure 11.7a. The value of q_{max} then is

$$q_{max} = \frac{4Q}{3L(B - 2e)} \tag{11.20}$$

2. Determine the effective dimensions of the foundation as

$B' = $ effective width $ = B - 2e$

$L' = $ effective length $ = L$

Note that, if the eccentricity were in the direction of the length of the foundation, then the value of L' would be equal to $L - 2e$. The value of B' would equal B. The smaller of the two dimensions (that is, L' and B') is the effective width of the foundation.

3. Use Eq. (11.7) for the ultimate bearing capacity as

$$q'_u = cN_cF_{cs}F_{cd}F_{ci} + qN_qF_{qs}F_{qd}F_{qi} + \frac{1}{2}\gamma B'N_\gamma F_{\gamma s}F_{\gamma d}F_{\gamma i} \tag{11.21}$$

To evaluate F_{cs}, F_{qs}, and $F_{\gamma s}$, use Table 11.2 with *effective length* and *effective width* dimensions instead of L and B, respectively. To determine F_{cd}, F_{qd}, and $F_{\gamma d}$, use Table 11.2 (*do not* replace B with B').

4. The total ultimate load that the foundation can sustain is

$$Q_{ult} = q'_u \overbrace{(B')(L')}^{A'} \tag{11.22}$$

where $A' = $ effective area.

5. The factor of safety against bearing capacity failure is

$$FS = \frac{Q_{ult}}{Q} \tag{11.23}$$

Foundations with Two-Way Eccentricity

Consider a situation in which a foundation is subjected to a vertical ultimate load Q_{ult} and a moment M, as shown in Figure 11.8a and b. For this case, the components of the moment, M, about the x and y axes can be determined as M_x and M_y, respectively (Figure 11.8c). This condition is equivalent to a load Q_{ult} placed eccentrically on the foundation with $x = e_B$ and $y = e_L$ (Figure 11.8d). Note that

$$e_B = \frac{M_y}{Q_{ult}} \tag{11.24}$$

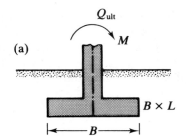

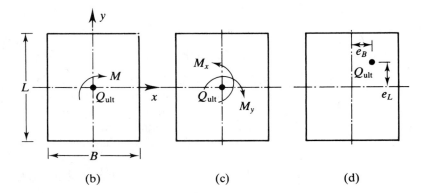

FIGURE 11.8 Analysis of foundation with two-way eccentricity

and

$$e_L = \frac{M_x}{Q_{ult}}$$ (11.25)

If Q_{ult} is needed, it can be obtained as follows [Eq. (11.22)]:

$$Q_{ult} = q'_u A'$$

where, from Eq. (11.21),

$$q'_u = cN_c F_{cs}F_{cd}F_{ci} + qN_q F_{qs}F_{qd}F_{qi} + \frac{1}{2}\gamma B' N_\gamma F_{\gamma s}F_{\gamma d}F_{\gamma i}$$

and

$$A' = \text{effective area} = B'L'$$

As before, to evaluate F_{cs}, F_{qs}, and $F_{\gamma s}$ (Table 11.2), we use the effective length (L') and effective width (B') dimensions instead of L and B, respectively. To calculate F_{cd}, F_{qd}, and $F_{\gamma d}$, we use Table 11.2; however, we do not replace B with B'. When we determine the effective area (A'), effective width (B'), and effective length (L'), four possible cases may arise (Higher and Anders, 1985).

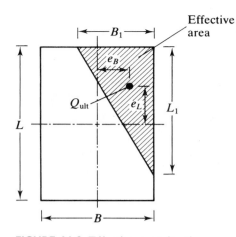

FIGURE 11.9 Effective area for the case of $e_L/L \geq \frac{1}{6}$ and $e_B/B \geq \frac{1}{6}$

Case I: $e_L/L \geq \frac{1}{6}$ and $e_B/B \geq \frac{1}{6}$. The effective area for this condition is shown in Figure 11.9, or

$$A' = \frac{1}{2} B_1 L_1 \tag{11.26}$$

where $B_1 = B \left(1.5 - \frac{3e_B}{B} \right)$ $\tag{11.27a}$

$$L_1 = L \left(1.5 - \frac{3e_L}{L} \right) \tag{11.27b}$$

The effective length, L', is the larger of the two dimensions—that is, B_1 or L_1. So, the effective width is

$$B' = \frac{A'}{L'} \tag{11.28}$$

Case II: $e_L/L < 0.5$ and $0 < e_B/B < \frac{1}{6}$. The effective area for this case is shown in Figure 11.10:

$$A' = \frac{1}{2} (L_1 + L_2) B \tag{11.29}$$

The magnitudes of L_1 and L_2 can be determined from Figure 11.10b. The effective width is

$$B' = \frac{A'}{L_1 \text{ or } L_2 \text{ (whichever is larger)}} \tag{11.30}$$

The effective length is

$$L' = L_1 \text{ or } L_2 \text{ (whichever is larger)} \tag{11.31}$$

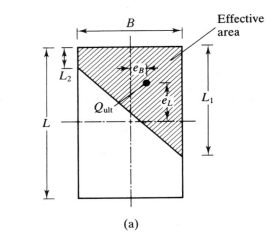

(a)

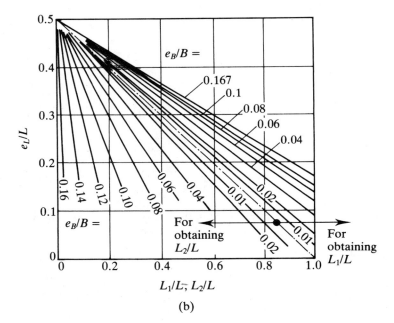

(b)

FIGURE 11.10 Effective area for the case of $e_L/L < 0.5$ and $0 < e_B/B < \frac{1}{6}$ (After Higher and Anders, 1985)

Case III: $e_L/L < \frac{1}{6}$ and $0 < e_B/B < 0.5$. The effective area is shown in Figure 11.11a:

$$A' = \frac{1}{2}(B_1 + B_2)L \tag{11.32}$$

The effective width is

$$B' = \frac{A'}{L} \tag{11.33}$$

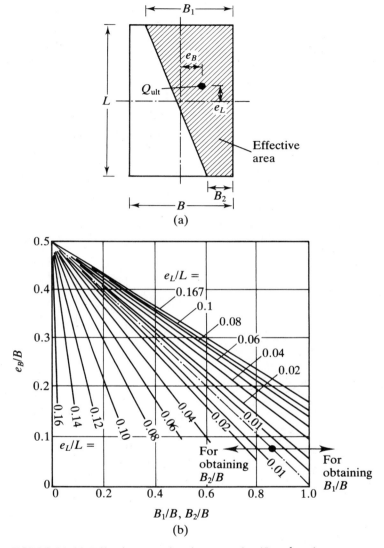

FIGURE 11.11 Effective area for the case of $e_L/L < \frac{1}{6}$ and $0 < e_B/B < 0.5$ (After Higher and Anders, 1985)

The effective length is

$$L' = L \qquad (11.34)$$

The magnitudes of B_1 and B_2 can be determined from Figure 11.11b.

Case IV: $e_L/L < \frac{1}{6}$ and $e_B/B < \frac{1}{6}$. Figure 11.12a shows the effective area for this case. The ratio B_2/B and thus B_2 can be determined by using the e_L/L curves that slope upward. Similarly, the ratio L_2/L and thus

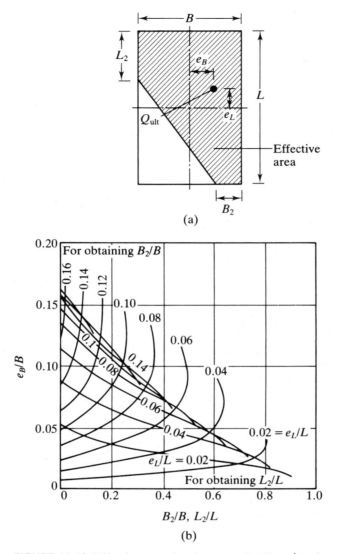

(a)

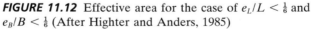

(b)

FIGURE 11.12 Effective area for the case of $e_L/L < \frac{1}{6}$ and $e_B/B < \frac{1}{6}$ (After Higher and Anders, 1985)

L_2 can be determined by using the e_L/L curves that slope downward. The effective area is then

$$A' = L_2 B + \frac{1}{2}(B + B_2)(L - L_2)$$
(11.35)

The effective width is

$$B' = \frac{A'}{L}$$
(11.36)

The effective length is

$$L' = L \tag{11.37}$$

EXAMPLE 11.2

A continuous foundation is shown in Figure 11.13. If the load eccentricity is 0.15 m, determine the ultimate load, Q_{ult}, per unit length of the foundation.

Solution For $c = 0$, Eq. (11.21) gives

$$q'_u = qN_qF_{qs}F_{qd}F_{qi} + \frac{1}{2}\gamma B'N_\gamma F_{\gamma s}F_{\gamma d}F_{\gamma i}$$

$$q = (17.3)(1.2) = 20.76 \text{ kN/m}^2$$

For $\phi = 35°$, from Table 11.1, we find $N_q = 33.3$ and $N_\gamma = 48.03$. We have

$$B' = 1.8 - (2)(0.15) = 1.5 \text{ m}$$

Because it is a strip foundation, B'/L' is 0. Hence, $F_{qs} = 1$ and $F_{\gamma s} = 1$, and

$$F_{qi} = F_{\gamma i} = 1$$

From Table 11.2, we have

$$F_{qd} = 1 + 2\tan\phi(1 - \sin\phi)^2\frac{D_f}{B} = 1 + 0.255\left(\frac{1.2}{1.8}\right) = 1.17$$

$$F_{\gamma d} = 1$$

$$q'_u = (20.76)(33.3)(1)(1.17)(1) + \left(\frac{1}{2}\right)(17.3)(1.5)(48.03)(1)(1)(1) = 1432 \text{ kN/m}^2$$

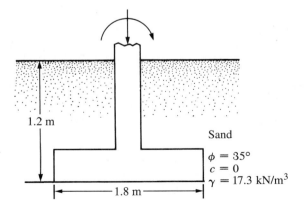

1.2 m

Sand

$\phi = 35°$
$c = 0$
$\gamma = 17.3 \text{ kN/m}^3$

1.8 m

FIGURE 11.13

Hence,

$$Q_{ult} = (B')(1)(q'_u) = (1.5)(1)(1432) = \mathbf{2148 \ kN/m} \qquad \blacksquare$$

**EXAMPLE
11.3**

A square foundation is shown in Figure 11.14, with $e_L = 0.3$ m and $e_B = 0.15$ m. Assume two-way eccentricity and determine the ultimate load, Q_{ult}.

Solution

$$\frac{e_L}{L} = \frac{0.3}{1.5} = 0.2$$

$$\frac{e_B}{B} = \frac{0.15}{1.5} = 0.1$$

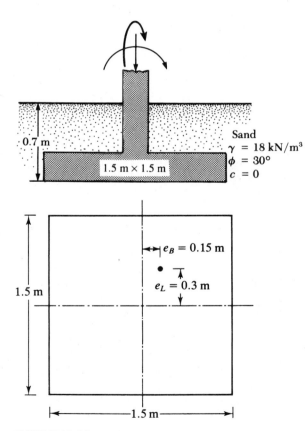

FIGURE 11.14

This case is similar to that shown in Figure 11.11a. From Figure 11.11b, for $e_L/L = 0.2$ and $e_B/B = 0.1$, we have

$$\frac{L_1}{L} \approx 0.85; \qquad L_1 = (0.85)(1.5) = 1.275 \text{ m}$$

and

$$\frac{L_2}{L} \approx 0.21; \qquad L_2 = (0.21)(1.5) = 0.315 \text{ m}$$

From Eq. (11.29),

$$A' = \frac{1}{2}(L_1 + L_2)B = \frac{1}{2}(1.275 + 0.315)(1.5) = 1.193 \text{ m}^2$$

From Eq. (11.31),

$$L' = L_1 = 1.275 \text{ m}$$

From Eq. (11.30),

$$B' = \frac{A'}{L_1} = \frac{1.193}{1.275} = 0.936 \text{ m}$$

Note, from Eq. (11.21), for $c = 0$, we have

$$q'_u = qN_qF_{qs}F_{qd}F_{qi} + \frac{1}{2}\gamma B'N_\gamma F_{\gamma s}F_{\gamma d}F_{\gamma i}$$

$$q = (0.7)(18) = 12.6 \text{ kN/m}^2$$

For $\phi = 30°$, from Table 11.1, $N_q = 18.4$ and $N_\gamma = 22.4$. Thus,

$$F_{qs} = 1 + \left(\frac{B'}{L'}\right)\tan\phi = 1 + \left(\frac{0.936}{1.275}\right)\tan 30° = 1.424$$

$$F_{\gamma s} = 1 - 0.4\left(\frac{B'}{L'}\right) = 1 - 0.4\left(\frac{0.936}{1.275}\right) = 0.706$$

$$F_{qd} = 1 + 2\tan\phi(1 - \sin\phi)^2\frac{D_f}{B} = 1 + \frac{(0.289)(0.7)}{1.5} = 1.135$$

$$F_{\gamma d} = 1$$

So

$$Q_{ult} = A'q'_u = A'\left(qN_qF_{qs}F_{qd} + \frac{1}{2}\gamma B'N_\gamma F_{\gamma s}F_{\gamma d}\right)$$

$$= (1.193)[(12.6)(18.4)(1.424)(1.135) + (0.5)(18)(0.936)(22.4)(0.706)(1)]$$

$$= \textbf{605.95 kN}$$ ∎

SETTLEMENT OF SHALLOW FOUNDATIONS

11.6   *Types of Foundation Settlement*

As was discussed in Chapter 6, foundation settlement is made up of *immediate* (or elastic) settlement, S_e, and consolidation settlement, S_c. The procedure for calculating the consolidation settlement of foundations was also explained in Chapter 6. The methods for estimating immediate settlement will be elaborated upon in the following sections.

It is important to point out that, theoretically at least, a foundation could be considered fully flexible or fully rigid. A uniformly loaded, perfectly flexible foundation resting on an elastic material such as saturated clay will have a sagging profile, as shown in Figure 11.15a, because of elastic settlement. However, if the foundation is rigid and is resting on an elastic material such as clay, it will undergo uniform settlement and the contact pressure will be redistributed (Figure 11.15b).

11.7   *Immediate Settlement*

Figure 11.16 shows a shallow foundation subjected to a net force per unit area equal to q_o. Let the Poisson's ratio and the modulus of elasticity of the soil supporting

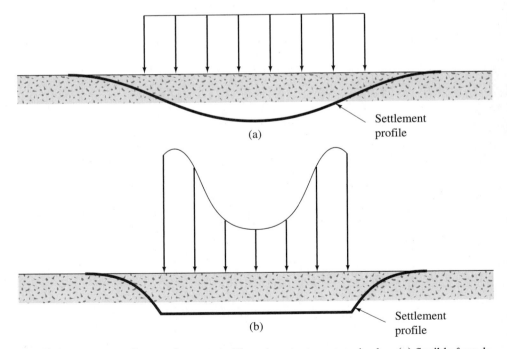

FIGURE 11.15 Immediate settlement profile and contact pressure in clay: (a) flexible foundation; (b) rigid foundation

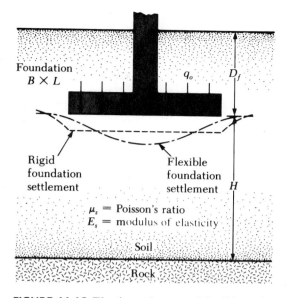

FIGURE 11.16 Elastic settlement of flexible and rigid foundations

it be μ_s and E_s, respectively. Theoretically, if $D_f = 0$, $H = \infty$, and the foundation is perfectly flexible, according to Harr (1966), the settlement may be expressed as

$$S_e = \frac{Bq_o}{E_s}(1 - \mu_s^2)\frac{\alpha}{2} \qquad \text{(corner of the flexible foundation)} \qquad (11.38)$$

$$S_e = \frac{Bq_o}{E_s}(1 - \mu_s^2)\alpha \qquad \text{(center of the flexible foundation)} \qquad (11.39)$$

where $\alpha = \frac{1}{\pi}\left[\ln\left(\frac{\sqrt{1 + m^2} + m}{\sqrt{1 + m^2} - m}\right) + m \ln\left(\frac{\sqrt{1 + m^2} + 1}{\sqrt{1 + m^2} - 1}\right)\right] \qquad (11.40)$

$$m = L/B \qquad (11.41)$$
$$B = \text{width of foundation}$$
$$L = \text{length of foundation}$$

The values of α for various length-to-width (L/B) ratios are shown in Figure 11.17. The average immediate settlement for a flexible foundation also may be expressed as

$$S_e = \frac{Bq_o}{E_s}(1 - \mu_s^2)\alpha_{av} \qquad \text{(average for flexible foundation)} \qquad (11.42)$$

Figure 11.17 also shows the values of α_{av} for various L/B ratios of foundation.

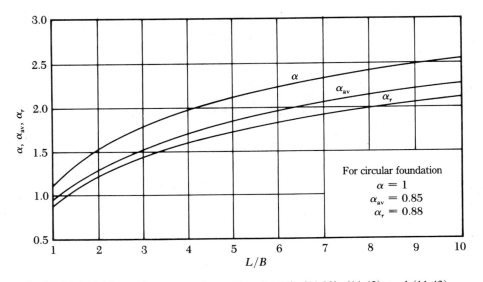

FIGURE 11.17 Values of α, α_{av}, and α_r—Eqs. (11.38), (11.39), (11.42), and (11.43)

If the foundation shown in Figure 11.16 is rigid, however, the immediate settlement will be different and may be expressed as

$$S_e = \frac{Bq_o}{E_s}(1 - \mu_s^2)\alpha_r \qquad \text{(rigid foundation)} \tag{11.43}$$

The values of α_r for various L/B ratios of foundation are shown in Figure 11.17.

The preceding equations for immediate settlement were obtained by integrating the strain at various depths below the foundations for limits of $z = 0$ to $z = \infty$. If an incompressible layer of rock is located at a limited depth, the actual settlement may be less than that calculated by the preceding equations. However, if the depth H in Figure 11.16 is greater than about $2B$ to $3B$, the actual settlement would not change considerably. Also note that the deeper the embedment, D_f, the less is the total elastic settlement.

11.8   *Immediate Settlement of Foundations on Saturated Clay*

Janbu et al. (1956) proposed an equation for evaluating the average settlement of flexible foundations on saturated clay soils (Poisson's ratio, $\mu_s = 0.5$). For the

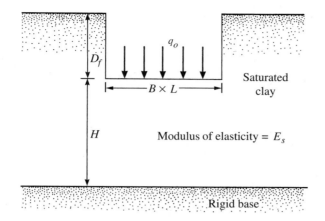

FIGURE 11.18 Foundation on saturated clay

notation used in Figure 11.18, this equation is

$$S_e = A_1 A_2 \frac{q_o B}{E_s}$$

(11.44)

where A_1 is a function of H/B and L/B and A_2 is a function of D_f/B.

Christian and Carrier (1978) modified the values of A_1 and A_2 somewhat and presented them in a graphical form. The interpolated values of A_1 and A_2 from those graphs are listed in Tables 11.3 and 11.4.

Table 11.3 Variation of A_1 with H/B

		A_1				
			L/B			
H/B	Circle	1	2	3	4	5
1	0.36	0.36	0.36	0.36	0.36	0.36
2	0.47	0.53	0.63	0.64	0.64	0.64
4	0.58	0.63	0.82	0.94	0.94	0.94
6	0.61	0.67	0.88	1.08	1.14	1.16
8	0.62	0.68	0.90	1.13	1.22	1.26
10	0.63	0.70	0.92	1.18	1.30	1.42
20	0.64	0.71	0.93	1.26	1.47	1.74
30	0.66	0.73	0.95	1.29	1.54	1.84

Table 11.4
Variation of A_2
with D_f/B

D_f/B	A_2
0	1.0
2	0.9
4	0.88
6	0.875
8	0.87
10	0.865
12	0.863
14	0.860
16	0.856
18	0.854
20	0.850

11.9   *Range of Material Parameters for Computing Immediate Settlement*

Section 11.7 presented the equations for calculating the immediate settlement of foundations. These equations contain the elastic parameters, such as E_s and μ_s. If the laboratory test results for these parameters are not available, certain realistic assumptions have to be made. Table 11.5 gives the approximate range of the elastic parameters for various soils.

Several investigators correlated the values of the modulus of elasticity, E_s, with the field standard penetration number, N_F, and the cone penetration resistance, q_c. Mitchell and Gardner (1975) compiled a comprehensive list of these correlations. Schmertmann (1970) proposed that the modulus of elasticity of sand may be given by

$$E_s \, (\text{kN/m}^2) = 766 N_F \qquad (11.45)$$

Table 11.5   Elastic parameters of various soils

Type of soil	Modulus of elasticity, E_s (MN/m²)	Poisson's ratio, μ_s
Loose sand	10–25	0.20–0.40
Medium dense sand	15–30	0.25–0.40
Dense sand	35–55	0.30–0.45
Silty sand	10–20	0.20–0.40
Sand and gravel	70–170	0.15–0.35
Soft clay	4–20	
Medium clay	20–40	0.20–0.50
Stiff clay	40–100	

where N_F = standard penetration number. Similarly,

$$E_s = 2q_c \tag{11.46}$$

where q_c = static cone penetration resistance. The modulus of elasticity of normally consolidated clays may be estimated as

$$E_s = 250c \text{ to } 500c \tag{11.47}$$

and for overconsolidated clays as

$$E_s = 750c \text{ to } 1000c \tag{11.48}$$

where c = undrained cohesion of clay soil.

11.10 Allowable Bearing Pressure in Sand Based on Settlement Consideration

Meyerhof (1956) proposed a correlation for the *net allowable bearing pressure* for foundations with the corrected standard penetration resistance, N_{cor}. The net allowable pressure can be defined as

$$q_{all(net)} = q_{all} - \gamma D_f \tag{11.49}$$

According to Meyerhof's theory, for 25 mm of estimated maximum settlement,

$$q_{all(net)} \text{ (kN/m}^2\text{)} = 11.98N_{cor} \qquad \text{(for } B \leq 1.22 \text{ m)} \tag{11.50}$$

$$q_{all(net)} \text{ (kN/m}^2\text{)} = 7.99N_{cor}\left(\frac{3.28B+1}{3.28B}\right)^2 \qquad \text{(for } B > 1.22 \text{ m)} \tag{11.51}$$

where N_{cor} = corrected standard penetration number.

Since Meyerhof proposed his original correlation, researchers have observed that its results are rather conservative. Later, Meyerhof (1965) suggested that the net allowable bearing pressure should be increased by about 50%. Bowles (1977) proposed that the modified form of the bearing pressure equations be expressed as

$$q_{all(net)} \text{ (kN/m}^2\text{)} = 19.16N_{cor}F_d\left(\frac{S_e}{25}\right) \qquad \text{(for } B \leq 1.22 \text{ m)} \tag{11.52}$$

$$q_{all(net)} \text{ (kN/m}^2\text{)} = 11.98N_{cor}\left(\frac{3.28B+1}{3.28B}\right)^2 F_d\left(\frac{S_e}{25}\right) \qquad \text{(for } B > 1.22 \text{ m)} \tag{11.53}$$

where F_d = depth factor = $1 + 0.33(D_f/B) \leq 1.33$ $\qquad$ (11.54)
$\qquad S_e$ = tolerable settlement (mm)

The empirical relations just presented may raise some questions. For example, which value of the standard penetration number should be used? What is the effect of the water table on the net allowable bearing capacity? The design value of N_{cor} should be determined by taking into account the N_{cor} values for a depth of $2B$ to $3B$, measured from the bottom of the foundation. Many engineers are also of the opinion that the N_{cor} value should be reduced somewhat if the water table is close to the foundation. However, the author believes that this reduction is not required because the penetration resistance reflects the location of the water table.

Meyerhof (1956) also prepared empirical relations for the net allowable bearing capacity of foundations based on the cone penetration resistance, q_c:

$$q_{all(net)} = \frac{q_c}{15} \qquad \text{(for } B \leq 1.22 \text{ m and settlement of 25 mm)}$$

(11.55)

and

$$q_{all(net)} = \frac{q_c}{25} \left(\frac{3.28B + 1}{3.28B} \right)^2 \qquad \text{(for } B > 1.22 \text{ m and settlement of 25 mm)}$$

(11.56)

Note that in Eqs. (11.55) and (11.56) the unit of B is meters and the units of $q_{all(net)}$ and q_c are kN/m².

The basic philosophy behind the development of these correlations is that, if the maximum settlement is no more than 25 mm for any foundation, the differential settlement would be no more than 19 mm. These are probably the allowable limits for most building foundation designs.

11.11 *Field Load Test*

The ultimate load-bearing capacity of a foundation, as well as the allowable bearing capacity based on tolerable settlement considerations, can be effectively determined from the field load test. It is generally referred to as the *plate load test* (ASTM, 1997, Test Designation D-1194-72). The plates that are used for tests in the field are usually made of steel and are 25 mm thick and 150 to 762 mm in diameter. Occasionally, square plates that are 305 mm × 305 mm are also used.

To conduct a plate load test, a hole is excavated with a minimum diameter of 4B (B = diameter of the test plate) to a depth of D_f (D_f = depth of the proposed foundation). The plate is placed at the center of the hole. Load is applied to the plate in steps—about one-fourth to one-fifth of the estimated ultimate load—by a jack. A schematic diagram of the test arrangement is shown in Figure 11.19a. During each step load application, the settlement of the plate is observed on dial gauges. At least 1 hour is allowed to elapse between each load application step. The test should be conducted until failure, or at least until the plate has gone through 25 mm of settlement. Figure 11.19b shows the nature of the load–settlement curve obtained from such tests, from which the ultimate load per unit area can be determined.

For tests in clay,

$$q_{u(F)} = q_{u(P)}$$

(11.57)

where $q_{u(F)}$ = ultimate bearing capacity of the proposed foundation
$\qquad q_{u(P)}$ = ultimate bearing capacity of the test plate

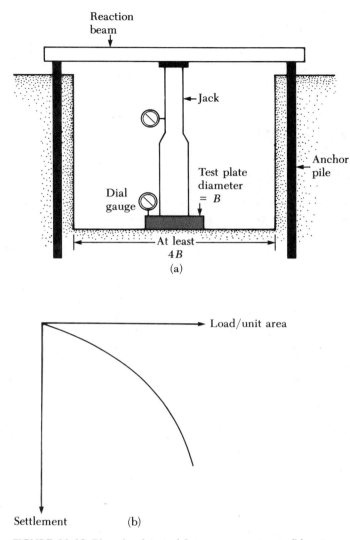

Reaction
beam

Jack

Anchor
pile

Test plate
diameter
= B

Dial
gauge

At least
4B

(a)

Load/unit area

Settlement   (b)

FIGURE 11.19 Plate load test: (a) test arrangement; (b) nature
of load–settlement curve

Equation (11.57) implies that the ultimate bearing capacity in clay is virtually
independent of the size of the plate.

For tests in sandy soils,

$$q_{u(F)} = q_{u(P)} \frac{B_F}{B_P}$$ (11.58)

where B_F = width of the foundation
B_P = width of the test plate

The allowable bearing capacity of a foundation, based on settlement considerations and for a given intensity of load, q_o, is

$$S_{e(F)} = S_{e(P)} \frac{B_F}{B_P} \quad \text{(for clayey soil)} \tag{11.59}$$

and

$$S_{e(F)} = S_{e(P)} \left(\frac{B_F}{B_P}\right)^2 \left(\frac{3.28 B_P + 1}{3.28 B_F + 1}\right)^2 \quad \text{(for sandy soil)} \tag{11.60}$$

In Eq. (11.60), the units of B_P and B_F are meters. Equation (11.60) is based on the works of Terzaghi and Peck (1967).

Housel (1929) proposed a different technique for determining the load-bearing capacity of shallow foundations based on settlement considerations:

1. Find the dimensions of a foundation that will carry a load of Q_o with a tolerable settlement of $S_{e(tol)}$.
2. Conduct two plate load tests with plates of diameters B_1 and B_2.
3. From the load–settlement curves obtained in Step 2, determine the total loads on the plates (Q_1 and Q_2) that correspond to the settlement of $S_{e(tol)}$. For plate no. 1, the total load can be expressed as

$$Q_1 = A_1 m + P_1 n \tag{11.61}$$

Similarly, for plate no. 2,

$$Q_2 = A_2 m + P_2 n \tag{11.62}$$

where A_1, A_2 = areas of plates no. 1 and no. 2, respectively
$\quad\quad P_1, P_2$ = perimeters of plates no. 1 and no. 2, respectively
$\quad\quad m, n$ = two constants that correspond to the bearing pressure and perimeter shear, respectively

The values of m and n can be determined by solving Eqs. (11.61) and (11.62).
4. For the foundation to be designed,

$$Q_o = A m + P n \tag{11.63}$$

where A = area of the foundation
$\quad\quad P$ = perimeter of the foundation

Because Q_o, m, and n are known, Eq. (11.63) can be solved to determine the foundation width. An application of this procedure is presented in Example 11.5.

EXAMPLE 11.4

The results of a plate load test in a sandy soil are shown in Figure 11.20. The size of the plate is 0.305 m × 0.305 m. Determine the size of a square column foundation that should carry a load of 2500 kN with a maximum settlement of 25 mm.

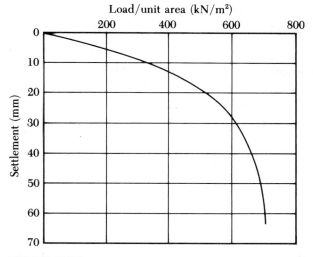

FIGURE 11.20

Solution The problem has to be solved by trial and error. Use the table and Eq. (11.60).

Q_o (kN) (1)	Assume width B_F (m) (2)	$q_o = \dfrac{Q_o}{B_F^2}$ (kN/m^2) (3)	$S_{e(P)}$ corresponding to q_o in col. 3 (mm) (4)	$S_{e(F)}$ from Eq. (11.60) (mm) (5)
2500	4.0	156.25	4.0	13.80
2500	3.0	277.80	8.0	26.35
2500	3.2	244.10	6.8	22.70
2500	3.1	260.10	7.2	23.86

So, a column footing with dimensions of **3.1 m × 3.1 m** will be appropriate. ■

EXAMPLE 11.5

The results of two plate load tests are given in the table:

Plate diameter, B (m)	Total load, Q (kN)	Settlement (mm)
0.305	32.2	20
0.610	71.8	20

A square column foundation has to be constructed to carry a total load of 715 kN. The tolerable settlement is 20 mm. Determine the size of the foundation.

Solution   Use Eqs. (11.61) and (11.62):

$$32.2 = \frac{\pi}{4}(0.305)^2 m + \pi(0.305)n \qquad \text{(a)}$$

$$71.8 = \frac{\pi}{4}(0.610)^2 m + \pi(0.610)n \qquad \text{(b)}$$

From Eqs. (a) and (b), we find

$$m = 50.68 \text{ kN/m}^2$$

$$n = 29.75 \text{ kN/m}$$

For the foundation to be designed [Eq. (11.63)],

$$Q_o = Am + Pn$$

or

$$Q_o = B_F^2 m + 4B_F n$$

For $Q_o = 715$ kN,

$$715 = B_F^2(50.68) + 4B_F(29.75)$$

or

$$50.68 B_F^2 + 119 B_F - 715 = 0$$

$$B_F \approx \textbf{2.8 m} \qquad \blacksquare$$

11.12   *Presumptive Bearing Capacity*

Several building codes (for example, Uniform Building Code, Chicago Building Code, New York City Building Code) specify the allowable bearing capacity of foundations on various types of soil. For minor construction, the codes often provide fairly acceptable guidelines. However, these bearing capacity values are based primarily on the *visual* classification of near-surface soils. They generally do not take into consideration factors such as the stress history of the soil, water table location, depth of the foundation, and tolerable settlement. So, for large construction projects, the codes' presumptive values should be used only as guides.

11.13   *Tolerable Settlement of Buildings*

As emphasized in this chapter, settlement analysis is an important part of the design and construction of foundations. Large settlements of various components of a structure may lead to considerable damage and/or may interfere with the proper functioning of the structure. Limited studies have been done to evaluate the conditions for tolerable settlement of various types of structure (for example, Bjerrum, 1963; Burland and Worth, 1974; Grant et al., 1974; Polshin and Tokar, 1957; and Wahls, 1981). Wahls (1981) has provided an excellent review of these studies.

Figure 11.21 gives the parameters for the definition of tolerable settlement. Figure 11.21a is for a structure that has undergone settlement without tilt; Figure 11.21b is for a structure that has undergone settlement with tilt.

The parameters are

ρ_i = total vertical displacement at point i

δ_{ij} = different settlement between points i and j

Δ = relative deflection

ω = tilt

$\eta_{ij} = \dfrac{\delta_{ij}}{l_{ij}} - \omega$ = angular distortion

$\dfrac{\Delta}{L}$ = deflection ratio

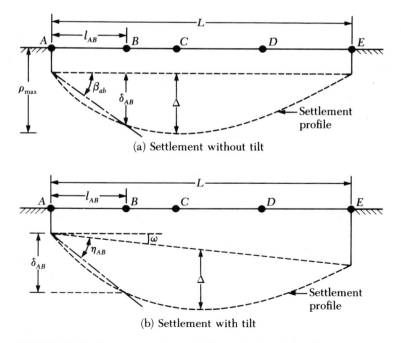

(a) Settlement without tilt

(b) Settlement with tilt

FIGURE 11.21 Parameters for definition of tolerable settlement (Redrawn after Wahls, 1981)

Table 11.6 Limiting angular distortion as recommended by Bjerrum (Compiled from Wahls, 1981)

Category of potential damage	η
Danger to machinery sensitive to settlement	1/750
Danger to frames with diagonals	1/600
Safe limit for no cracking of buildings*	1/500
First cracking of panel walls	1/300
Difficulties with overhead cranes	1/300
Tilting of high rigid buildings becomes visible	1/250
Considerable cracking of panel and brick walls	1/150
Danger of structural damage to general buildings	1/150
Safe limit for flexible brick walls, $L/H > 4$*	1/150

*Safe limits include a factor of safety.

$$L = \text{lateral dimension of the structure}$$

Bjerrum (1963) provided the conditions of *limiting* angular distortion, η, for various structures (see Table 11.6).

Polshin and Tokar (1957) presented the settlement criteria of the 1955 U.S.S.R. Building Code. These criteria were based on experience gained from observations of foundation settlement over 25 years. Tables 11.7 and 11.8 present the criteria.

Table 11.7 Allowable settlement criteria: 1955 U.S.S.R. Building Code (Compiled from Wahls, 1981)

Type of structure	Sand and hard clay	Plastic clay
(a) η		
Civil- and industrial-building column foundations:		
For steel and reinforced concrete structures	0.002	0.002
For end rows of columns with brick cladding	0.007	0.001
For structures where auxiliary strain does not arise during nonuniform settlement of foundations	0.005	0.005
Tilt of smokestacks, towers, silos, and so on	0.004	0.004
Craneways	0.003	0.003
(b) Δ/L		
Plain brick walls:		
For multistory dwellings and civil buildings		
at $L/H \leq 3$	0.0003	0.0004
at $L/H \geq 5$	0.0005	0.0007
For one-story mills	0.0010	0.0010

Table 11.8 Allowable average settlement for different building types (Compiled from Wahls, 1981)

Type of building	Allowable average settlement (mm)
Building with plain brick walls	
$L/H \geq 2.5$	80
$L/H \leq 1.5$	100
Building with brick walls, reinforced with reinforced concrete	
or reinforced brick	150
Framed building	100
Solid reinforced concrete foundations of smokestacks, silos, towers,	
and so on	300

MAT FOUNDATIONS

11.14 *Combined Footing and Mat Foundation*

Mat foundations are primarily shallow foundations. They are one of four major types of *combined footing* (see Figure 11.22a). A brief overview of combined footings and the methods used to calculate their dimensions follows.

1. *Rectangular combined footing:* In several instances, the load to be carried by a column and the soil bearing capacity are such that the standard spread footing design will require extension of the column foundation beyond the property line. In such a case, two or more columns can be supported on a single rectangular foundation, as shown in Figure 11.22b. If the net allowable soil pressure is known, the size of the foundation ($B \times L$) can be determined in the following manner:

   a. Determine the area of the foundation, A:

   $$A = \frac{Q_1 + Q_2}{q_{\text{all(net)}}} \tag{11.64}$$

   where $Q_1, Q_2 = $ column loads

   $q_{\text{all(net)}} = $ net allowable soil bearing capacity

   b. Determine the location of the resultant of the column loads. From Figure 11.22b, we see

   $$X = \frac{Q_2 L_3}{Q_1 + Q_2} \tag{11.65}$$

   c. For uniform distribution of soil pressure under the foundation, the resultant of the column loads should pass through the centroid of the foundation. Thus,

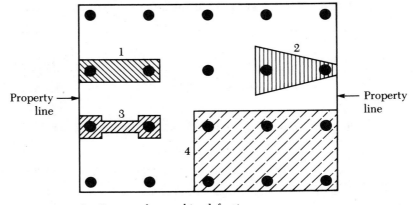

1   Rectangular combined footing
2   Trapezoidal combined footing
3   Cantilever footing
4   Mat foundation

(a)

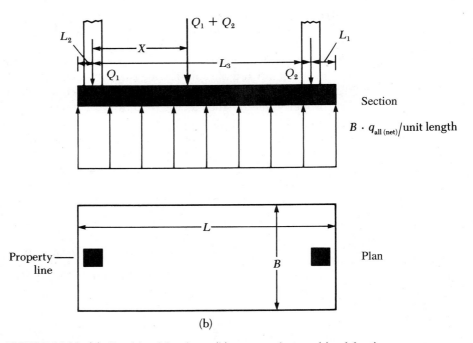

(b)

FIGURE 11.22 (a) Combined footings; (b) rectangular combined footing; (c) trapezoidal combined footing; (d) cantilever footing

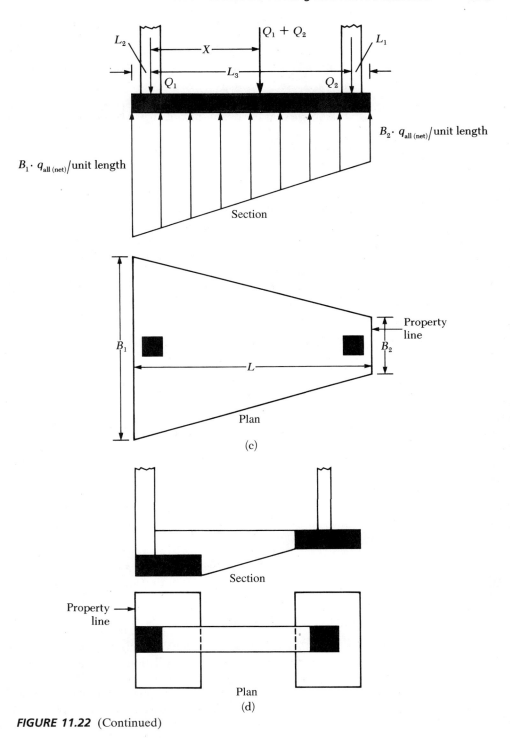

FIGURE 11.22 (Continued)

$$L = 2(L_2 + X) \tag{11.66}$$

where L = length of the foundation.

d. Once the length L is determined, obtain the value of L_1:

$$L_1 = L - L_2 - L_3 \tag{11.67}$$

Note that the magnitude of L_2 will be known and depends on the location of the property line.

e. The width of the foundation then is

$$B = \frac{A}{L} \tag{11.68}$$

2. *Trapezoidal combined footing:* This type of combined footing (Figure 11.22c) is sometimes used as an isolated spread foundation of a column carrying a large load where space is tight. The size of the foundation that will uniformly distribute pressure on the soil can be obtained in the following manner:

a. If the net allowable soil pressure is known, determine the area of the foundation:

$$A = \frac{Q_1 + Q_2}{q_{all(net)}}$$

From Figure 11.22c, we see

$$A = \frac{B_1 + B_2}{2} L \tag{11.69}$$

b. Determine the location of the resultant for the column loads:

$$X = \frac{Q_2 L_3}{Q_1 + Q_2}$$

c. From the property of a trapezoid, we have

$$X + L_2 = \left(\frac{B_1 + 2B_2}{B_1 + B_2}\right) \frac{L}{3} \tag{11.70}$$

With known values of A, L, X, and L_2, solve Eqs. (11.69) and (11.70) to obtain B_1 and B_2. Note that for a trapezoid,

$$\frac{L}{3} < X + L_2 < \frac{L}{2}$$

3. *Cantilever footing:* This type of combined footing construction uses a *strap beam* to connect an eccentrically loaded column foundation to the foundation of an interior column (Figure 11.22d). Cantilever footings may be used in place of trapezoidal or rectangular combined footings when the allowable soil bearing capacity is high and the distances between the columns are large.

4. *Mat foundation:* This type of foundation, which is sometimes referred to as a *raft foundation,* is a combined footing that may cover the entire area under a structure supporting several columns and walls (Figure 11.22a). Mat foundations are sometimes preferred for soils that have low load-bearing capacities but that will have to support high column and/or wall loads. Under some conditions, spread footings would have to cover more than half the building area, and mat foundations might be more economical.

11.15   *Common Types of Mat Foundations*

Several types of mat foundations are used currently. Some of the common types are shown schematically in Figure 11.23 and include:

1. Flat plate (Figure 11.23a). The mat is of uniform thickness.
2. Flat plate thickened under columns (Figure 11.23b).

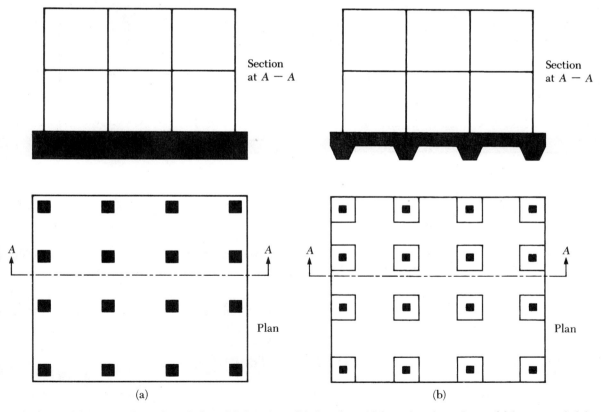

FIGURE 11.23 Types of mat foundation: (a) flat plate; (b) flat plate thickened under column; (c) beams and slab; (d) slab with basement walls

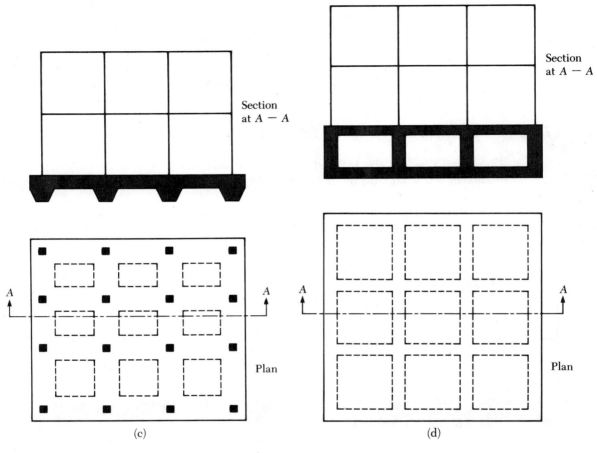

FIGURE 11.23 (Continued)

3. Beams and slab (Figure 11.23c). The beams run both ways, and the columns are located at the intersection of the beams.
4. Slab with basement walls as a part of the mat (Figure 11.23d). The walls act as stiffeners for the mat.

Mats may be supported by piles. The piles help in reducing the settlement of a structure built over highly compressible soil. Where the water table is high, mats are often placed over piles to control buoyancy.

11.16 Bearing Capacity of Mat Foundations

The *gross ultimate bearing capacity* of a mat foundation can be determined by the same equation used for shallow foundations, or

$$q_u = cN_cF_{cs}F_{cd}F_{ci} + qN_qF_{qs}F_{qd}F_{qi} + \frac{1}{2}\gamma BN_\gamma F_{\gamma s}F_{\gamma d}F_{\gamma i} \tag{11.7}$$

Tables 11.1 and 11.2 give the proper values of the bearing capacity factors and the shape, depth, and load inclination factors. The term B in Eq. (11.7) is the smallest dimension of the mat.

The *net ultimate bearing capacity* is

$$q_{net(u)} = q_u - q \tag{11.8}$$

A suitable factor of safety should be used to calculate the net *allowable* bearing capacity. For rafts on clay, the factor of safety should not be less than 3 under dead load and maximum live load. However, under the most extreme conditions, the factor of safety should be at least 1.75 to 2. For rafts constructed over sand, a factor of safety of 3 should normally be used. Under most working conditions, the factor of safety against bearing capacity failure of rafts on sand is very large.

For saturated clays with $\phi = 0$ and vertical loading condition, Eq. (11.7) gives

$$q_u = c_u N_c F_{cs} F_{cd} + q \tag{11.71}$$

where c_u = undrained cohesion. (*Note:* $N_c = 5.14$, $N_q = 1$, and $N_\gamma = 0$.) From Table 11.2, for $\phi = 0$,

$$F_{cs} = 1 + \left(\frac{B}{L}\right)\left(\frac{N_q}{N_c}\right) = 1 + \left(\frac{B}{L}\right)\left(\frac{1}{5.14}\right) = 1 + \frac{0.195B}{L}$$

and

$$F_{cd} = 1 + 0.4\left(\frac{D_f}{B}\right)$$

Substitution of the preceding shape and depth factors into Eq. (11.71) yields

$$q_u = 5.14c_u\left(1 + \frac{0.195B}{L}\right)\left(1 + 0.4\frac{D_f}{B}\right) + q \tag{11.72}$$

Hence, the net ultimate bearing capacity is

$$q_{net(u)} = q_u - q = 5.14c_u\left(1 + \frac{0.195B}{L}\right)\left(1 + 0.4\frac{D_f}{B}\right) \tag{11.73}$$

For $FS = 3$, the net allowable soil bearing capacity becomes

$$q_{all(net)} = \frac{q_{net(u)}}{FS} = 1.713c_u\left(1 + \frac{0.195B}{L}\right)\left(1 + 0.4\frac{D_f}{B}\right) \tag{11.74}$$

The net allowable bearing capacity for mats constructed over granular soil deposits can be adequately determined from the standard penetration resistance numbers. From Eq. (11.53), for shallow foundations, we have

$$q_{all(net)} \ (kN/m^2) = 11.98 N_{cor} \left(\frac{3.28B + 1}{3.28B}\right)^2 F_d \left(\frac{S_e}{25}\right)$$

where N_{cor} = corrected standard penetration resistance
B = width (m)
$F_d = 1 + 0.33(D_f/B) \le 1.33$
S_e = settlement (mm)

When the width, B, is large, the preceding equation can be approximated (assuming $3.28B + 1 \approx 3.28B$) as

$$q_{all(net)} \ (kN/m^2) \approx 11.98 N_{cor} F_d \left(\frac{S_e}{25}\right)$$

$$= 11.98 N_{cor} \left[1 + 0.33 \left(\frac{D_f}{B}\right)\right] \left[\frac{S_e \ (mm)}{25}\right]$$

$$\le 15.93 N_{cor} \left[\frac{S_e \ (mm)}{25}\right] \tag{11.75}$$

Note that the original Eq. (11.53) was for a settlement of 25 mm, with a differential settlement of about 19 mm. However, the widths of the raft foundations are larger than the isolated spread footings. The depth of significant stress increase in the soil below a foundation depends on the foundation width. Hence, for a raft foundation, the depth of the zone of influence is likely to be much larger than that of a spread footing. Thus, the loose soil pockets under a raft may be more evenly distributed, resulting in a smaller differential settlement. Hence, the customary assumption is that, for a maximum raft settlement of 50 mm, the differential settlement would be 19 mm. Using this logic and conservatively assuming that F_d equals 1, we can approximate Eq. (11.75) as

$$q_{all(net)} \ (kN/m^2) \approx 23.96 N_{cor} \tag{11.76}$$

The net pressure applied on a foundation (Figure 11.24) may be expressed as

$$q = \frac{Q}{A} - \gamma D_f \tag{11.77}$$

where Q = dead weight of the structure and the live load
A = area of the raft

Hence, in all cases, q should be less than or equal to $q_{all(net)}$.

EXAMPLE 11.6

Determine the net ultimate bearing capacity of a mat foundation measuring 13 m × 9 m on a saturated clay with $c_u = 94 \ kN/m^2$, $\phi = 0$, and $D_f = 2$ m.

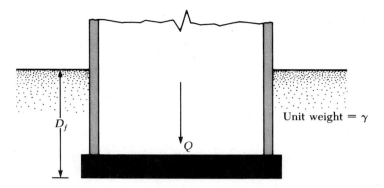

FIGURE 11.24 Definition of net pressure on soil caused by a mat foundation

Solution From Eq. (11.73), we have

$$q_{net(u)} = 5.14c_u \left[1 + \left(\frac{0.195B}{L} \right) \right] \left[1 + 0.4 \left(\frac{D_f}{B} \right) \right]$$

$$= (5.14)(94) \left[1 + \left(\frac{0.195 \times 9}{13} \right) \right] \left[1 + 0.4 \left(\frac{2}{9} \right) \right]$$

$$= \mathbf{597\ kN/m^2} \qquad \blacksquare$$

EXAMPLE 11.7

What will be the net allowable bearing capacity of a mat foundation with dimensions of 13 m × 9 m constructed over a sand deposit? Here, $D_f = 2$ m, allowable settlement = 25 mm, and corrected average penetration number $N_{cor} = 10$.

Solution From Eq. (11.75), we have

$$q_{all(net)} = 11.98N_{cor} \left[1 + 0.33 \left(\frac{D_f}{B} \right) \right] \left[\frac{S_e}{25} \right] \le 15.93N_{cor} \left[\frac{S_e}{25} \right]$$

$$= (11.98)(10) \left[1 + \frac{(0.33)(2)}{9} \right] \left(\frac{25}{25} \right) \approx \mathbf{128.6\ kN/m^2} \qquad \blacksquare$$

11.17 *Compensated Foundations*

The settlement of a mat foundation can be reduced by decreasing the net pressure increase on soil and by increasing the depth of embedment, D_f. This increase is particularly important for mats on soft clays, where large consolidation settlements are expected. From Eq. (11.77), the net average applied pressure on soil is

$$q = \frac{Q}{A} - \gamma D_f$$

For no increase of the net soil pressure on soil below a raft foundation, q should be 0. Thus,

$$D_f = \frac{A}{A\gamma} \tag{11.78}$$

This relation for D_f is usually referred to as the depth of embedment of a *fully compensated foundation*.

The factor of safety against bearing capacity failure for partially compensated foundations (that is, $D_f < Q/A\gamma$) may be given as

$$FS = \frac{q_{net(u)}}{q} = \frac{q_{net(u)}}{\dfrac{Q}{A} - \gamma D_f} \tag{11.79}$$

For saturated clays, the factor of safety against bearing capacity failure can thus be obtained by substituting Eq. (11.73) into Eq. (11.79):

$$FS = \frac{5.14c_u\left(1 + \dfrac{0.195B}{L}\right)\left(1 + 0.4\dfrac{D_f}{B}\right)}{\dfrac{Q}{A} - \gamma D_f} \tag{11.80}$$

EXAMPLE 11.8

Refer to Figure 11.24. The mat has dimensions of 30 m × 40 m, and the live load and dead load on the mat are 200 MN. The mat is placed over a layer of soft clay that has a unit weight of 18.8 kN/m³. Find D_f for a fully compensated foundation.

Solution From Eq. (11.78), we have

$$D_f = \frac{Q}{A\gamma} = \frac{200 \times 10^3 \text{ kN}}{(30 \times 40)(18.8)} = \textbf{8.87 m} \qquad \blacksquare$$

EXAMPLE 11.9

Refer to Example 11.8. For the clay, $c_u = 12.5$ kN/m². If the required factor of safety against bearing capacity failure is 3, determine the depth of the foundation.

Solution From Eq. (11.80), we have

$$FS = \frac{5.14c_u\left(1 + \dfrac{0.195B}{L}\right)\left(1 + 0.4\dfrac{D_f}{B}\right)}{\dfrac{Q}{A} - \gamma D_f}$$

Here, $FS = 3$, $c_u = 12.5$ kN/m², $B/L = 30/40 = 0.75$, and $Q/A = (200 \times 10^3)/(30 \times 40) = 166.67$ kN/m². Substituting these values into Eq. (11.80) yields

$$3 = \frac{(5.14)(12.5)[1 + (0.195)(0.75)]\left[1 + 0.4\left(\dfrac{D_f}{30}\right)\right]}{166.67 - (18.8)D_f}$$

$$500.01 - 56.4D_f = 73.65 + 0.982D_f$$

$$426.36 = 57.382D_f$$

or

$$D_f \approx \mathbf{7.5\ m} \qquad\blacksquare$$

**EXAMPLE
11.10**

Consider a mat foundation 27 m × 37 m in plan, as shown in Figure 11.25. The total dead load and live load on the raft is 200 MN. Estimate the consolidation settlement at the center of the foundation.

Solution For $Q = 200$ MN, the load per unit area is

$$q = \frac{Q}{A} - \gamma D_f = \frac{200 \times 10^3}{27 \times 37} - (15.7)(2) \approx 168.8\ \text{kN/m}^2$$

From Chapter 6, we know that the average pressure increase on the clay layer below the center of the foundation is

$$\Delta\sigma_{av} = \frac{1}{6}(\Delta\sigma_t + 4\,\Delta\sigma_m + \Delta\sigma_b)$$

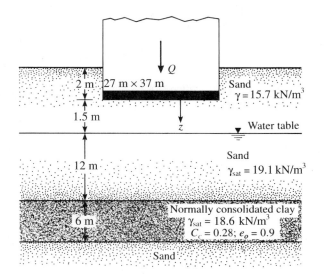

FIGURE 11.25

Refer to Figure 5.21 to find the values of $\Delta\sigma_t$, $\Delta\sigma_m$, and $\Delta\sigma_b$. At the *top of the clay layer*,

$$\frac{z}{B} = \frac{13.5}{27} = 0.5$$

$$\frac{L}{B} = \frac{37}{27} = 1.37$$

So, for $z/B = 0.5$ and $L/B = 1.37$, we have

$$\frac{\Delta\sigma_t}{q} = 0.75 \quad \text{and} \quad \Delta\sigma_t = (0.75)(168.8) = 126.6 \text{ kN/m}^2$$

Similarly, for the *middle of the clay layer*,

$$\frac{z}{B} = \frac{16.5}{27} = 0.61$$

$$\frac{L}{B} = 1.37$$

So $\Delta\sigma_m/q = 0.66$ and $\Delta\sigma_m = 114.4 \text{ kN/m}^2$. At the *bottom of the clay layer*,

$$\frac{z}{B} = \frac{19.5}{27} = 0.72$$

$$\frac{L}{B} = 1.37$$

So, $\Delta\sigma_b/q = 0.58$ and $\Delta\sigma_b = 97.9 \text{ kN/m}^2$. Hence,

$$\Delta\sigma_{av} = \frac{1}{6}[126.6 + (4)(114.4) + 97.9] = 113.7 \text{ kN/m}^2$$

From Eq. (6.14), the consolidation settlement is

$$S = \frac{C_c H_c}{1 + e_0} \log \frac{\sigma_o' + \Delta\sigma_{av}'}{\sigma_o'}$$

$$\sigma_o' = (3.5)(15.7) + (12)(19.1 - 9.81) + \frac{6}{2}(18.6 - 9.81) \approx 192.8 \text{ kN/m}^2$$

$$S = \frac{(0.28)(6)}{1.9} \log\left(\frac{192.8 + 113.7}{192.8}\right) = 0.178 \text{ m} = \textbf{178 mm} \qquad\blacksquare$$

Problems

11.1   A continuous foundation is 1.5 m wide. The design conditions are $D_f = 1.1$ m, $\gamma = 17.2 \text{ kN/m}^3$, $\phi = 26°$, and $c = 28 \text{ kN/m}^2$. Determine the allowable gross vertical load-bearing capacity ($FS = 4$).

11.2 A square column foundation is 2 m × 2 m in plan. The design conditions are $D_f = 1.5$ m, $\gamma = 15.9$ kN/m³, $\phi = 34°$, and $c = 0$. Determine the allowable gross vertical load that the column could carry ($FS = 3$).

11.3 For the foundation given in Problem 11.2, what will be the gross allowable load-bearing capacity if the load is inclined at an angle 10° to the vertical?

11.4 A square foundation ($B \times B$), has to be constructed as shown in Figure 11.26. Assume that $\gamma = 16.5$ kN/m³, $\gamma_{sat} = 18.6$ kN/m³, $D_f = 1.2$ m, and $D_1 = 0.6$ m. The gross allowable load, Q_{all}, with $FS = 3$ is 670 kN. The field standard penetration resistance, N_F, values are given in the table.

Depth (m)	N_F
1.5	4
3.0	6
4.5	6
6.0	10
7.5	5

Determine the size of the footing.

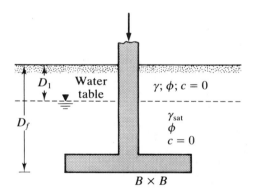

FIGURE 11.26

11.5 A column foundation is 4 m × 2 m in plan. For $D_f = 1.4$ m, $c = 153$ kN/m², $\phi = 0$, and $\gamma = 18.4$ kN/m³, what is the net ultimate load that the column could carry?

11.6 For a square foundation that is $B \times B$ in plan, $D_f = 0.9$ m, vertical gross allowable load, $Q_{all} = 667$ kN, $\gamma = 18.1$ kN/m³, $\phi = 40°$, $c = 0$, and $FS = 3$. Determine the size of the foundation.

11.7 A square footing is shown in Figure 11.27. Use an FS of 6 and determine the size of the footing.

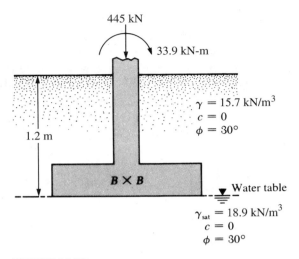

FIGURE 11.27

11.8   An eccentrically loaded foundation is shown in Figure 11.28. Use an *FS* of 4 and determine the maximum allowable load that the foundation can carry.

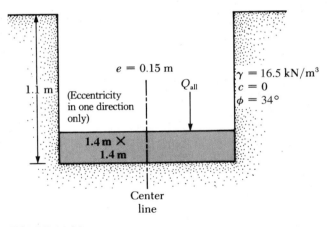

FIGURE 11.28

11.9   An eccentrically loaded foundation is shown in Figure 11.29. Determine the ultimate load, Q_u, that the foundation can carry.

11.10   Refer to Figure 11.8 for a foundation with a two-way eccentricity. The soil conditions are $\gamma = 18$ kN/m^3, $\phi = 35°$, and $c = 0$. The design criteria are $D_f = 1$ m, $B = 1.5$ m, $L = 2$ m, $e_B = 0.3$ m, and $e_L = 0.364$ m. Determine the gross ultimate load that the foundation could carry.

11.11   Repeat Problem 11.10 for $e_L = 0.4$ m and $e_B = 0.19$ m.

11.12   Refer to Figure 11.16. A foundation that is 3 m × 2 m in plan is resting on

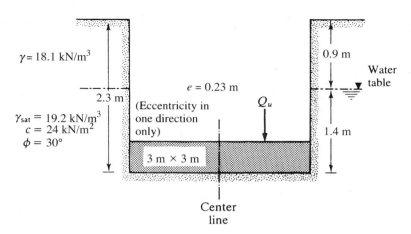

$\gamma = 18.1$ kN/m^3

$e = 0.23$ m

0.9 m

Water table

2.3 m

(Eccentricity in one direction only)

Q_u

$\gamma_{sat} = 19.2$ kN/m^3
$c = 24$ kN/m^2
$\phi = 30°$

1.4 m

3 m × 3 m

Center line

FIGURE 11.29

a sand deposit. The net load per unit area at the level of the foundation, q_o, is 153 kN/m^2. For the sand, $\mu_s = 0.3$, $E_s = 22$ MN/m^2, $D_f = 0.9$ m, and $H = 12$ m. Assume that the foundation is rigid and determine the elastic settlement that the foundation would undergo. Use Eq. (11.43).

11.13 Repeat Problem 11.12 for foundation criteria of size = 1.8 m × 1.8 m, $q_o = 190$ kN/m^2, $D_f = 1$ m, and $H = 8$ m; and soil conditions of $\mu_s = 0.35$, $E_s = 16,500$ kN/m^2, and $\gamma = 16.5$ kN/m^3.

11.14 A square column foundation is shown in Figure 11.30. Determine the average

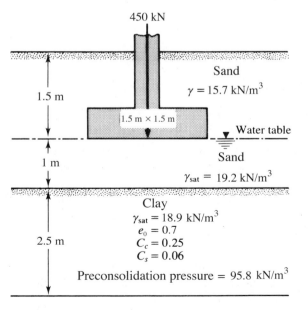

450 kN

Sand
$\gamma = 15.7$ kN/m^3

1.5 m

1.5 m × 1.5 m

Water table

Sand
$\gamma_{sat} = 19.2$ kN/m^3

1 m

Clay
$\gamma_{sat} = 18.9$ kN/m^3
$e_0 = 0.7$
$C_c = 0.25$
$C_s = 0.06$

2.5 m

Preconsolidation pressure = 95.8 kN/m^3

FIGURE 11.30

increase of pressure in the clay layer below the center of the foundation. Use Eqs. (6.43) and (6.44).

11.15   Estimate the consolidation settlement of the clay layer shown in Figure 11.30 from the results of Problem 11.14.

11.16   Two plate load tests with square plates were conducted in the field. At 25-mm settlement, the results were as given in the table.

Width of plate (mm)	Load (kN)
305	35.9
610	114.8

What size of square footing is required to carry a net load of 1050 kN at a settlement of 25 mm?

11.17   A mat foundation measuring 14 m × 9 m has to be constructed on a saturated clay. For the clay, $c_u = 93$ kN/m^2 and $\phi = 0$. The depth, D_f, for the mat foundation is 2 m. Determine the net ultimate bearing capacity.

11.18   Repeat Problem 11.17 with the following:

- Mat foundation: $B = 10$ m, $L = 20$ m, and $D_f = 3$ m
- Clay: $\phi = 0$ and $c_u = 100$ kN/m^2

11.19   The table gives the results of a standard penetration test in the field (sandy soil):

Depth (m)	Field value of N_F
2	8
4	10
6	12
8	9
10	14

Estimate the net allowable bearing capacity of a mat foundation 6 m × 5 m in plan. Here, $D_f = 1.5$ m and allowable settlement = 50 mm. Assume that the unit weight of soil, $\gamma = 17.5$ kN/m^3.

11.20   Repeat Problem 11.19 for an allowable settlement of 30 mm.

11.21   Consider a mat foundation with dimensions of 18 m × 12 m. The combined dead and live load on the mat is 44.5 MN. The mat is to be placed on a clay with $c_u = 40.7$ kN/m^2 and $\gamma = 17.6$ kN/m^3. Find the depth, D_f, of the mat for a fully compensated foundation.

11.22   For the mat in Problem 11.21, what will be the depth, D_f, of the mat for $FS = 3$ against bearing capacity failure?

11.23   Repeat Problem 11.22 for an undrained cohesion of the clay of 60 kN/m^2.

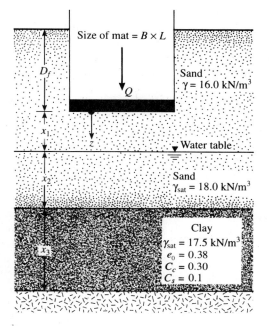

Size of mat = $B \times L$

Sand $\gamma = 16.0$ kN/m^3

Q

Water table

Sand $\gamma_{sat} = 18.0$ kN/m^3

Clay
$\gamma_{sat} = 17.5$ kN/m^3
$e_0 = 0.38$
$C_c = 0.30$
$C_s = 0.1$

FIGURE 11.31

11.24 A mat foundation is shown in Figure 11.31. The design considerations are $L = 15$ m, $B = 7.5$ m, $D_f = 3$ m, $Q = 35$ MN, $x_1 = 2.5$ m, $x_2 = 2.75$ m, $x_3 = 4$ m, and $\sigma'_c = 105$ kN/m^2. Calculate the consolidation settlement under the center of the mat.

References

American Society for Testing and Materials (1997). *Annual Book of ASTM Standards,* Vol. 04.08, West Conshohocken, PA.

Bjerrum, L. (1963). "Allowable Settlement of Structures," *Proceedings,* European Conference on Soil Mechanics and Foundation Engineering, Wiesbaden, Germany, Vol. III, 135–137.

Bowles, J. E. (1977). *Foundation Analysis and Design,* 2nd ed., McGraw-Hill, New York.

Burland, J. B., and Worth, C. P. (1974). "Allowable and Differential Settlement of Structures Including Damage and Soil-Structure Interaction," *Proceedings,* Conference on Settlement of Structures, Cambridge University, England, 611–654.

Caquot, A., and Kerisel, J. (1953). "Sur le terme de surface dans le calcul des fondations en milieu pulverulent," *Proceedings,* Third International Conference on Soil Mechanics and Foundation Engineering, Zürich, Vol. I, 336–337.

Christian, J. T., and Carrier, W. D. (1978). "Janbu, Bjerrum, and Kjaernsli's Chart Reinterpreted," *Canadian Geotechnical Journal,* Vol. 15, 124–128.

De Beer, E. E. (1970). "Experimental Determination of the Shape Factors and Bearing Capacity Factors of Sand," *Geotechnique,* Vol. 20, No. 4, 387–411.

Grant, R. J., Christian, J. T., and Vanmarcke, E. H. (1974). "Differential Settlement of Buildings," *Journal of the Geotechnical Engineering Division,* American Society of Civil Engineers, Vol. 100, No. GT9, 973–991.

Hanna, A. M., and Meyerhof, G. G. (1981). "Experimental Evaluation of Bearing Capacity of Footings Subjected to Inclined Loads," *Canadian Geotechnical Journal,* Vol. 18, No. 4, 599–603.

Hansen, J. B. (1970). "A Revised and Extended Formula for Bearing Capacity," Danish Geotechnical Institute, *Bulletin 28,* Copenhagen.

Harr, M. E. (1966). *Fundamentals of Theoretical Soil Mechanics,* McGraw-Hill, New York.

Highter, W. H., and Anders, J. C. (1985). "Dimensioning Footings Subjected to Eccentric Loads," *Journal of Geotechnical Engineering,* American Society of Civil Engineers, Vol. 111, No. GT5, 659–665.

Housel, W. S. (1929). "A Practical Method for the Selection of Foundations Based on Fundamental Research in Soil Mechanics," *Research Bulletin No. 13,* University of Michigan, Ann Arbor.

Janbu, N., Bjerrum, L., and Kjaernsli, B. (1956). "Veiledning ved losning av fundamentering—soppgaver," *Publication No. 16,* Norwegian Geotechnical Institute, 30–32.

Meyerhof, G. G. (1953). "The Bearing Capacity of Foundations Under Eccentric and Inclined Loads," *Proceedings,* Third International Conference on Soil Mechanics and Foundation Engineering, Zürich, Vol. 1, 440–445.

Meyerhof, G. G. (1956). "Penetration Tests and Bearing Capacity of Cohesionless Soils," *Journal of the Soil Mechanics and Foundations Division,* American Society of Civil Engineers, Vol. 82, No. SM1, 1–19.

Meyerhof, G. G. (1963). "Some Recent Research on the Bearing Capacity of Foundations," *Canadian Geotechnical Journal,* Vol. 1, No. 1, 16–26.

Meyerhof, G. G. (1965). "Shallow Foundations," *Journal of the Soil Mechanics and Foundations Division,* ASCE, Vol. 91, No. SM2, 21–31.

Mitchell, J. K., and Gardner, W. S. (1975). "*In Situ* Measurement of Volume Change Characteristics," *Proceedings,* Specialty Conference, American Society of Civil Engineers, Vol. 2, 279–345.

Polshin, D. E., and Tokar, R. A. (1957). "Maximum Allowable Nonuniform Settlement of Structures," *Proceedings,* Fourth International Conference on Soil Mechanics and Foundation Engineering, London, Vol. 1, 402–405.

Prandtl, L. (1921). "Über die Eindringungsfestigkeit (Härte) plastischer Baustoffe und die Festigkeit von Schneiden," *Zeitschrift für angewandte Mathematik und Mechanik,* Vol. 1, No. 1, 15–20.

Reissner, H. (1924). "Zum Erddruckproblem," *Proceedings,* First International Congress of Applied Mechanics, Delft, 295–311.

Schmertmann, J. H. (1970). "Static Cone to Compute Settlement Over Sand," *Journal of the Soil Mechanics and Foundations Division,* American Society of Civil Engineers, Vol. 96, No. SM3, 1011–1043.

Terzaghi, K. (1943). *Theoretical Soil Mechanics,* Wiley, New York.

Terzaghi, K., and Peck, R. B. (1967). *Soil Mechanics in Engineering Practice,* 2nd ed., Wiley, New York.

Vesic, A. S. (1963). "Bearing Capacity of Deep Foundations in Sand," *Highway Research Record No. 39,* National Academy of Sciences, 112–153.

Vesic, A. S. (1973). "Analysis of Ultimate Loads of Shallow Foundations," *Journal of the Soil Mechanics and Foundations Division,* American Society of Civil Engineers, Vol. 99, No. SM1, 45–73.

Wahls, H. E. (1981). "Tolerable Settlement of Buildings," *Journal of the Geotechnical Engineering Division,* American Society of Civil Engineers, Vol. 107, No. GT11, 1489–1504.

Supplementary References for Further Study

Adams, M. T., and Collin, J. G. (1997). "Large Model Spread Footing Load Tests on Geosynthetic Reinforced Soil Foundations," *Journal of Geotechnical and Geoenvironmental Engineering,* American Society of Civil Engineers, Vol. 123, No. 1, 66–72.

American Concrete Institute Committee 336 (1988). "Suggested Design Procedures for Combined Footings and Mats," *Journal of the American Concrete Institute,* Vol. 63, No. 10, 1041–1077.

Kumbhojkar, A. S. (1993). "Numerical Evaluation of Terzaghi's N_γ," *Journal of Geotechnical Engineering,* American Society of Civil Engineers, Vol. 119, No. 3, 598–607.

Meyerhof, G. G. (1974). "Ultimate Bearing Capacity of Footings on Sand Layer Overlying Clay," *Canadian Geotechnical Journal,* Vol. 11, No. 2, 224–229.

Meyerhof, G. G., and Hanna, A. M. (1978). "Ultimate Bearing Capacity of Foundations on Layered Soil Under Inclined Load," *Canadian Geotechnical Journal,* Vol. 15, No. 4, 565–572.

Omar, M. T., Das, B. M., Yen, S. C., Puri, V. K., and Cook, E. E. (1993a). "Ultimate Bearing Capacity of Rectangular Foundations on Geogrid-Reinforced Sand," *Geotechnical Testing Journal,* American Society for Testing and Materials, Vol. 16, No. 2, 246–252.

Omar, M. T., Das, B. M., Yen, S. C., Puri, V. K., and Cook, E. E. (1993b). "Shallow Foundations on Geogrid-Reinforced Sand," *Transportation Research Record No. 1414,* National Academy of Sciences, National Research Council, 59–64.

Richards, R., Jr., Elms, D. G., and Budhu, M. (1993). "Seismic Bearing Capacity and Settlement of Foundations," *Journal of Geotechnical Engineering,* American Society of Civil Engineers, Vol. 119, No. 4, 662–674.

12

Retaining Walls and Braced Cuts

The general principles of lateral earth pressure were presented in Chapter 9. Those principles can be extended to the analysis and design of earth-retaining structures such as retaining walls and braced cuts. *Retaining walls* provide permanent lateral support to *vertical* or *near-vertical* slopes of soil. Also, at times, construction work requires ground excavations with vertical or near-vertical faces—for example, basements of buildings in developed areas or underground transportation facilities at shallow depths below the ground surface (cut-and-cover type of construction). The vertical faces of the cuts should be protected by *temporary bracing systems* to avoid failure that may be accompanied by considerable settlement or by bearing capacity failure of nearby foundations. These cuts are called *braced cuts*. This chapter is divided into two parts: The first part discusses the analysis of retaining walls, and the second part presents the analysis of braced cuts.

RETAINING WALLS

12.1 Retaining Walls—General

Retaining walls are commonly used in construction projects and may be grouped into four classifications:

1. Gravity retaining walls
2. Semigravity retaining walls
3. Cantilever retaining walls
4. Counterfort retaining walls

Gravity retaining walls (Figure 12.1a) are constructed with plain concrete or stone masonry. They depend on their own weight and any soil resting on the masonry for stability. This type of construction is not economical for high walls.

445

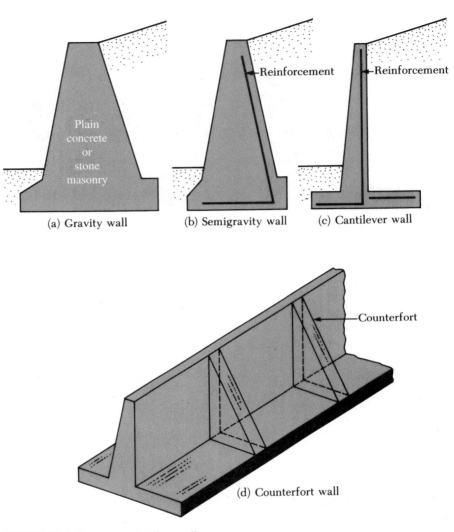

(a) Gravity wall   (b) Semigravity wall   (c) Cantilever wall

Plain
concrete
or
stone
masonry

Reinforcement   Reinforcement

Counterfort

(d) Counterfort wall

FIGURE 12.1 Types of retaining wall

In many cases, a small amount of steel may be used for the construction of gravity walls, thereby minimizing the size of wall sections. Such walls are generally referred to as *semigravity retaining walls* (Figure 12.1b).

Cantilever retaining walls (Figure 12.1c) are made of reinforced concrete that consists of a thin stem and a base slab. This type of wall is economical to a height of about 8 m.

Counterfort retaining walls (Figure 12.1d) are similar to cantilever walls. At regular intervals, however, they have thin vertical concrete slabs known as *counterforts* that tie the wall and the base slab together. The purpose of the counterforts is to reduce the shear and the bending moments.

To design retaining walls properly, an engineer must know the basic soil parameters—that is, the *unit weight, angle of friction*, and *cohesion*—of the soil retained behind the wall and the soil below the base slab. Knowing the properties of the soil behind the wall enables the engineer to determine the lateral pressure distribution that has to be considered in the design.

The design of a retaining wall proceeds in two phases. First, with the lateral earth pressure known, the structure as a whole is checked for *stability,* including checking for possible *overturning, sliding,* and *bearing capacity* failures. Second, each component of the structure is checked for *adequate strength*, and the *steel reinforcement* of each component is determined.

12.2 *Proportioning Retaining Walls*

When designing retaining walls, an engineer must assume some of the dimensions, called *proportioning,* to check trial sections for stability. If the stability checks yield undesirable results, the sections can be changed and rechecked. Figure 12.2 shows

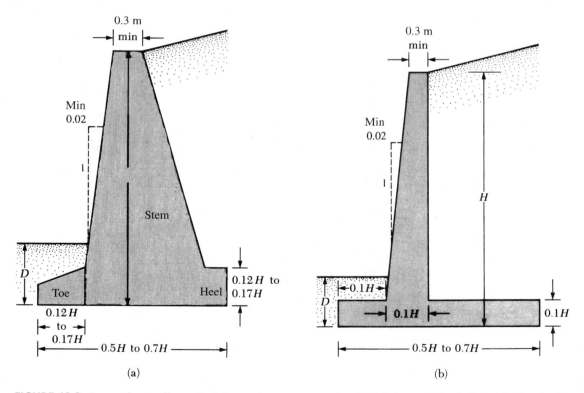

FIGURE 12.2 Approximate dimensions for various components of retaining wall for initial stability checks: (a) gravity wall; (b) cantilever wall [*Note:* minimum dimension of D is 0.6m.]

the general proportions of various retaining wall components that can be used for initial checks.

Note that the top of the stem of any retaining wall should be no less than about 0.3 m wide for proper placement of concrete. The depth, D, to the bottom of the base slab should be a minimum of 0.6 m. However, the bottom of the base slab should be positioned below the seasonal frost-line.

For counterfort retaining walls, the general proportion of the stem and the base slab is the same as for cantilever walls. However, the counterfort slabs may be about 0.3 m thick and spaced at center-to-center distances of $0.3H$ to $0.7H$.

12.3 Application of Lateral Earth Pressure Theories to Design

Chapter 9 presented the fundamental theories for calculating lateral earth pressure. To use these theories in design, an engineer must make several simple assumptions. In the case of cantilever walls, using the Rankine earth pressure theory for stability checks involves drawing a vertical line AB through point A, as shown in Figure 12.3a (located at the edge of the heel of the base slab). The Rankine active condition

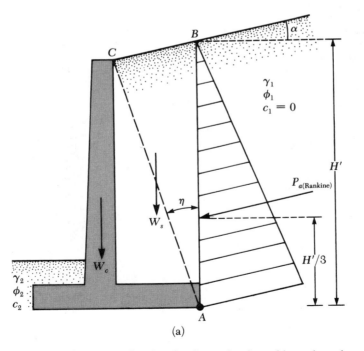

FIGURE 12.3 Assumption for the determination of lateral earth pressure: (a) cantilever wall; (b) and (c) gravity wall

is assumed to exist along the vertical plane AB. Rankine's active earth pressure equations may then be used to calculate the lateral pressure on the face AB. In the analysis of stability for the wall, the force $P_{a(\text{Rankine})}$, the weight of soil above the heel, W_s, and the weight of the concrete, W_c, all should be taken into consideration. The assumption for the development of Rankine's active pressure along the soil face AB is theoretically correct if the shear zone bounded by the line AC is not

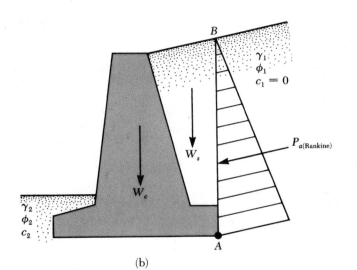

(b)

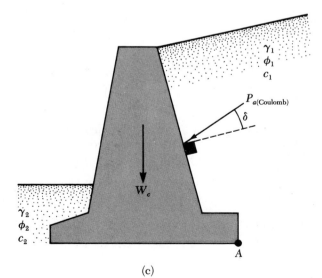

(c)

FIGURE 12.3 (Continued)

obstructed by the stem of the wall. The angle, η, that the line AC makes with the vertical is

$$\eta = 45 + \frac{\alpha}{2} - \frac{\phi_1}{2} - \sin^{-1}\left(\frac{\sin \alpha}{\sin \phi_1}\right)$$ (12.1)

For gravity walls, a similar type of analysis may be used, as shown in Figure 12.3b. However, Coulomb's theory also may be used, as shown in Figure 12.3c. If *Coulomb's active pressure theory* is used, the only forces to be considered are $P_{a(Coulomb)}$ and the weight of the wall, W_c.

In the case of ordinary retaining walls, water table problems and hence hydrostatic pressure are not encountered. Facilities for drainage from the soils retained are always provided.

To check the stability of a retaining wall, the following steps are taken:

1. Check for *overturning* about its toe.
2. Check for *sliding failure* along its base.
3. Check for *bearing capacity failure* of the base.
4. Check for *settlement*.
5. Check for *overall stability*.

The following sections describe the procedure for checking for overturning, sliding, and bearing capacity failure. The principles of investigation for settlement were covered in Chapters 6 and 11 and will not be repeated here.

12.4   *Check for Overturning*

Figure 12.4 shows the forces that act on a cantilever and a gravity-retaining wall, based on the assumption that the Rankine active pressure is acting along a vertical plane AB drawn through the heel. P_p is the Rankine passive pressure; recall that its magnitude is [from Eq. (9.42) with $\gamma = \gamma_2$, $c = c_2$, and $H = D$]

$$P_p = \frac{1}{2} K_p \gamma_2 D^2 + 2c_2 \sqrt{K_p} D$$ (12.2)

where γ_2 = unit weight of soil in front of the heel and under the base slab
K_p = Rankine's passive earth pressure coefficient = $\tan^2(45 + \phi_2/2)$
c_2, ϕ_2 = cohesion and soil friction angle, respectively

The factor of safety against overturning about the toe—that is, about point C in Figure 12.4—may be expressed as

$$FS_{(overturning)} = \frac{\Sigma M_R}{\Sigma M_O}$$ (12.3)

where ΣM_O = sum of the moments of forces tending to overturn about point C
ΣM_R = sum of the moments of forces tending to resist overturning about point C

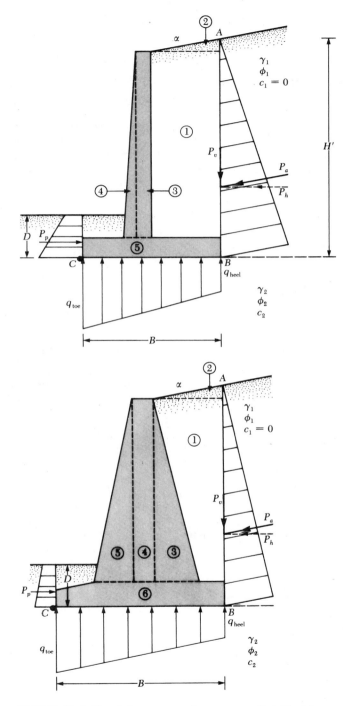

FIGURE 12.4 Check for overturning; assume that Rankine
pressure is valid

The overturning moment is

$$\Sigma M_O = P_h \left(\frac{H'}{3}\right) \tag{12.4}$$

where $P_h = P_a \cos \alpha$.

When calculating the resisting moment, ΣM_R (neglecting P_p), we can prepare a table such as Table 12.1. The weight of the soil above the heel and the weight of the concrete (or masonry) are both forces that contribute to the resisting moment. Note that the force P_v also contributes to the resisting moment. P_v is the vertical component of the active force P_a, or

$$P_v = P_a \sin \alpha \tag{12.5}$$

The moment of the force P_v about C is

$$M_v = P_v B = P_a \sin \alpha B \tag{12.6}$$

where B = width of the base slab.

Once ΣM_R is known, the factor of safety can be calculated as

$$FS_{(overturning)} = \frac{M_1 + M_2 + M_3 + M_4 + M_5 + M_6 + M_v}{P_a \cos \alpha (H'/3)} \tag{12.7}$$

The usual minimum desirable value of the factor of safety with respect to overturning is 1.5 to 2.

Some designers prefer to determine the factor of safety against overturning with

$$FS_{(overturning)} = \frac{M_1 + M_2 + M_3 + M_4 + M_5 + M_6}{P_a \cos \alpha (H'/3) - M_v} \tag{12.8}$$

Table 12.1 Procedure for calculation of ΣM_R

Section (1)	Area (2)	Weight/unit length of wall (3)	Moment arm measured from C (4)	Moment about C (5)
1	A_1	$W_1 = \gamma_1 \times A_1$	X_1	M_1
2	A_2	$W_2 = \gamma_2 \times A_2$	X_2	M_2
3	A_3	$W_3 = \gamma_c \times A_3$	X_3	M_3
4	A_4	$W_4 = \gamma_c \times A_4$	X_4	M_4
5	A_5	$W_5 = \gamma_c \times A_5$	X_5	M_5
6	A_6	$W_6 = \gamma_c \times A_6$	X_6	M_6
		P_v	B	M_v
		ΣV		ΣM_R

Note: γ_1 = unit weight of backfill
γ_c = unit weight of concrete

12.5 Check for Sliding Along the Base

The factor of safety against sliding may be expressed by the equation

$$FS_{(sliding)} = \frac{\Sigma F_{R'}}{\Sigma F_d} \tag{12.9}$$

where $\Sigma F_{R'}$ = sum of the horizontal resisting forces
      ΣF_d = sum of the horizontal driving forces

Figure 12.5 indicates that the shear strength of the soil below the base slab may be represented as

$$\tau_f = \sigma' \tan \phi_2 + c_2$$

Thus, the maximum resisting force that can be derived from the soil per unit length of the wall along the bottom of the base slab is

$$R' = \tau_f(\text{area of cross section}) = s(B \times 1) = B\sigma' \tan \phi_2 + Bc_2$$

However,

$$B\sigma' = \text{sum of the vertical force} = \Sigma V \text{ (see Table 12.1)}$$

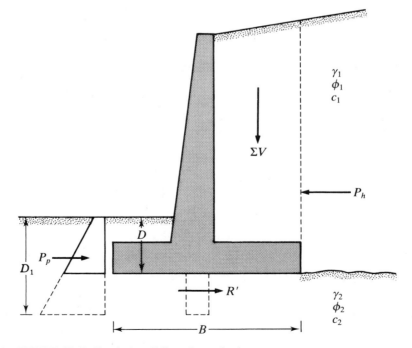

FIGURE 12.5 Check for sliding along the base

so

$$R' = (\Sigma V) \tan \phi_2 + Bc_2$$

Figure 12.5 shows that the passive force, P_p, is also a horizontal resisting force. The expression for P_p is given in Eq. (12.2). Hence,

$$\Sigma F_{R'} = (\Sigma V) \tan \phi_2 + Bc_2 + P_p \tag{12.10}$$

The only horizontal force that will tend to cause the wall to slide (*driving force*) is the horizontal component of the active force P_a, so

$$\Sigma F_d = P_a \cos \alpha \tag{12.11}$$

Combining Eqs. (12.9), (12.10), and (12.11) yields

$$FS_{(sliding)} = \frac{(\Sigma V) \tan \phi_2 + Bc_2 + P_p}{P_a \cos \alpha} \tag{12.12}$$

A minimum factor of safety of 1.5 against sliding is generally required.

In many cases, the passive force, P_p, is ignored when calculating the factor of safety with respect to sliding. The friction angle, ϕ_2, is also reduced in several instances for safety. The reduced soil friction angle may be on the order of one-half to two-thirds of the angle ϕ_2. In a similar manner, the cohesion c_2 may be reduced to the value of $0.5c_2$ to $0.67c_2$. Thus,

$$FS_{(sliding)} = \frac{(\Sigma V) \tan (k_1 \phi_2) + Bk_2c_2 + P_p}{P_a \cos \alpha} \tag{12.13}$$

where k_1 and k_2 are in the range of $\frac{1}{2}$ to $\frac{2}{3}$.

In some instances, certain walls may not yield a desired factor of safety of 1.5. To increase their resistance to sliding, a base key may be used. Base keys are illustrated by broken lines in Figure 12.5. The passive force at the toe *without the key* is

$$P_p = \frac{1}{2}\gamma_2 D^2 K_p + 2c_2 D\sqrt{K_p}$$

However, if a key is included, the passive force per unit length of the wall becomes (*note:* $D = D_1$)

$$P_p = \frac{1}{2}\gamma_2 D_1^2 K_p + 2c_2 D_1 \sqrt{K_p}$$

where $K_p = \tan^2(45 + \phi_2/2)$. Because $D_1 > D$, a key obviously will help increase the passive resistance at the toe and hence the factor of safety against sliding. Usually the base key is constructed below the stem, and some main steel is run into the key.

Another way to increase the value of $FS_{(sliding)}$ is to consider reducing the value of P_a [see Eq. (12.13)]. One possible way to do so is to use the method developed by Elman and Terry (1988). The discussion here is limited to the case in which the retaining wall has a horizontal granular backfill (Figure 12.6). In Figure 12.6a, the active force, P_a, is horizontal ($\alpha = 0$) so that

$$P_a \cos \alpha = P_h = P_a$$

and

$$P_a \sin \alpha = P_v = 0$$

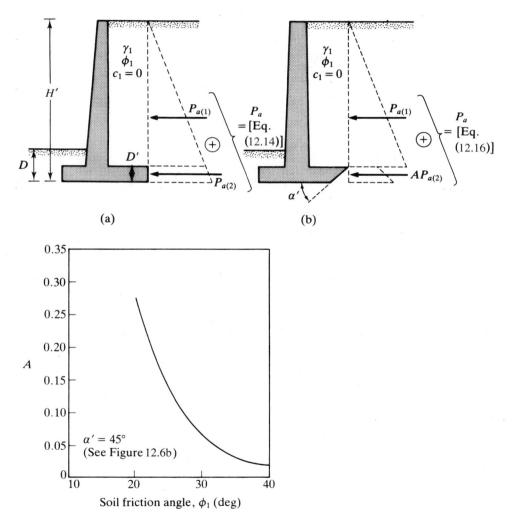

(a)

(b)

(c)

FIGURE 12.6 (a) Retaining wall with vertical heel; (b) retaining wall with sloped heel; (c) variation of A with friction angle of backfill [Eq. (12.16)] (Based on Elman and Terry, 1988)

However,

$$P_a = P_{a(1)} + P_{a(2)} \tag{12.14}$$

The magnitude of $P_{a(2)}$ can be reduced if the heel of the retaining wall is sloped as shown in Figure 12.6b. For this case,

$$P_a = P_{a(1)} + AP_{a(2)} \tag{12.15}$$

The magnitude of A, as shown in Figure 12.6c, is valid for $\alpha' = 45°$. However, note that in Figure 12.6a,

$$P_{a(1)} = \frac{1}{2} \gamma_1 K_a (H' - D')^2$$

and

$$P_a = \frac{1}{2} \gamma_1 K_a H'^2$$

Hence,

$$P_{a(2)} = \frac{1}{2} \gamma_1 K_a [H'^2 - (H' - D')^2]$$

So, for the active pressure diagram shown in Figure 12.6b, we have

$$P_a = \frac{1}{2} \gamma_1 K_a (H' - D')^2 + \frac{A}{2} \gamma_1 K_a [H'^2 - (H' - D')^2] \tag{12.16}$$

Sloping the heel of a retaining wall can thus be extremely helpful in some cases.

12.6 Check for Bearing Capacity Failure

The vertical pressure transmitted to the soil by the base slab of the retaining wall should be checked against the ultimate bearing capacity of the soil. The nature of variation of the vertical pressure transmitted by the base slab into the soil is shown in Figure 12.7. Note that q_{toe} and q_{heel} are the *maximum* and the *minimum* pressures occurring at the ends of the toe and heel sections, respectively. The magnitudes of q_{toe} and q_{heel} can be determined in the following manner.

The sum of the vertical forces acting on the base slab is ΣV (see col. 3, Table 12.1), and the horizontal force is $P_a \cos \alpha$. Let R be the resultant force, or

$$\overrightarrow{R} = \overrightarrow{\Sigma V} + \overrightarrow{(P_a \cos \alpha)} \tag{12.17}$$

The net moment of these forces about point C (Figure 12.7) is

$$M_{net} = \Sigma M_R - \Sigma M_O \tag{12.18}$$

The values of ΣM_R and ΣM_O were previously determined [see col. 5, Table 12.1,

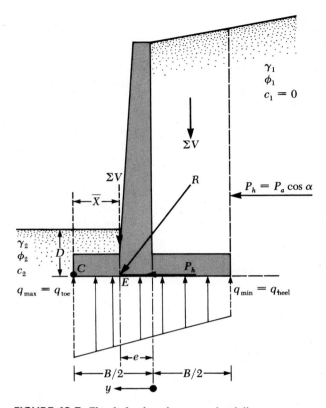

FIGURE 12.7 Check for bearing capacity failure

and Eq. (12.4).] Let the line of action of the resultant, R, intersect the base slab at E, as shown in Figure 12.7. The distance CE then is

$$\overline{CE} = \overline{X} = \frac{M_{net}}{\Sigma V} \qquad (12.19)$$

Hence, the eccentricity of the resultant, R, may be expressed as

$$e = \frac{B}{2} - \overline{CE} \qquad (12.20)$$

The pressure distribution under the base slab may be determined by using the simple principles of mechanics of materials:

$$q = \frac{\Sigma V}{A} \pm \frac{M_{net} y}{I} \qquad (12.21)$$

where M_{net} = moment = $(\Sigma V)e$
$\quad\quad I$ = moment of inertia per unit length of the base section = $\frac{1}{12}(1)(B^2)$

For maximum and minimum pressures, the value of y in Eq. (12.21) equals $B/2$. Substituting the preceding values into Eq. (12.21) gives

$$q_{max} = q_{toe} = \frac{\Sigma V}{(B)(1)} + \frac{e(\Sigma V)\frac{B}{2}}{\left(\frac{1}{12}\right)(B^3)} = \frac{\Sigma V}{B}\left(1 + \frac{6e}{B}\right) \tag{12.22}$$

Similarly,

$$q_{min} = q_{heel} = \frac{\Sigma V}{B}\left(1 - \frac{6e}{B}\right) \tag{12.23}$$

Note that ΣV includes the soil weight, as shown in Table 12.1, and that, when the value of the eccentricity, e, becomes greater than $B/6$, q_{min} becomes negative [Eq. (12.23)]. Thus, there will be some tensile stress at the end of the heel section. This stress is not desirable because the tensile strength of soil is very small. If the analysis of a design shows that $e > B/6$, the design should be reproportioned and calculations redone.

The relationships for the ultimate bearing capacity of a shallow foundation were discussed in Chapter 11. Recall that

$$q_u = c_2 N_c F_{cd} F_{ci} + q N_q F_{qd} F_{qi} + \frac{1}{2}\gamma_2 B' N_\gamma F_{\gamma d} F_{\gamma i} \tag{12.24}$$

where $q = \gamma_2 D$
$$B' = B - 2e$$
$$F_{cd} = 1 + 0.4\frac{D}{B'}$$

$$F_{qd} = 1 + 2\tan\phi_2(1 - \sin\phi_2)^2\frac{D}{B'}$$

$$F_{\gamma d} = 1$$
$$F_{ci} = F_{qi} = \left(1 - \frac{\psi°}{90°}\right)^2$$

$$F_{\gamma i} = \left(1 - \frac{\psi°}{\phi_2°}\right)^2$$

$$\psi° = \tan^{-1}\left(\frac{P_a\cos\alpha}{\Sigma V}\right)$$

Note that the shape factors F_{cs}, F_{qs}, and $F_{\gamma s}$ given in Chapter 11 are all equal to 1 because they can be treated as a continuous foundation. For this reason, the shape factors are not shown in Eq. (12.24).

Once the ultimate bearing capacity of the soil has been calculated using Eq. (12.24), the factor of safety against bearing capacity failure can be determined:

$$FS_{(\text{bearing capacity})} = \frac{q_u}{q_{\max}} \qquad (12.25)$$

Generally, a factor of safety of 3 is required. In Chapter 11, we noted that the ultimate bearing capacity of shallow foundations occurs at a settlement of about 10% of the foundation width. In the case of retaining walls, the width B is large. Hence, the ultimate load q_u will occur at a fairly large foundation settlement. A factor of safety of 3 against bearing capacity failure may not ensure, in all cases, that settlement of the structure will be within the tolerable limit. Thus, this situation needs further investigation.

EXAMPLE 12.1

The cross section of a cantilever retaining wall is shown in Figure 12.8. Calculate the factors of safety with respect to overturning, sliding, and bearing capacity.

Solution   Referring to Figure 12.8, we find

$$H' = H_1 + H_2 + H_3 = 2.6 \tan 10° + 6 + 0.7$$

$$= 0.458 + 6 + 0.7 = 7.158 \text{ m}$$

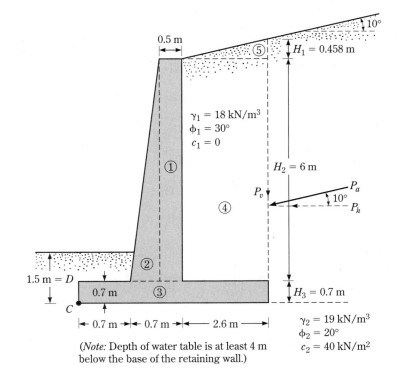

(Note: Depth of water table is at least 4 m below the base of the retaining wall.)

FIGURE 12.8

The Rankine active force per unit length of wall is

$$P_a = \frac{1}{2}\gamma_1 H'^2 K_a$$

For $\phi_1 = 30°$ and $\alpha = 10°$, K_a is equal to 0.350 (Table 9.6). Thus,

$$P_a = \frac{1}{2}(18)(7.158)^2(0.35) = 161.4 \text{ kN/m}$$

$$P_v = P_a \sin 10° = 161.4(\sin 10°) = 28.03 \text{ kN/m}$$

$$P_h = P_a \cos 10° = 161.4(\cos 10°) = 158.95 \text{ kN/m}$$

Factor of Safety Against Overturning   The following table can now be prepared to determine the resisting moment.

Section no.*	Area (m²)	Weight/unit length (kN/m)†	Moment arm from point C (m)	Moment (kN-m/m)
1	$6 \times 0.5 = 3$	70.74	1.15	81.35
2	$\frac{1}{2}(0.2)6 = 0.6$	14.15	0.833	11.79
3	$4 \times 0.7 = 2.8$	66.02	2.0	132.04
4	$6 \times 2.6 = 15.6$	280.80	2.7	758.16
5	$\frac{1}{2}(2.6)(0.458) = 0.595$	10.71	3.13	33.52
		$P_v = 28.03$	4.0	112.12
		$\Sigma V = 470.45$		$\Sigma 1128.98 = \Sigma M_R$

*For section numbers, refer to Figure 12.8.
†$\gamma_{\text{concrete}} = 23.58 \text{ kN/m}^3$

For the overturning moment, we get

$$M_O = P_h\left(\frac{H'}{3}\right) = 158.95\left(\frac{7.158}{3}\right) = 379.25 \text{ kN-m/m}$$

Hence,

$$FS_{(\text{overturning})} = \frac{\Sigma M_R}{M_O} = \frac{1128.98}{379.25} = \textbf{2.98} > 2\text{—OK}$$

Factor of Safety Against Sliding   From Eq. (12.13), we have

$$FS_{(\text{sliding})} = \frac{(\Sigma V)\tan(k_1\phi_1) + Bk_2c_2 + P_p}{P_a \cos \alpha}$$

Let $k_1 = k_2 = \frac{2}{3}$. Also,

$$P_p = \frac{1}{2}K_p\gamma_2 D^2 + 2c_2\sqrt{K_p}D$$

$$K_p = \tan^2\left(45 + \frac{\phi_2}{2}\right) = \tan^2(45 + 10) = 2.04$$

$$D = 1.5 \text{ m}$$

So

$$P_p = \frac{1}{2}(2.04)(19)(1.5)^2 + 2(40)(\sqrt{2.04})(1.5)$$

$$= 43.61 + 171.39 = 215 \text{ kN/m}$$

Hence,

$$FS_{(\text{sliding})} = \frac{(470.45)\tan\left(\dfrac{2 \times 20}{3}\right) + (4)\left(\dfrac{2}{3}\right)(40) + 215}{158.95}$$

$$= \frac{111.5 + 106.67 + 215}{158.95} = \textbf{2.73} > 1.5\text{—OK}$$

Note: For some designs, the depth, D, for passive pressure calculation may be taken to be *equal to the thickness of the base slab.*

Factor of Safety Against Bearing Capacity Failure Combining Eqs. (12.18), (12.19), and (12.20), we have

$$e = \frac{B}{2} - \frac{\Sigma M_R - \Sigma M_O}{\Sigma V} = \frac{4}{2} - \frac{1128.98 - 379.25}{470.45}$$

$$= 0.406 \text{ m} < \frac{B}{6} = \frac{4}{6} = 0.666 \text{ m}$$

Again, from Eqs. (12.22) and (12.23),

$$q_{\substack{\text{toe} \\ \text{heel}}} = \frac{\Sigma V}{B}\left(1 \pm \frac{6e}{B}\right) = \frac{470.45}{4}\left(1 \pm \frac{6 \times 0.406}{4}\right) = 189.2 \text{ kN/m}^2 \text{ (toe)}$$

$$= 45.99 \text{ kN/m}^2 \text{ (heel)}$$

The ultimate bearing capacity of the soil can be determined from Eq. (12.24):

$$q_u = c_2 N_c F_{cd} F_{ci} + q N_q F_{qd} F_{qi} + \frac{1}{2}\gamma_2 B' N_\gamma F_{\gamma d} F_{\gamma i}$$

For $\phi_2 = 20°$, we find $N_c = 14.83$, $N_q = 6.4$, and $N_\gamma = 5.39$ (Table 11.1). Also,

$$q = \gamma_2 D = (19)(1.5) = 28.5 \text{ kN/m}^2$$

$$B' = B - 2e = 4 - 2(0.406) = 3.188 \text{ m}$$

$$F_{cd} = 1 + 0.4\left(\frac{D}{B'}\right) = 1 + 0.4\left(\frac{1.5}{3.188}\right) = 1.188$$

$$F_{qd} = 1 + 2 \tan \phi_2 (1 - \sin \phi_2)^2 \left(\frac{D}{B'} \right) = 1 + 0.315 \left(\frac{1.5}{3.188} \right) = 1.148$$

$$F_{\gamma d} = 1$$

$$F_{ci} = F_{qi} = \left(1 - \frac{\psi^\circ}{90^\circ} \right)^2$$

$$\psi = \tan^{-1} \left(\frac{P_a \cos \alpha}{\Sigma V} \right) = \tan^{-1} \left(\frac{158.95}{470.45} \right) = 18.67^\circ$$

So

$$F_{ci} = F_{qi} = \left(1 - \frac{18.67}{90} \right)^2 = 0.628$$

$$F_{\gamma i} = \left(1 - \frac{\psi}{\phi} \right)^2 = \left(1 - \frac{18.67}{20} \right)^2 \approx 0$$

Hence,

$$q_u = (40)(14.83)(1.188)(0.628) + (28.5)(6.4)(1.148)(0.628)$$

$$+ \frac{1}{2}(19)(5.93)(3.188)(1)(0)$$

$$= 442.57 + 131.50 + 0 = 574.07 \text{ kN/m}^2$$

$$FS_{(\text{bearing capacity})} = \frac{q_u}{q_{toe}} = \frac{574.07}{189.2} = \textbf{3.03} > 3 \text{—OK} \qquad \blacksquare$$

EXAMPLE 12.2   A gravity retaining wall is shown in Figure 12.9. Use $\delta = \frac{2}{3}\phi_1$ and Coulomb's active earth pressure theory. Determine these values:

    a. The factor of safety against overturning
    b. The factor of safety against sliding
    c. The pressure on the soil at the toe and heel

Solution

$$H' = 5 + 1.5 = 6.5 \text{ m}$$

Coulomb's active force

$$P_a = \frac{1}{2}\gamma_1 H'^2 K_a$$

With $\alpha = 0^\circ$, $\theta = 15^\circ$, $\delta = \frac{2}{3}\phi_1$, and $\phi_1 = 32^\circ$, we find $K_a = 0.4023$ (Table 9.3). So

$$P_a = \frac{1}{2}(18.5)(6.5)^2(0.4023) = 157.22 \text{ kN/m}$$

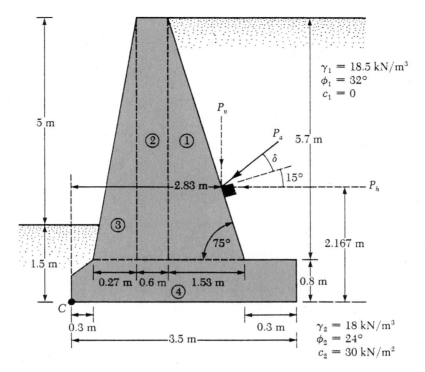

FIGURE 12.9

$$P_h = P_a \cos\left(15 + \frac{2}{3}\phi_1\right) = 157.22\cos 36.33 = 126.65 \text{ kN/m}$$

$$P_v = P_a \sin\left(15 + \frac{2}{3}\phi_1\right) = 157.22\sin 36.33 = 93.14 \text{ kN/m}$$

Part a: Factor of Safety Against Overturning Referring to Figure 12.9, we can prepare the following table:

Area no.	Area (m²)	Weight/unit length (kN/m)*	Moment arm from point C (m)	Moment (kN-m/m)
1	$\frac{1}{2}(5.7)(1.53) = 4.36$	102.81	2.18	224.13
2	$(0.6)(5.7) = 3.42$	80.64	1.37	110.48
3	$\frac{1}{2}(0.27)(5.7) = 0.77$	18.16	0.98	17.80
4	$\approx(3.5)(0.8) = 2.8$	66.02	1.75	115.54
		$P_v = 93.14$	2.83	263.59
		$\Sigma V = 360.77$ kN/m		$\Sigma M_R = 731.54$ kN-m/m

*$\gamma_{\text{concrete}} = 23.58$ kN/m³

We have for the overturning moment

$$M_o = P_h \left(\frac{H'}{3} \right) = 126.65(2.167) = 274.45 \text{ kN-m/m}$$

Hence,

$$FS_{\text{(overturning)}} = \frac{\Sigma M_R}{\Sigma M_O} = \frac{731.54}{274.45} = \textbf{2.665} > 2 \text{—OK}$$

Part b: Factor of Safety Against Sliding

$$FS_{\text{(sliding)}} = \frac{(\Sigma V) \tan \left(\frac{2}{3} \phi_2 \right) + \frac{2}{3} c_2 B + P_p}{P_h}$$

$$P_p = \tfrac{1}{2} K_p \gamma_2 D^2 + 2c_2 \sqrt{K_p} D$$

$$K_p = \tan^2 \left(45 + \frac{24}{2} \right) = 2.37$$

Hence,

$$P_p = \frac{1}{2} (2.37)(18)(1.5)^2 + 2(30)(1.54)(1.5) = 186.59 \text{ kN/m}$$

So

$$FS_{\text{(sliding)}} = \frac{360.77 \tan \left(\frac{2}{3} \times 24 \right) + \frac{2}{3} (30)(3.5) + 186.59}{126.65}$$

$$= \frac{103.45 + 70 + 186.59}{126.65} = \textbf{2.84}$$

If P_p is ignored, the factor of safety would be **1.37**.

Part c: Pressure on Soil at Toe and Heel From Eqs. (12.18), (12.19), and (12.20), we have

$$e = \frac{B}{2} - \frac{\Sigma M_R - \Sigma M_O}{\Sigma V} = \frac{3.5}{2} - \frac{731.54 - 274.45}{360.77} = 0.483 < \frac{B}{6} = 0.583$$

$$q_{\text{toe}} = \frac{\Sigma V}{B} \left[1 + \frac{6e}{B} \right] = \frac{360.77}{3.5} \left[1 + \frac{(6)(0.483)}{3.5} \right] = \textbf{188.43 kN/m}^2$$

$$q_{\text{heel}} = \frac{V}{B} \left[1 - \frac{6e}{B} \right] = \frac{360.77}{3.5} \left[1 - \frac{(6)(0.483)}{3.5} \right] = \textbf{17.73 kN/m}^2 \quad \blacksquare$$

12.7 Comments Relating to Stability

When a weak soil layer is located at a shallow depth—that is, within a depth of about 1.5 times the height of the retaining wall—the bearing capacity of the weak layer should be carefully investigated. The possibility of excessive settlement also should be considered. In some cases, the use of lightweight backfill material behind the retaining wall may solve the problem.

In many instances, piles are used to transmit the foundation load to a firmer layer. However, often the thrust of the sliding wedge of soil, in the case of deep shear failure, bends the piles and eventually causes them to fail. Careful attention should be given to this possibility when considering the option of using pile foundations for retaining walls. (Pile foundations may be required for bridge abutments to avoid the problem of scouring.)

As illustrated in Examples 12.1 and 12.2, the *active earth pressure coefficient* is used to determine the lateral force of the backfill. The active state of the backfill can be established only if the wall yields sufficiently, which does not happen in all cases. The degree of wall yielding will depend on its height and the section modulus. Furthermore, the lateral force of the backfill will depend on many factors, as identified by Casagrande (1973):

1. Effect of temperature
2. Groundwater fluctuation
3. Readjustment of the soil particles due to creep and prolonged rainfall
4. Tidal changes
5. Heavy wave action
6. Traffic vibration
7. Earthquakes

Insufficient wall yielding when combined with other unforeseen factors may generate a large lateral force on the retaining structure compared to that obtained from the active earth pressure theory.

12.8   Drainage from the Backfill of the Retaining Wall

As the result of rainfall or other wet conditions, the backfill material for a retaining wall may become saturated. Saturation will increase the pressure on the wall and may create an unstable condition. For this reason, adequate drainage must be provided by using *weepholes* and/or *perforated drainage pipes* (see Figure 12.10).

The *weepholes,* if provided, should have a minimum diameter of about 0.1 m and be adequately spaced. Note that there is always a possibility that the backfill material may be washed into weepholes or drainage pipes and ultimately clog them. Thus, a filter material should be placed behind the weepholes or around the drainage pipes, as the case may be; geotextiles now serve that purpose.

12.9   Provision of Joints in Retaining-Wall Construction

A retaining wall may be constructed with one or more of the following joints:

1. *Construction joints* (Figure 12.11a) are vertical and horizontal joints that are placed between two successive pours of concrete. To increase the shear

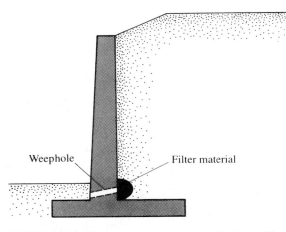

FIGURE 12.10 Drainage provisions for the backfill of a retaining wall

at the joints, keys may be used. If keys are not used, the surface of the first pour is cleaned and roughened before the next pour of concrete.

2. *Contraction joints* (Figure 12.11b) are vertical joints (grooves) placed in the face of a wall (from the top of the base slab to the top of the wall) that allow the concrete to shrink without noticeable harm. The grooves may be about 6 to 8 mm wide and ≈12 to 16 mm deep.

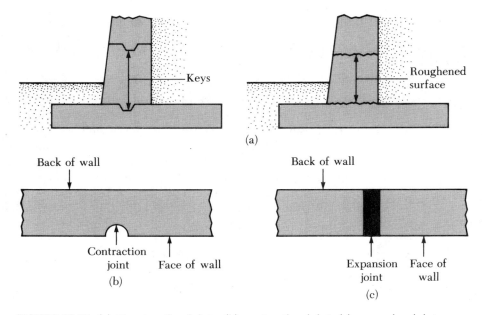

FIGURE 12.11 (a) Construction joints; (b) contraction joint; (c) expansion joint

3. *Expansion joints* (Figure 12.11c) allow for the expansion of concrete caused by temperature changes; vertical expansion joints from the base to the top of the wall may also be used. These joints may be filled with flexible joint fillers. In most cases, horizontal reinforcing steel bars running across the stem are continuous through all joints. The steel is greased to allow the concrete to expand.

BRACED CUTS

12.10 *Braced Cuts—General*

Figure 12.12 shows two types of braced cut commonly used in construction work. One type uses the *soldier beam* (Figure 12.12a), which is a vertical steel or timber beam driven into the ground before excavation. *Laggings,* which are horizontal timber planks, are placed between soldier beams as the excavation proceeds. When the excavation reaches the desired depth, *wales* and *struts* (horizontal steel beams) are installed. The struts are horizontal compression members. Figure 12.12b shows another type of braced excavation. In this case, interlocking *sheet piles* are driven into the soil before excavation. Wales and struts are inserted immediately after excavation reaches the appropriate depth. A majority of braced cuts use sheet piles.

Steel sheet piles in the United States are about 10 to 13 mm thick. European sections may be thinner and wider. Sheet pile sections may be *Z, deep arch, low arch,* or *straight web* sections. The interlocks of the sheet pile sections are shaped like a *thumb and finger* or a *ball and socket* for watertight connections. Figure 12.13a shows schematic diagrams of the thumb-and-finger type of interlocking for straight web sections. The ball-and-socket type of interlocking for Z section piles is shown in Figure 12.13b. Table 12.2 shows the properties of the sheet pile sections produced by Bethlehem Steel Corporation. The allowable design flexural stress for the steel sheet piles is as follows:

Type of steel	Allowable stress (MN/m^2)
ASTM A-328	170
ASTM A-572	210
ASTM A-690	210

Steel sheet piles are convenient to use because of their resistance to high driving stress developed when being driven into hard soils. They are also lightweight and reusable.

To design braced excavations (that is, to select wales, struts, sheet piles, and soldier beams), an engineer must estimate the lateral earth pressure to which the braced cuts will be subjected. This topic is discussed in Section 12.11; subsequent sections cover the procedures of analysis and design of braced cuts.

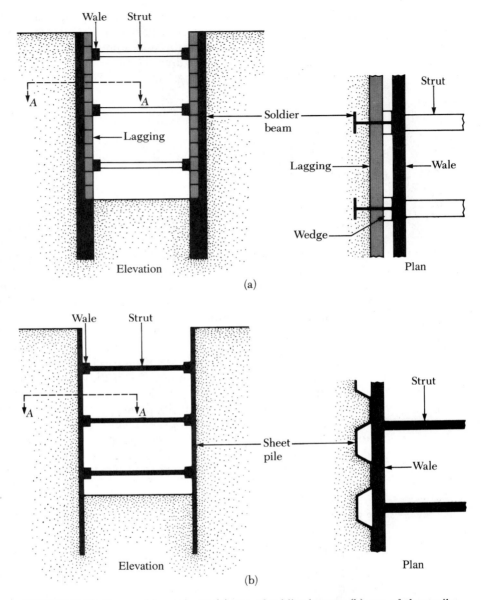

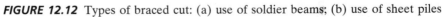

FIGURE 12.12 Types of braced cut: (a) use of soldier beams; (b) use of sheet piles

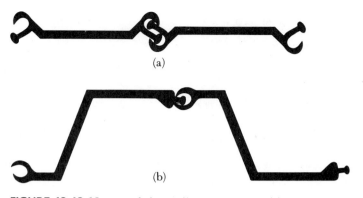

FIGURE 12.13 Nature of sheet pile connections: (a) thumb-and-finger type; (b) ball-and-socket type

12.11   *Lateral Earth Pressure in Braced Cuts*

Chapter 9 explained that a retaining wall rotates about its bottom (Figure 12.14a). With sufficient yielding of the wall, the lateral earth pressure is approximately equal to that obtained by Rankine's theory or Coulomb's theory. In contrast to retaining walls, braced cuts show a different type of wall yielding (see Figure 12.14b). In this case, deformation of the wall gradually increases with the depth of excavation. The variation of the amount of deformation depends on several factors, such as the type of soil, the depth of excavation, and the workmanship. However, with very little wall yielding at the top of the cut, the lateral earth pressure will be close to the at-rest pressure. At the bottom of the wall, with a much larger degree of yielding, the lateral earth pressure will be substantially lower than the Rankine active earth pressure. As a result, the distribution of lateral earth pressure will vary substantially in comparison to the linear distribution assumed in the case of retaining walls.

The total lateral force, P, imposed on a wall may be evaluated theoretically by using Terzaghi's (1943a) general wedge theory (Figure 12.14c). The failure surface is assumed to be the arc of a logarithmic spiral, defined as

$$r = r_o e^{\theta \tan \phi} \tag{12.26}$$

where ϕ = angle of friction of soil. A detailed outline for the evaluation of P is beyond the scope of this text; those interested should check a soil mechanics text for more information.

A comparison of the lateral earth pressure for braced cuts in sand (with angle of wall friction $\delta = 0$) with that for a retaining wall ($\delta = 0$) is shown in Figure 12.14c. If $\delta = 0$, a retaining wall of height H will be subjected to a Rankine active earth pressure, and the resultant active force will intersect the wall at distance nH from the bottom of the wall. For this case, $n = \frac{1}{3}$. In contrast, the value of n for a braced cut may vary from 0.33 to 0.5 or 0.6. The general wedge theory may also be used to analyze braced cuts in saturated clay (for example, see Das and Seeley, 1975).

In any event, when choosing a lateral soil pressure distribution for the design

Table 12.2 Properties of some sheet pile sections (Produced by Bethlehem Steel Corporation)

Section designation	Sketch of section	Section modulus (m³/m of wall)	Moment of inertia (m⁴/m of wall)
PZ-40		326.4×10^{-5}	670.5×10^{-6}
PZ-35		260.5×10^{-5}	493.4×10^{-6}
PZ-27		162.3×10^{-5}	251.5×10^{-6}
PZ-22		97×10^{-5}	115.2×10^{-6}
PSA-31		10.8×10^{-5}	4.41×10^{-6}
PSA-23		12.8×10^{-5}	5.63×10^{-6}

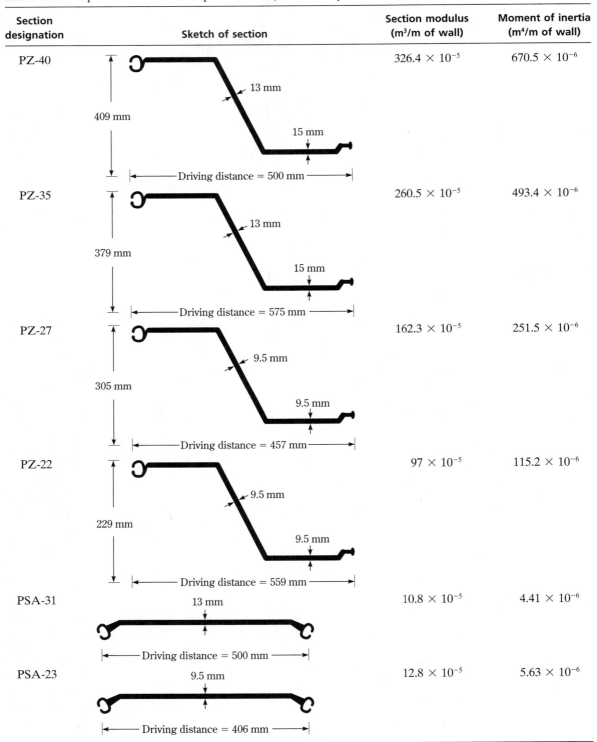

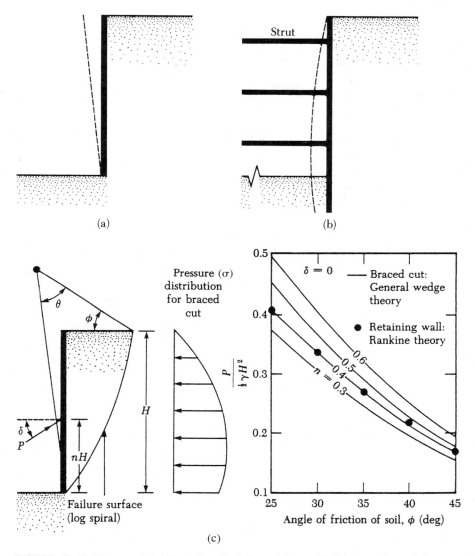

FIGURE 12.14 Nature of yielding of walls: (a) retaining wall; (b) braced cut; (c) comparison of lateral earth pressure for braced cuts and retaining walls in sand ($\delta = 0$)

of braced cuts, the engineer must keep in mind that the nature of failure in braced cuts is much different from that in retaining walls. After observing several braced cuts, Peck (1969) suggested using *design pressure envelopes* for braced cuts in sand and clay. Figures 12.15, 12.16, and 12.17 show Peck's pressure envelopes, to which the following guidelines apply.

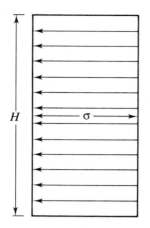

FIGURE 12.15 Peck's (1969)
apparent pressure envelope
for cuts in sand

Cuts in Sand

Figure 12.15 shows the pressure envelope for cuts in sand. This pressure, σ, may be expressed as

$$\sigma = 0.65\gamma HK_a \tag{12.27}$$

where γ = unit weight
H = height of the cut
K_a = Rankine's active pressure coefficient = $\tan^2 (45 - \phi/2)$

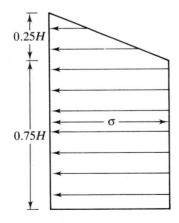

FIGURE 12.16 Peck's (1969) apparent
pressure envelope for cuts in soft
to medium clay

Cuts in Soft and Medium Clay

The pressure envelope for soft to medium clay is shown in Figure 12.16. It is applicable for the condition

$$\frac{\gamma H}{c} > 4$$

where c = undrained cohesion ($\phi = 0$). The pressure, σ, is the larger of

$$\sigma = \gamma H \left[1 - \left(\frac{4c}{\gamma H} \right) \right]$$

or   (12.28)

$$\sigma = 0.3\gamma H$$

where γ = unit weight of clay.

Cuts in Stiff Clay

The pressure envelope shown in Figure 12.17, in which

$$\sigma = 0.2\gamma H \text{ to } 0.4\gamma H \qquad (\text{with an average of } 0.3\gamma H)$$ (12.29)

is applicable to the condition $\gamma H/c \leq 4$.

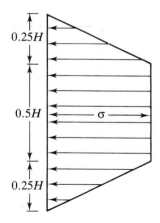

FIGURE 12.17 Peck's (1969) apparent pressure envelope for cuts in stiff clay

Limitations for the Pressure Envelopes

When using the pressure envelopes just described, keep the following points in mind:

1. The pressure envelopes are sometimes referred to as *apparent pressure envelopes*. However, the actual pressure distribution is a function of the construction sequence and the relative flexibility of the wall.
2. They apply to excavations with depths greater than about 6 m.
3. They are based on the assumption that the water table is below the bottom of the cut.
4. Sand is assumed to be drained with 0 pore water pressure.
5. Clay is assumed to be undrained, and pore water pressure is not considered.

12.12   *Soil Parameters for Cuts in Layered Soil*

Sometimes, layers of both sand and clay are encountered when a braced cut is being constructed. In this case, Peck (1943) proposed that an equivalent value of cohesion ($\phi = 0$ concept) should be determined in the following manner (refer to Figure 12.18a):

$$c_{av} = \frac{1}{2H}\left[\gamma_s K_s H_s^2 \tan \phi_s + (H - H_s)n'q_u\right] \tag{12.30}$$

where H = total height of the cut
γ_s = unit weight of sand
H_s = height of the sand layer

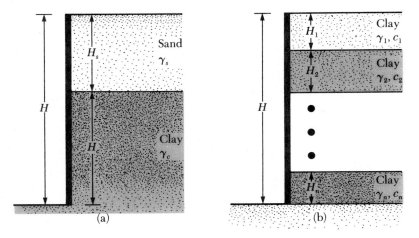

FIGURE 12.18 Layered soils in braced cuts

K_s = a lateral earth pressure coefficient for the sand layer (≈ 1)

ϕ_s = angle of friction of sand

q_u = unconfined compression strength of clay

n' = a coefficient of progressive failure (ranges from 0.5 to 1.0; average value 0.75)

The average unit weight, γ_a, of the layers may be expressed as

$$\gamma_a = \frac{1}{H}[\gamma_s H_s + (H - H_s)\gamma_c] \tag{12.31}$$

where γ_c = saturated unit weight of clay layer. Once the average values of cohesion and unit weight are determined, the pressure envelopes in clay can be used to design the cuts.

Similarly, when several clay layers are encountered in the cut (Figure 12.18b), the average undrained cohesion becomes

$$c_{av} = \frac{1}{H}(c_1 H_1 + c_2 H_2 + \cdots + c_n H_n) \tag{12.32}$$

where $c_1, c_2, \ldots, c_n$ = undrained cohesion in layers $1, 2, \ldots, n$

$H_1, H_2, \ldots, H_n$ = thickness of layers $1, 2, \ldots, n$

The average unit weight, γ_a, is

$$\gamma_a = \frac{1}{H}(\gamma_1 H_1 + \gamma_2 H_2 + \gamma_3 H_3 + \cdots + \gamma_n H_n) \tag{12.33}$$

12.13 *Design of Various Components of a Braced Cut*

Struts

In construction work, struts should have a minimum vertical spacing of about 3 m or more. The struts are actually horizontal columns subject to bending. The load-carrying capacity of columns depends on the *slenderness ratio*, l/r. The slenderness ratio can be reduced by providing vertical and horizontal supports at intermediate points. For wide cuts, splicing the struts may be necessary. For braced cuts in clayey soils, the depth of the first strut below the ground surface should be less than the depth of tensile crack, z_o. From Eq. (9.15), we have

$$\sigma'_a = \gamma z K_a - 2c\sqrt{K_a}$$

where K_a = coefficient of Rankine's active pressure. For determining the depth of tensile crack, we use

$$\sigma'_a = 0 = \gamma z_o K_a - 2c\sqrt{K_a}$$

or

$$z_o = \frac{2c}{\sqrt{K_a \gamma}}$$

With $\phi = 0$, $K_a = \tan^2(45 - \phi/2) = 1$. So

$$z_o = \frac{2c}{\gamma} \qquad (\textit{Note: } c = c_u)$$

A simplified conservative procedure may be used to determine the strut loads. Although this procedure will vary depending on the engineers involved in the project, the following is a step-by-step outline of the general procedure (refer to Figure 12.19):

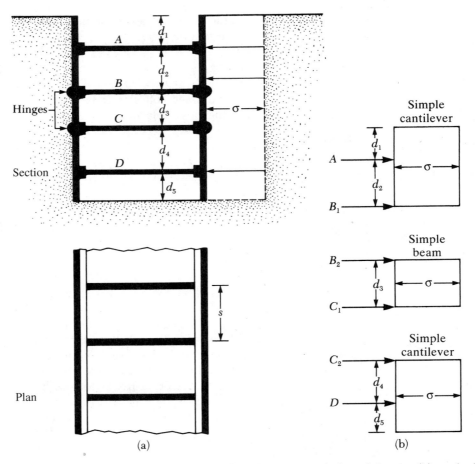

FIGURE 12.19 Determination of strut loads; (a) section and plan of the cut; (b) method for determining strut loads

1. Draw the pressure envelope for the braced cut (see Figures 12.15, 12.16, and 12.17). Also show the proposed strut levels. Figure 12.19a shows a pressure envelope for a sandy soil; however, it could also be for a clay. The strut levels are marked *A*, *B*, *C*, and *D*. The sheet piles (or soldier beams) are assumed to be hinged at the strut levels, except for the top and bottom ones. In Figure 12.19a, the hinges are at the level of struts *B* and *C*. (Many designers also assume the sheet piles, or soldier beams, to be hinged at all strut levels, except for the top.)

2. Determine the reactions for the two simple cantilever beams (top and bottom) and all the simple beams between. In Figure 12.19b, these reactions are *A*, B_1, B_2, C_1, C_2, and *D*.

3. Calculate the strut loads in Figure 12.19 as follows:

$$P_A = (A)(s)$$

$$P_B = (B_1 + B_2)(s)$$

$$P_C = (C_1 + C_2)(s)$$

$$P_D = (D)(s) \tag{12.34}$$

where P_A, P_B, P_C, P_D = loads to be taken by the individual struts at levels *A*, *B*, *C*, and *D*, respectively

A, B_1, B_2, C_1, C_2, D = reactions calculated in step 2 (note unit: force/unit length of the braced cut)

s = horizontal spacing of the struts (see plan in Figure 12.19a)

4. Knowing the strut loads at each level and the intermediate bracing conditions allows selection of the proper sections from the steel construction manual.

Sheet Piles

The following steps are taken in designing the sheet piles:

1. For each of the sections shown in Figure 12.19b, determine the maximum bending moment.

2. Determine the maximum value of the maximum bending moments (M_{max}) obtained in step 1. Note that the unit of this moment will be, for example, kN-m/m length of the wall.

3. Obtain the required section modulus of the sheet piles:

$$S = \frac{M_{max}}{\sigma_{all}} \tag{12.35}$$

where σ_{all} = allowable flexural stress of the sheet pile material.

4. Choose a sheet pile that has a section modulus greater than or equal to the required section modulus from a table such as Table 12.2.

Wales

Wales may be treated as continuous horizontal members if they are spliced properly. Conservatively, they may also be treated as though they are pinned at the struts. For the section shown in Figure 12.19a, the maximum moments for the wales (assuming that they are pinned at the struts) are:

At level A, $M_{max} = \dfrac{(A)(s^2)}{8}$

At level B, $M_{max} = \dfrac{(B_1 + B_2)s^2}{8}$

At level C, $M_{max} = \dfrac{(C_1 + C_2)s^2}{8}$

At level D, $M_{max} = \dfrac{(D)(s^2)}{8}$

where A, B_1, B_2, C_1, C_2, and D are the reactions under the struts per unit length of the wall (step 2 of strut design).

We can determine the section modulus of the wales with

$$S = \frac{M_{max}}{\sigma_{all}}$$

The wales are sometimes fastened to the sheet piles at points that satisfy the lateral support requirements.

EXAMPLE 12.3

Figure 12.20 shows the cross section of a long braced cut.

a. Draw the pressure envelope.
b. Determine the strut loads at levels A, B, and C.

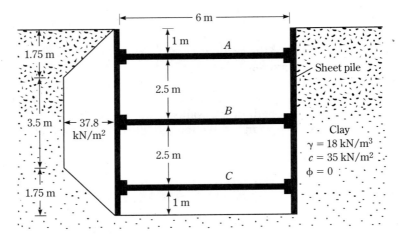

FIGURE 12.20

c. Determine the section modulus of the sheet pile required.
d. Determine the section modulus for the wales at level B.

The struts are placed 3 m center-to-center.

Solution

Part a   We are given $\gamma = 18$ kN/m³, $c = 35$ kN/m², and $H = 7$ m, we find

$$\frac{\gamma H}{c} = \frac{(18)(7)}{35} = 3.6 < 4$$

So, the pressure envelope will be like the one in Figure 12.17. This is plotted in Figure 12.20 with maximum pressure intensity, σ, equal to

$$0.3\gamma H = 0.3(18)(7) = 37.8 \text{ kN/m}^2.$$

Part b   For determination of the strut loads, refer to Figure 12.21. Taking the

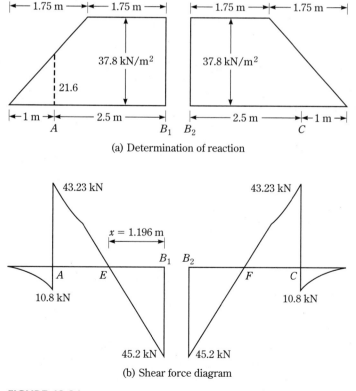

(a) Determination of reaction

(b) Shear force diagram

FIGURE 12.21

moment about B_1, $\Sigma\ M_{B1} = 0$, we have

$$A(2.5) = \left(\frac{1}{2}\right)(37.8)(1.75)\left(1.75 + \frac{1.75}{3}\right) - (1.75)(37.8)\left(\frac{1.75}{2}\right) = 0$$

or

$$A = 54.02 \text{ kN/m}$$

Also, Σ vertical forces $= 0$. Thus,

$$\frac{1}{2}(1.75)(37.8) + (37.8)(1.75) = A + B_1$$

$$33.08 + 66.15 - A = B_1$$

So

$$B_1 = 45.2 \text{ kN/m}$$

Due to symmetry, we have

$$B_2 = 45.2 \text{ kN/m}$$

$$C = 54.02 \text{ kN/m}$$

The strut loads at levels A, B, and C are

$$P_A = 54.02 \times \text{horizontal spacing}, s = 54.02 \times 3 = \textbf{162.06 kN}$$

$$P_B = (B_1 + B_2)3 = (45.2 + 45.2)3 = \textbf{271.2 kN}$$

$$P_c = 54.02 \times 3 = \textbf{162.06 kN}$$

Part c Refer to the left side of Figure 12.21a. For the maximum moment, the shear force should be 0. The nature of variation of the shear force is shown in Figure 12.21b. The location of point E can be given as

$$x = \frac{\text{reaction at } B_1}{37.8} = \frac{45.2}{37.8} = 1.196 \text{ m}$$

We have

$$\text{magnitude of moment at } A = \frac{1}{2}(1)\left(\frac{37.8}{1.75} \times 1\right)\left(\frac{1}{3}\right)$$

$$= 3.6 \text{ kN-m/m of wall}$$

$$\text{magnitude of moment at } E = (45.2 \times 1.196) - (37.8 \times 1.196)\left(\frac{1.196}{2}\right)$$

$$= 54.06 - 27.03 = 27.03 \text{ kN-m/m of wall}$$

Because the loading on the left and right sections of Figure 12.21a are the same, the magnitude of moments at F and C (Figure 12.21b) will be the same as at E and A, respectively. Hence, the maximum moment is 27.03 kN-m/m of wall.

The section modulus of the sheet pile is

$$S = \frac{M_{max}}{\sigma_{all}} = \frac{27.03 \text{ kN-m}}{170 \times 10^3 \text{ kN/m}^2} = \mathbf{15.9 \times 10^{-5} \text{ m}^3/\text{m of the wall}}$$

Part d   The reaction at level B was calculated in Part b. Hence,

$$M_{max} = \frac{(B_1 + B_2)s^2}{8} = \frac{(45.2 + 45.2)3^2}{8} = 101.7 \text{ kN-m}$$

The section modulus for the wales at level B is

$$S = \frac{101.7}{\sigma_{all}} = \frac{101.7}{0.6F_y} = \frac{101.7}{0.6(248.4 \times 1000)}$$

$$= \mathbf{0.682 \times 10^{-3} \text{ m}^3}$$ ∎

12.14   *Heave of the Bottom of a Cut in Clay*

Braced cuts in clay may become unstable as a result of heaving of the bottom of the excavation. Terzaghi (1943b) analyzed the factor of safety of braced excavations against bottom heave. The failure surface for such a case is shown in Figure 12.22. The vertical load per unit length of the cut at the bottom of the cut along line *bd* and *af* is

$$Q = \gamma H B_1 - cH \tag{12.36}$$

where $B_1 = 0.7B$
$\quad c = $ cohesion ($\phi = 0$ concept)

This load Q may be treated as a load per unit length on a continuous foundation

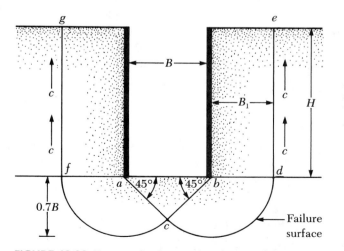

Note: cd and *cf* are arcs of circles with centers at *b* and *a*, respectively

FIGURE 12.22 Factor of safety against bottom heave

at the level of bd (and af) and having a width of $B_1 = 0.7B$. Based on Terzaghi's bearing capacity theory, the net ultimate load-carrying capacity per unit length of this foundation (Chapter 11) is

$$Q_u = cN_cB_1 = 5.7cB_1$$

Hence, from Eq. (12.36), the factor of safety against bottom heave is

$$FS_{(\text{heave})} = \frac{Q_u}{Q} = \frac{5.7cB_1}{\gamma HB_1 - cH} = \frac{1}{H}\left(\frac{5.7c}{\gamma - \dfrac{c}{0.7B}}\right) \tag{12.37}$$

This factor of safety is based on the assumption that the clay layer is homogeneous, at least to a depth of $0.7B$ below the bottom of the cut. However, a *hard layer of rock or rocklike material at a depth of $D < 0.7B$* will modify the failure surface to some extent. In such a case, the factor of safety becomes

$$FS_{(\text{heave})} = \frac{1}{H}\left(\frac{5.7c}{\gamma - c/D}\right) \tag{12.38}$$

Bjerrum and Eide (1956) also studied the problem of bottom heave for braced cuts in clay. For the factor of safety, they proposed

$$FS_{(\text{heave})} = \frac{cN_c}{\gamma H} \tag{12.39}$$

The bearing capacity factor, N_c, varies with the ratios H/B and L/B (where $L =$ length of the cut). For infinitely long cuts ($B/L = 0$), $N_c = 5.14$ at $H/B = 0$ and increases to $N_c = 7.6$ at $H/B = 4$. Beyond that—that is, for $H/B > 4$—the value of N_c remains constant. For cuts square in plan ($B/L = 1$), $N_c = 6.3$ at $H/B = 0$, and $N_c = 9$ for $H/B \geq 4$. In general, for any H/B,

$$N_{c(\text{rectangle})} = N_{c(\text{square})}\left(0.84 + 0.16\frac{B}{L}\right) \tag{12.40}$$

Figure 12.23 shows the variation in the values of N_c for $L/B = 1, 2, 3,$ and ∞.

When Eqs. (12.39) and (12.40) are combined, the factor of safety against heave becomes

$$FS_{(\text{heave})} = \frac{cN_{c(\text{square})}\left(0.84 + 0.16\dfrac{B}{L}\right)}{\gamma H} \tag{12.41}$$

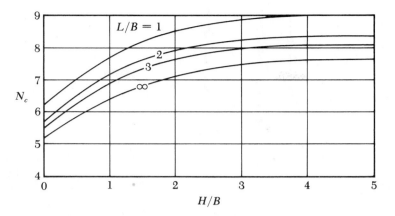

FIGURE 12.23 Variation of N_c with L/B and H/B [Based on Bjerrum and Eide's equation, Eq. (12.40)]

Equation (12.41) and the variation of the bearing capacity factor, N_c, as shown in Figure 12.23 are based on the assumptions that the clay layer below the bottom of the cut is homogeneous and that the magnitude of the undrained cohesion in the soil that contains the failure surface is equal to c (Figure 12.24). However, if a stronger clay layer is encountered at a shallow depth, as shown in Figure 12.25a, the failure surface below the cut will be controlled by the undrained cohesions c_1

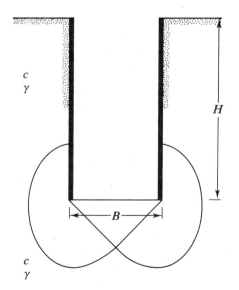

FIGURE 12.24 Derivation of Eq. (12.42)

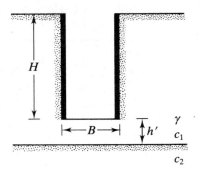

(Note: $c_2 > c_1$)

(a)

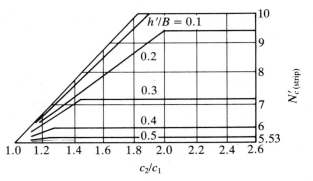

(b)

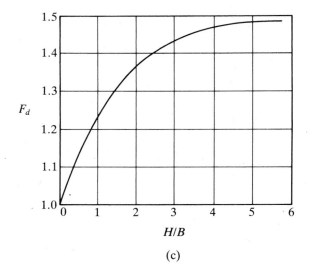

(c)

FIGURE 12.25 (a) Layered clay below the bottom of the cut; (b) variation of $N'_{c(strip)}$ with c_2/c_1 and h'/B (redrawn after Reddy and Srinivasan, 1967); and (c) variation of F_d with H/B

and c_2. For this type of condition, the factor of safety is

$$FS_{(heave)} = \frac{c_1 [N'_{c(strip)} F_d] F_s}{\gamma H}$$

(12.42)

where $N'_{c(strip)}$ = bearing capacity factor of an infinitely long cut ($B/L = 0$), which
is a function of h'/B and c_2/c_1
F_d = depth factor, which is a function of H/B
F_s = shape factor

The variation of $N'_{c(strip)}$ is shown in Figure 12.25b, and the variation of F_d as a
function of H/B is given in Figure 12.25c. The shape factor, F_s, is

$$F_s = 1 + 0.2 \frac{B}{L}$$

(12.43)

In most cases, a factor of safety of about 1.5 is generally recommended. If F_s
becomes less than about 1.5, the sheet pile is driven deeper (Figure 12.26). Usually
the depth, d, is kept less than or equal to $B/2$. In that case, the force, P, per unit
length of the buried sheet pile (aa' and bb') may be expressed as follows (U.S.
Department of the Navy, 1971):

$$P = 0.7(\gamma H B - 1.4 c H - \pi c B) \qquad \text{for } d \geq 0.47B$$

(12.44)

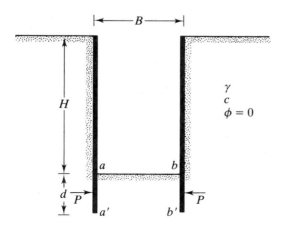

FIGURE 12.26 Force on the buried length of sheet pile

and

$$P = 1.5d\left(\gamma H - \frac{1.4cH}{B} - \pi c\right) \qquad \text{for } d \leq 0.47B \qquad (12.45)$$

EXAMPLE 12.4

A long braced cut in saturated clay has a width of cut, B, of 4.5 m and a depth of cut, H, of 8 m. For the clay, $\gamma = 17.2$ kN/m^3 and $c = 42$ kN/m^2. Determine the factor of safety against bottom heave by using each of these equations:

    a. Eq. (12.37)
    b. Eq. (12.39)

Assume that the clay extends to a great depth below the bottom of the cut.

Solution

    a.

$$FS_{(heave)} = \frac{1}{H}\left(\frac{5.7c}{\gamma - \dfrac{c}{0.7B}}\right) = \frac{1}{8}\left(\frac{5.7 \times 42}{17.2 - \dfrac{42}{0.7 \times 4.5}}\right) = \mathbf{7.74}$$

    b.

$$FS_{(heave)} = \frac{cN_c}{\gamma H}$$

From Figure 12.23 for $H/B = 8/4.5 = 1.78$ and $L/B \approx \infty$, the magnitude of $N_c \approx$ 7. So

$$FS_{(heave)} = \frac{(42)(7)}{(17.2)(8)} = \mathbf{2.14} \qquad \blacksquare$$

EXAMPLE 12.5

Refer to Example 12.4. If a stiffer clay layer ($c = 55$ kN/m^2) is encountered 1.5 m below the bottom of the cut, what will be the factor of safety against bottom heave?

Solution For layered clay encountered below the bottom of the cut, use Eq. (12.42). Refer to Figure 12.25a. First,

$$\frac{c_2}{c_1} = \frac{55}{42} = 1.31$$

$$\frac{h'}{B} = \frac{1.5}{4.5} = 0.33$$

From Figure 12.25b for $c_2/c_1 = 1.31$ and $h'/B \approx 0.33$, the magnitude of $N'_{c(\text{strip})} \approx$ 6.5. Again,

$$\frac{H}{B} = \frac{8}{4.5} = 1.78$$

So, from Figure 12.25c, $F_d \approx 1.34$. Also,

$$F_s = 1 + 0.2 \left(\frac{B}{L}\right) = 1 + 0.2 \left(\frac{10}{\infty}\right) = 1.0$$

So

$$FS_{(\text{heave})} = \frac{c_1[N'_{c(\text{strip})} F_d] F_s}{\gamma H} = \frac{(42)[(6.5)(1.34)](1.0)}{(17.2)(8)} = \mathbf{2.66} \qquad \blacksquare$$

12.15   *Lateral Yielding of Sheet Piles and Ground Settlement*

In braced cuts, some lateral movement of sheet pile walls may be expected (Figure 12.27). The amount of lateral yield depends on several factors, the most important of which is the elapsed time between excavation and placement of wales and struts. Mana and Clough (1981) analyzed the field records of several braced cuts in clay from the San Francisco, Oslo (Norway), Boston, Chicago, and Bowline Point (New York) areas. Under ordinary construction conditions, they found that the maximum

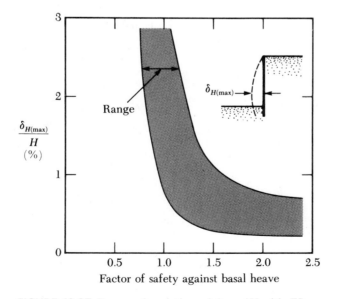

FIGURE 12.27 Range of variation of $\delta_{H(\text{max})}/H$ with FS against basal heave from field observation (Redrawn after Mana and Clough, 1981)

lateral wall yield, $\delta_{H(max)}$, has a definite relationship with the factor of safety against heave, as shown in Figure 12.27. Note that the factor of safety against heave plotted in Figure 12.27 was calculated by using Eqs. (12.37) and (12.38).

As discussed before, in several instances the sheet piles (or the soldier piles, as the case may be) are driven to a certain depth below the bottom of the excavation. The reason is to reduce the lateral yielding of the walls during the last stages of excavation. Lateral yielding of the walls will cause the ground surface surrounding the cut to settle. The degree of lateral yielding, however, depends mostly on the soil type below the bottom of the cut. If clay below the cut extends to a great depth and $\gamma H/c$ is less than about 6, extension of the sheet piles or soldier piles below the bottom of the cut will help considerably in reducing the lateral yield of the walls.

However, under similar circumstances, if $\gamma H/c$ is about 8, the extension of sheet piles into the clay below the cut does not help greatly. In such circumstances, we may expect a great degree of wall yielding that may result in the total collapse of the bracing systems. If a hard soil layer lies below a clay layer at the bottom of the cut, the piles should be embedded in the stiffer layer. This action will greatly reduce lateral yield.

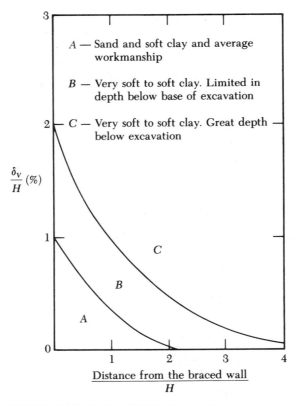

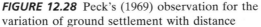

FIGURE 12.28 Peck's (1969) observation for the variation of ground settlement with distance

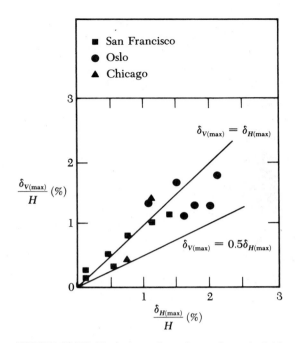

FIGURE 12.29 Variation of maximum lateral yield with maximum ground settlement (After Mana and Clough, 1981)

The lateral yielding of walls will generally induce ground settlement, δ_V, around a braced cut, which is generally referred to as *ground loss*. Based on several field observations, Peck (1969) provided curves for predicting ground settlement in various types of soil (see Figure 12.28). The magnitude of ground loss varies extensively; however, Figure 12.28 may be used as a general guide.

Based on the field data obtained from various cuts in the areas of San Francisco, Oslo, and Chicago, Mana and Clough (1981) provided a correlation between the maximum lateral yield of sheet piles, $\delta_{H(max)}$, and the maximum ground settlement, $\delta_{V(max)}$. The variation is shown in Figure 12.29. Note that

$$\delta_{V(max)} \approx 0.5\delta_{H(max)} \quad \text{to} \quad 1.0\delta_{H(max)} \tag{12.46}$$

Problems

For Problems 12.1–12.5, use unit weight of concrete, $\gamma_c = 23.58$ kN/m^3.

12.1 For the cantilever retaining wall shown in Figure 12.30, the wall dimensions are $H = 8$ m, $x_1 = 0.4$ m, $x_2 = 0.6$ m, $x_3 = 1.5$ m, $x_4 = 3.5$ m, $x_5 = 0.96$ m, $D = 1.75$ m, and $\alpha = 10°$; and the soil properties are $\gamma_1 = 16.8$ kN/m^3, $\phi_1 = 32°$, $\gamma_2 = 17.6$ kN/m^3, $\phi_2 = 28°$, and $c_2 = 30$ kN/m^2. Calculate the factors of safety with respect to overturning, sliding, and bearing capacity.

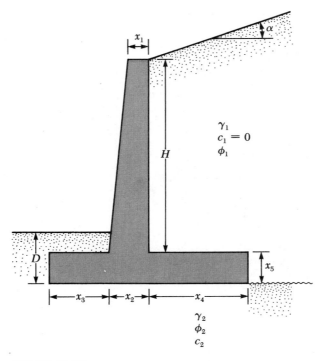

γ_1
$c_1 = 0$
ϕ_1

H

D

x_5

γ_2
ϕ_2
c_2

FIGURE 12.30

12.2 Repeat Problem 12.1 for the wall dimensions $H = 6$ m, $x_1 = 0.3$ m, $x_2 = 0.7$ m, $x_3 = 1.4$ m, $x_4 = 2.3$ m, $x_5 = 0.85$ m, $D = 1.25$ m, and $\alpha = 5°$; and the soil properties $\gamma_1 = 18.4$ kN/m³, $\phi_1 = 34°$, $\gamma_2 = 16.8$ kN/m³, $\phi_2 = 18°$, and $c_2 = 50$ kN/m².

12.3 Repeat Problem 12.1 with wall dimensions of $H = 5.49$ m, $x_1 = 0.46$ m, $x_2 = 0.58$ m, $x_3 = 0.92$ m, $x_4 = 1.55$ m, $x_5 = 0.61$ m, $D = 1.22$ m, and $\alpha = 0°$; and soil properties of $\gamma_1 = 18.08$ kN/m³, $\phi_1 = 36°$, $\gamma_2 = 19.65$ kN/m³, $\phi_2 = 15°$, and $c_2 = 44$ kN/m².

12.4 A gravity retaining wall is shown in Figure 12.31. Calculate the factors of safety with respect to overturning and sliding. We have wall dimensions $H = 6$ m, $x_1 = 0.6$ m, $x_2 = 0.2$ m, $x_3 = 2$ m, $x_4 = 0.5$ m, $x_5 = 0.75$ m, $x_6 = 0.8$ m, and $D = 1.5$ m; and soil properties $\gamma_1 = 16.5$ kN/m³, $\phi_1 = 32°$, $\gamma_2 = 18$ kN/m³, $\phi_2 = 22°$, and $c_2 = 40$ kN/m². Use Rankine's active pressure for calculation.

12.5 Repeat Problem 12.4 using Coulomb's active pressure for calculation and $\delta = \frac{2}{3}\phi_1$.

12.6 Refer to the braced cut in Figure 12.32, for which $\gamma = 17.6$ kN/m³, $\phi = 32°$, and $c = 0$. The struts are located at 4 m on center in the plan. Draw the earth pressure envelope and determine the strut loads at levels A, B, and C.

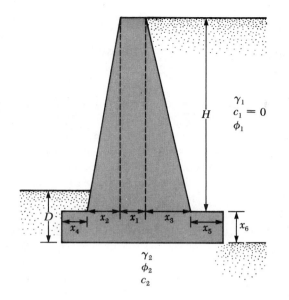

FIGURE 12.31

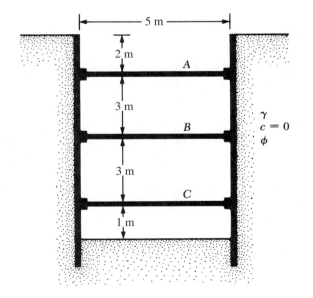

FIGURE 12.32

12.7 For the braced cut described in Problem 12.6, assume that $\sigma_{all} = 172 \text{ MN/m}^2$.
   a. Determine the sheet pile section.
   b. What is the section modulus of the wales at level A?

12.8 Redo Problem 12.6 for $\gamma = 18.2 \text{ kN/m}^3$, $\phi = 35°$, $c = 0$, and center-to-center strut spacing in plan of 3 m.

12.9 Determine the sheet pile section required for the braced cut described in Problem 12.8 for $\sigma_{all} = 172 \text{ MN/m}^2$.

12.10 Refer to Figure 12.18a. For the braced cut, $H = 6$ m, $H_s = 2$ m, $\gamma_s = 16.2$ kN/m^3, angle of friction of sand, $\phi_s = 34°$, $H_c = 4$ m, $\gamma_c = 17.5$ kN/m^3, and the unconfined compression strength of the clay layer, $q_u = 68$ kN/m^2.
   a. Estimate the average cohesion, c_{av}, and average unit weight, γ_{av}, for development of the earth pressure envelope.
   b. Plot the earth pressure envelope.

12.11 Refer to Figure 12.18b, which shows a braced cut in clay. Here, $H = 7$ m, $H_1 = 2$ m, $c_1 = 102$ kN/m^2, $\gamma_1 = 17.5$ kN/m^3, $H_2 = 2.5$ m, $c_2 = 75$ kN/m^2, $\gamma_2 = 16.8$ kN/m^3, $H_3 = 2.5$ m, $c_3 = 80$ kN/m^2, and $\gamma_3 = 17$ kN/m^3.
   a. Determine the average cohesion, c_{av}, and the average unit weight, γ_{av}, for development of the earth pressure envelope.
   b. Plot the earth pressure envelope.

12.12 Refer to Figure 12.33 in which $\gamma = 17.5$ kN/m^3, $c = 30$ kN/m^2, and center-to-center spacing of struts is 5 m. Draw the earth pressure envelope and determine the strut loads at levels A, B, and C.

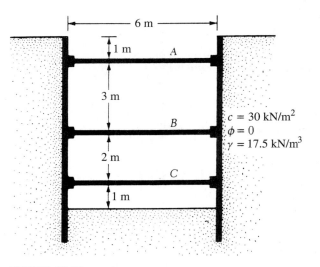

FIGURE 12.33

12.13 For the braced cut described in Problem 12.12, determine the sheet pile section. Use $\sigma_{all} = 170 \text{ MN/m}^2$.

12.14 Redo Problem 12.12 for $c = 60$ kN/m^2.

12.15 Determine the factor of safety against bottom heave for the braced cut described in Problem 12.12. Use Eqs. (12.37) and (12.41). For Eq. (12.41), assume the length of the cut, $L = 18$ m.

12.16 Determine the factor of safety against bottom heave for the braced cut described in Problem 12.14. Use Eq. (12.41). The length of the cut is 12.5 m.

References

Bjerrum, L., and Eide, O. (1956). "Stability of Strutted Excavation in Clay," *Geotechnique,* Vol. 6, No. 1, 32–47.

Casagrande, L. (1973). "Comments on Conventional Design of Retaining Structure," *Journal of the Soil Mechanics and Foundations Division,* ASCE, Vol. 99, No. SM2, 181–198.

Das, B. M., and Seeley, G. R. (1975). "Active Thrust on Braced Cut in Clay," *Journal of the Construction Division,* American Society of Civil Engineers, Vol. 101, No. CO4, 945–949.

Elman, M. T., and Terry, C. F. (1988). "Retaining Walls with Sloped Heel," *Journal of Geotechnical Engineering,* American Society of Civil Engineers, Vol. 114, No. GT10, 1194–1199.

Mana, A. I., and Clough, G. W. (1981). "Prediction of Movements for Braced Cuts in Clay," *Journal of the Geotechnical Engineering Division,* American Society of Civil Engineers, Vol. 107, No. GT8, 759–777.

Peck, R. B. (1943). "Earth Pressure Measurements in Open Cuts, Chicago (Ill.) Subway," *Transactions,* American Society of Civil Engineers, Vol. 108, 1008–1058.

Peck, R. B. (1969). "Deep Excavation and Tunneling in Soft Ground," *Proceedings,* Seventh International Conference on Soil Mechanics and Foundation Engineering, Mexico City, State-of-the-Art Volume, 225–290.

Reddy, A. S., and Srinivasan, R. J. (1967). "Bearing Capacity of Footing on Layered Clay," *Journal of the Soil Mechanics and Foundations Division,* American Society of Civil Engineers, Vol. 93, No. SM2, 83–99.

Terzaghi, K. (1943a). "General Wedge Theory of Earth Pressure," *Transactions,* American Society of Civil Engineers, Vol. 106, 69–97.

Terzaghi, K. (1943b). *Theoretical Soil Mechanics,* Wiley, New York.

U.S. Department of the Navy (1971). "Design Manual—Soil Mechanics, Foundations, and Earth Structures," NAVFAC DM-7, Washington, D.C.

Supplementary References for Further Study

Bell, J. R., Stilley, A. N., and Vandre, B. (1975). "Fabric Retaining Earth Walls," *Proceedings,* Thirteenth Engineering Geology and Soils Engineering Symposium, Moscow, Idaho.

Casagrande, L. (1973). "Comments on Conventional Design of Retaining Structure," *Journal of the Soil Mechanics and Foundations Division,* ASCE, Vol. 99, No. SM2, 181–198.

Federal Highway Administration (1996). *Mechanically Stabilized Earth Walls and Reinforced Soil Slopes Design and Construction Guidelines, Publication No. FHWA-SA-96-071,* Washington, D.C.

Goh, A. T. C. (1993). "Behavior of Cantilever Retaining Walls," *Journal of Geotechnical Engineering,* ASCE, Vol. 119, No. 11, 1751–1770.

Hijab, W. (1956). "A Note on the Centroid of a Logarithmic Spiral Sector," *Geotechnique,* Vol. 4, No. 2, 96–99.

Kim, J. S., and Preber, T. (1969). "Earth Pressure Against Braced Excavations," *Journal of the Soil Mechanics and Foundations Division,* ASCE, Vol. 96, No. 6, 1581–1584.

Richards, R., and Elms, D. G. (1979). "Seismic Behavior of Gravity Retaining Walls," *Journal of the Geotechnical Engineering Division,* American Society of Civil Engineers, Vol. 105, No. GT4, 449–464.

13

Deep Foundations—Piles and Drilled Shafts

Piles are structural members made of steel, concrete, and/or timber. They are used to build pile foundations, which are deep and more costly than shallow foundations (see Chapter 11). Despite the cost, the use of piles is often necessary to ensure structural safety. Drilled shafts are cast-in-place piles that generally have a diameter greater than 750 mm with or without steel reinforcement and with or without an enlarged bottom. The first part of this chapter considers pile foundations, and the second part presents a detailed discussion on drilled shafts.

PILE FOUNDATIONS

13.1 Need for Pile Foundations

Pile foundations are needed in special circumstances. The following are some situations in which piles may be considered for the construction of a foundation.

1. When the upper soil layer(s) is (are) highly compressible and too weak to support the load transmitted by the superstructure, piles are used to transmit the load to underlying bedrock or a stronger soil layer, as shown in Figure 13.1a. When bedrock is not encountered at a reasonable depth below the ground surface, piles are used to transmit the structural load to the soil gradually. The resistance to the applied structural load is derived mainly from the frictional resistance developed at the soil–pile interface (Figure 13.1b).

2. When subjected to horizontal forces (see Figure 13.1c), pile foundations resist by bending while still supporting the vertical load transmitted by the superstructure. This situation is generally encountered in the design and construction of earth-retaining structures and foundations of tall structures that are subjected to strong wind and/or earthquake forces.

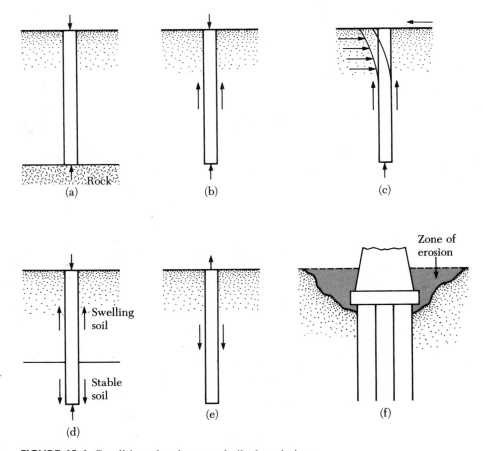

FIGURE 13.1 Conditions for the use of pile foundations

3. In many cases, the soils at the site of a proposed structure may be expansive and collapsible. These soils may extend to a great depth below the ground surface. Expansive soils swell and shrink as the moisture content increases and decreases, and the swelling pressure of such soils can be considerable. If shallow foundations are used, the structure may suffer considerable damage. However, pile foundations may be considered as an alternative when piles are extended beyond the active zone, which swells and shrinks (Figure 13.1d). Soils such as loess are collapsible. When the moisture content of these soils increases, their structures may break down. A sudden decrease in the void ratio of soil induces large settlements of structures supported by shallow foundations. In such cases, pile foundations may be used, in which piles are extended into stable soil layers beyond the zone of possible moisture change.

4. The foundations of some structures, such as transmission towers, offshore platforms, and basement mats below the water table, are subjected to

uplifting forces. Piles are sometimes used for these foundations to resist the uplifting force (Figure 13.1e).

5. Bridge abutments and piers are usually constructed over pile foundations to avoid the possible loss of bearing capacity that a shallow foundation might suffer because of soil erosion at the ground surface (Figure 13.1f).

Although numerous investigations, both theoretical and experimental, have been conducted to predict the behavior and the load-bearing capacity of piles in granular and cohesive soils, the mechanisms are not yet entirely understood and may never be clear. The design of pile foundations may be considered somewhat of an "art" as a result of the uncertainties involved in working with some subsoil conditions.

13.2   *Types of Piles and Their Structural Characteristics*

Different types of piles are used in construction work, depending on the type of load to be carried, the subsoil conditions, and the water table. Piles can be divided into these categories: (a) steel piles, (b) concrete piles, (c) wooden (timber) piles, and (d) composite piles.

Steel Piles

Steel piles generally are either *pipe piles* or *rolled steel H-section piles*. Pipe piles can be driven into the ground with their ends open or closed. Wide-flange and I-section steel beams can also be used as piles; however, H-section piles are usually preferred because their web and flange thicknesses are equal. In wide-flange and I-section beams, the web thicknesses are smaller than the thicknesses of the flange. Table 13.1 gives the dimensions of some standard H-section steel piles used in the United States. Table 13.2 shows selected pipe sections frequently used for piling purposes. In many cases, the pipe piles are filled with concrete after they are driven.

When necessary, steel piles are spliced by welding or by riveting. Figure 13.2a shows a typical splicing by welding for an H-pile. A typical splicing by welding for a pipe pile is shown in Figure 13.2b. Figure 13.2c shows a diagram of splicing an H-pile by rivets or bolts.

When hard driving conditions are expected, such as driving through dense gravel, shale, and soft rock, steel piles can be fitted with driving points or shoes. Figures 13.2d and e are diagrams of two types of shoe used for pipe piles.

Concrete Piles

Concrete piles may be divided into two basic types: precast piles and cast-*in-situ* piles. *Precast piles* can be prepared using ordinary reinforcement, and they can be square or octagonal in cross section (Figure 13.3). Reinforcement is provided to enable the pile to resist the bending moment developed during pickup and transpor-

Table 13.1 Common H-section piles used in the United States

Designation, size (mm) × weight (kN/m)	Depth, d_1 (mm)	Section area ($m^2 \times 10^{-3}$)	Flange and web thickness, w (mm)	Flange width (mm)	Moment of inertia ($m^4 \times 10^{-6}$)	
					I_{xx}	I_{yy}
HP 200 × 0.52	204	6.84	11.3	207	49.4	16.8
HP 250 × 0.834	254	10.8	14.4	260	123	42
× 0.608	246	8.0	10.6	256	87.5	24
HP 310 × 1.226	312	15.9	17.5	312	271	89
× 1.079	308	14.1	15.5	310	237	77.5
× 0.912	303	11.9	13.1	308	197	63.7
× 0.775	299	10.0	11.1	306	164	62.9
HP 330 × 1.462	334	19.0	19.5	335	370	123
× 1.264	329	16.5	16.9	333	314	104
× 1.069	324	13.9	14.5	330	263	86
× 0.873	319	11.3	11.7	328	210	69
HP 360 × 1.707	361	22.2	20.5	378	508	184
× 1.491	356	19.4	17.9	376	437	158
× 1.295	351	16.8	15.6	373	374	136
× 1.060	346	13.8	12.8	371	303	109

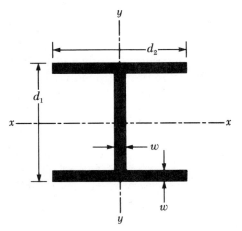

tation, the vertical load, and the bending moment caused by lateral load. The piles are cast to desired lengths and cured before being transported to the work sites.

Precast piles can also be prestressed by using high-strength steel prestressing cables. The ultimate strength of these steel cables is about 1800 MN/m². During casting of the piles, the cables are pretensioned to 900 to 1300 MN/m², and concrete is poured around them. After curing, the cables are cut, thus producing a compressive force on the pile section. Table 13.3 gives additional information about prestressed concrete piles with square and octagonal cross sections.

Cast-in-situ, or *cast-in-place, piles* are built by making a hole in the ground

Table 13.2 Selected pipe pile sections

Outside diameter (mm)	Wall thickness (mm)	Area of steel (cm²)
219	3.17	21.5
	4.78	32.1
	5.56	37.3
	7.92	52.7
254	4.78	37.5
	5.56	43.6
	6.35	49.4
305	4.78	44.9
	5.56	52.3
	6.35	59.7
406	4.78	60.3
	5.56	70.1
	6.35	79.8
457	5.56	80
	6.35	90
	7.92	112
508	5.56	88
	6.35	100
	7.92	125
610	6.35	121
	7.92	150
	9.53	179
	12.70	238

and then filling it with concrete. Various types of cast-in-place concrete pile are currently used in construction, and most of them have been patented by their manufacturers. These piles may be divided into two broad categories: cased and uncased. Both types may have a pedestal at the bottom.

Cased piles are made by driving a steel casing into the ground with the help of a mandrel placed inside the casing. When the pile reaches the proper depth, the mandrel is withdrawn and the casing is filled with concrete. Figures 13.4a, b, c, and d show some examples of cased piles without a pedestal. Figure 13.4e shows a cased pile with a pedestal. The pedestal is an expanded concrete bulb that is formed by dropping a hammer on fresh concrete.

Figures 13.4f and 13.4g are two types of *uncased pile,* one without a pedestal and the other with one. The uncased piles are made by first driving the casing to the desired depth and then filling it with fresh concrete. The casing is then gradually withdrawn.

Additional information about cast-in-place piles is given in Table 13.4.

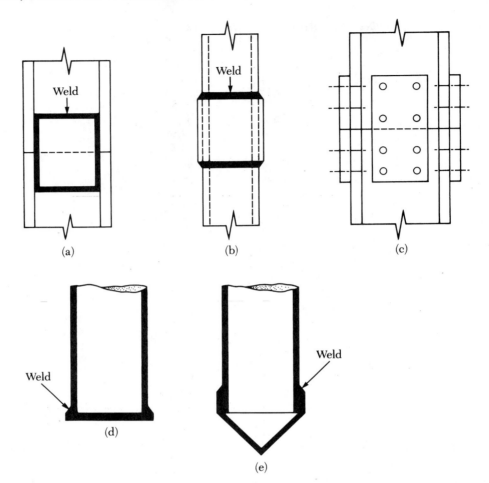

FIGURE 13.2   Steel piles: (a) splicing of H-pile by welding; (b) splicing of pipe pile by welding; (c) splicing of H-pile by rivets or bolts; (d) flat driving point of pipe pile; (e) conical driving point of pipe pile.

Timber Piles

Timber piles are tree trunks that have had their branches and bark carefully trimmed off. The maximum length of most timber piles is 10 to 20 m. To qualify for use as a pile, the timber should be straight, sound, and without any defects. The American Society of Civil Engineers' *Manual of Practice,* No. 17 (1959), divided timber piles into three classifications:

1. *Class A piles* carry heavy loads. The minimum diameter of the butt should be 356 mm.
2. *Class B piles* are used to carry medium loads. The minimum butt diameter should be 305 to 330 mm.

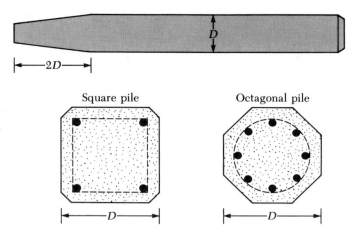

FIGURE 13.3 Precast piles with ordinary reinforcement

Table 13.3 Typical prestressed concrete piles

Pile shape*	D (mm)	Area of cross section (cm²)	Perimeter (mm)	Number of strands 12.7-mm diameter	Number of strands 11.1-mm diameter	Minimum effective prestress force (kN)	Section modulus (m³ × 10⁻³)	Design bearing capacity (kN) Concrete strength (MN/m²) 34.5	Design bearing capacity (kN) Concrete strength (MN/m²) 41.4
S	254	645	1016	4	4	312	2.737	556	778
O	254	536	838	4	4	258	1.786	462	555
S	305	929	1219	5	6	449	4.719	801	962
O	305	768	1016	4	5	369	3.097	662	795
S	356	1265	1422	6	8	610	7.489	1091	1310
O	356	1045	1168	5	7	503	4.916	901	1082
S	406	1652	1626	8	11	796	11.192	1425	1710
O	406	1368	1346	7	9	658	7.341	1180	1416
S	457	2090	1829	10	13	1010	15.928	1803	2163
O	457	1729	1524	8	11	836	10.455	1491	1790
S	508	2581	2032	12	16	1245	21.844	2226	2672
O	508	2136	1677	10	14	1032	14.355	1842	2239
S	559	3123	2235	15	20	1508	29.087	2694	3232
O	559	2587	1854	12	16	1250	19.107	2231	2678
S	610	3658	2438	18	23	1793	37.756	3155	3786
O	610	3078	2032	15	19	1486	34.794	2655	3186

* S = square section; O = octagonal section

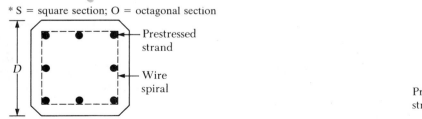

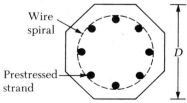

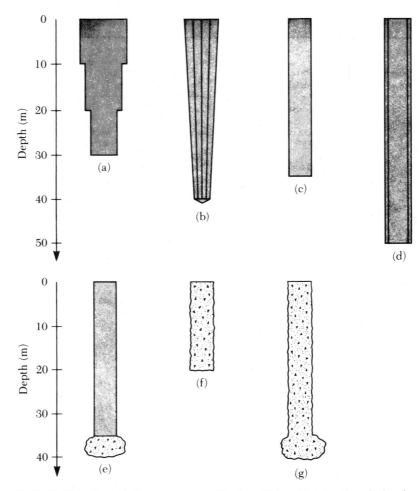

FIGURE 13.4 Cast-in-place concrete piles (see Table 13.4 for descriptions)

3. *Class C piles* are used in temporary construction work. They can be used permanently for structures when the entire pile is below the water table. The minimum butt diameter should be 305 mm.

In any case, a pile tip should have a diameter not less than 150 mm.

Timber piles cannot withstand hard driving stress; therefore, the pile capacity is generally limited to about 220 to 270 kN. Steel shoes may be used to avoid damage at the pile tip (bottom). The tops of timber piles may also be damaged during the driving operation. To avoid damage to the pile top, a metal band or cap may be used. The crushing of the wooden fibers caused by the impact of the hammer is referred to as *brooming*.

Splicing of timber piles should be avoided, particularly when they are expected to carry tensile load or lateral load. However, if splicing is necessary, it can be done

Table 13.4 Descriptions of the cast-in-place piles shown in Figure 13.4

Part in Figure 13.4	Name of pile	Type of casing	Maximum usual depth of pile (m)
a	Raymond Step-Taper	Corrugated, thin cylindrical casing	30
b	Monotube of Union Metal	Thin fluted, tapered steel casing driven without mandrel	40
c	Western cased	Thin sheet casing	30–40
d	Seamless pipe or Armco	Straight steel pipe casing	50
e	Franki cased pedestal	Thin sheet casing	30–40
f	Western uncased without pedestal	—	15–20
g	Franki uncased pedestal	—	30–40

by using *pipe sleeves* (Figure 13.5a) or *metal straps* and *bolts* (Figure 13.5b). The length of the pipe sleeve should be at least five times the diameter of the pile. The butting ends should be cut square so that full contact can be maintained. The spliced portions should be carefully trimmed so that they fit tightly to the inside of the pipe sleeve. In the case of metal straps and bolts, the butting ends should also be

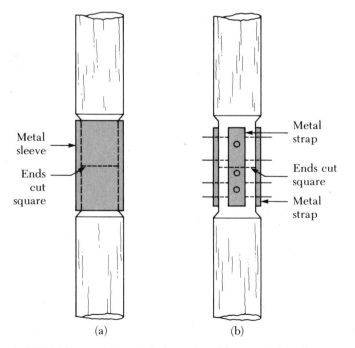

FIGURE 13.5 Splicing of timber piles: (a) use of pipe sleeves; (b) use of metal straps and bolts

Table 13.5 Comparisons of piles made of different materials

Pile type	Usual length of piles (m)	Maximum length of pile (m)	Usual load (kN)	Approximate maximum load (kN)	Comments
Steel	15–60	Practically unlimited	300–1200		*Advantages* a. Easy to handle with respect to cutoff and extension to the desired length b. Can stand high driving stresses c. Can penetrate hard layers such as dense gravel, soft rock d. High load-carrying capacity *Disadvantages* a. Relatively costly material b. High level of noise during pile driving c. Subject to corrosion d. H-piles may be damaged or deflected from the vertical during driving through hard layers or past major obstructions
Precast concrete	*Precast:* 10–15 *Prestressed:* 10–35	*Precast:* 30 *Prestressed:* 60	300–3000	*Precast:* 800–900 *Prestressed:* 7500–8500	*Advantages* a. Can be subjected to hard driving b. Corrosion resistant c. Can be easily combined with concrete superstructure *Disadvantages* a. Difficult to achieve proper cutoff b. Difficult to transport
Cased cast-in-place concrete	5–15	15–40	200–500	800	*Advantages* a. Relatively cheap b. Possibility of inspection before pouring concrete c. Easy to extend *Disadvantages* a. Difficult to splice after concreting b. Thin casings may be damaged during driving

Table 13.5 (Continued)

Pile type	Usual length of piles (m)	Maximum length of pile (m)	Usual load (kN)	Approximate maximum load (kN)	Comments
Uncased cast-in-place concrete	5–15	30–40	300–500	700	*Advantages* a. Initially economical b. Can be finished at any elevation *Disadvantages* a. Voids may be created if concrete is placed rapidly b. Difficult to splice after concreting c. In soft soils, the sides of the hole may cave in, thus squeezing the concrete
Wood	10–15	30	100–200	270	*Advantages* a. Economical b. Easy to handle c. Permanently submerged piles are fairly resistant to decay *Disadvantages* a. Decay above water table b. Can be damaged in hard driving c. Low load-bearing capacity d. Low resistance to tensile load when spliced

cut square. Also, the sides of the spliced portion should be trimmed plane for putting the straps on.

Timber piles can stay undamaged indefinitely if they are surrounded by saturated soil. However, in a marine environment, timber piles are subject to attack by various organisms and can be damaged extensively in a few months. When located above the water table, the piles are subject to attack by insects. The life of the piles may be increased by treating them with preservatives such as creosote.

Composite Piles

The upper and lower portions of *composite piles* are made of different materials. For example, composite piles may be made of steel and concrete or timber and concrete. Steel and concrete piles consist of a lower portion of steel and an upper portion of cast-in-place concrete. This type of pile is used when the length of the pile required for adequate bearing exceeds the capacity of simple cast-in-place concrete piles. Timber and concrete piles usually consist of a lower portion of timber pile below the permanent water table and an upper portion of concrete. In any case, forming proper joints between two dissimilar materials is difficult, and, for that reason, composite piles are not widely used.

Comparison of Pile Types

Several factors determine the selection of piles for a particular structure at a specific site. Table 13.5 gives a brief comparison of the advantages and disadvantages of the various types of pile based on the pile material.

13.3   Estimation of Pile Length

Selecting the type of pile to be used and estimating its necessary length are fairly difficult tasks that require good judgment. In addition to the classifications given in Section 13.2, piles can be divided into two major categories, depending on their lengths and the mechanisms of load transfer to the soil: (a) point bearing piles, and (b) friction piles.

Point Bearing Piles

If soil-boring records establish the presence of bedrock or rocklike material at a site within a reasonable depth, piles can be extended to the rock surface (Figure 13.6a). In this case, the ultimate capacity of the piles depends entirely on the load-bearing capacity of the underlying material; thus, the piles are called *point bearing piles*. In most of these cases, the necessary length of the pile can be fairly well established.

Instead of bedrock, if a fairly compact and hard stratum of soil is encountered at a reasonable depth, piles can be extended a few meters into the hard stratum

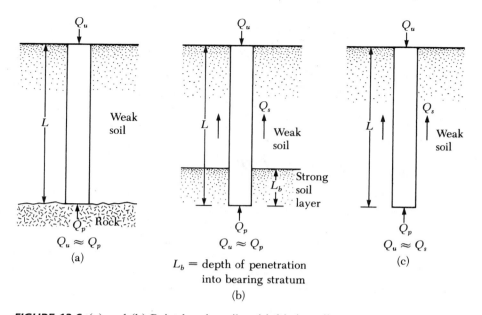

FIGURE 13.6 (a) and (b) Point bearing piles; (c) friction piles

(Figure 13.6b). Piles with pedestals can be constructed on the bed of the hard stratum, and the ultimate pile load may be expressed as

$$Q_u = Q_p + Q_s \tag{13.1}$$

where Q_p = load carried at the pile point
        Q_s = load carried by skin friction developed at the side of the pile (caused by shearing resistance between the soil and the pile)

If Q_s is very small, then

$$Q_u \approx Q_p \tag{13.2}$$

In this case, the required pile length may be estimated accurately if proper subsoil exploration records are available.

Friction Piles

When no layer of rock or rocklike material is present at a reasonable depth at a site, point bearing piles become very long and uneconomical. For this type of subsoil condition, piles are driven through the softer material to specified depths (Figure 13.6c). The ultimate load of these piles may be expressed by Eq. (13.1). However, if the value of Q_p is relatively small,

$$Q_u \approx Q_s \tag{13.3}$$

These piles are called *friction piles* because most of the resistance is derived from skin friction. However, the term *friction pile,* although used often in literature, is a misnomer in clayey soils, the resistance to applied load is also caused by *adhesion*.

The length of friction piles depends on the shear strength of the soil, the applied load, and the pile size. To determine the necessary lengths of these piles, an engineer needs a good understanding of soil–pile interaction, good judgment, and experience. Theoretical procedures for calculating the load-bearing capacity of piles are presented in Section 13.6.

13.4 Installation of Piles

Most piles are driven into the ground by *hammers* or *vibratory drivers*. In special circumstances, piles can also be inserted by *jetting* or *partial augering*. The types of hammer used for pile driving include the (a) drop hammer, (b) single-acting air or steam hammer, (c) double-acting and differential air or steam hammer, and (d) diesel hammer. In the driving operation, a cap is attached to the top of the pile. A cushion may be used between the pile and the cap. This cushion has the effect of reducing the impact force and spreading it over a longer time; however, its use is optional. A hammer cushion is placed on the pile cap. The hammer drops on the cushion.

Table 13.6 Partial list of typical air and steam hammers

Maker of hammer*	Model no.	Type of hammer	Rated energy (kN-m)	Blows per minute	Ram weight (kN)
V	3100	Single acting	407	58	449
V	540	Single acting	271	48	182
V	060	Single acting	244	62	267
MKT	OS-60	Single acting	244	55	267
V	040	Single acting	163	60	178
V	400C	Differential	154	100	178
R	8/0	Single acting	110	35	111
MKT	S-20	Single acting	82	60	89
R	5/0	Single acting	77	44	78
V	200-C	Differential	68	98	89
R	150-C	Differential	66	95–105	67
MKT	S-14	Single acting	51	60	62
V	140C	Differential	49	103	62
V	08	Single acting	35	50	36
MKT	S-8	Single acting	35	55	36
MKT	11B3	Double acting	26	95	22
MKT	C-5	Double acting	22	110	22
V	30-C	Double acting	10	133	13

*V—Vulcan Iron Works, Florida
MKT—McKiernan-Terry, New Jersey
R—Raymond International, Inc., Texas

Table 13.7 Partial list of typical diesel hammers

Maker of hammer*	Model no.	Rated energy (kN-m)	Blows per minute	Piston weight (kN)
K	K150	379.7	45–60	147
M	MB70	191.2–86	38–60	71
K	K-60	143.2	42–60	59
K	K-45	123.5	39–60	44
M	M-43	113.9–51.3	40–60	42
K	K-35	96	39–60	34
MKT	DE70B	85.4–57	40–50	31
K	K-25	68.8	39–60	25
V	N-46	44.1	50–60	18
L	520	35.7	80–84	23
M	M-14S	35.3–16.1	42–60	13
V	N-33	33.4	50–60	13
L	440	24.7	86–90	18
MKT	DE20	24.4–16.3	40–50	9
MKT	DE-10	11.9	40–50	5
L	180	11.0	90–95	8

*V—Vulcan Iron Works, Florida
M—Mitsubishi International Corporation
MKT—McKiernan-Terry, New Jersey
L—Link Belt, Cedar Rapids, Iowa
K—Kobe Diesel

Tables 13.6 and 13.7 list some of the commercially available single-acting, double-acting, differential, and diesel hammers.

In pile driving, when the pile needs to penetrate a thin layer of hard soil (such as sand and gravel) overlying a softer soil layer, a technique called *jetting* is sometimes used. In jetting, water is discharged at the pile point by a pipe 50 to 75 mm in diameter to wash and loosen the sand and gravel.

Based on the nature of their placement, piles may be divided into two categories: *displacement piles* and *nondisplacement piles*. Driven piles are displacement piles because they move some soil laterally; hence, there is a tendency for the densification of soil surrounding them. Concrete piles and closed-ended pipe piles are high-displacement piles. However, steel H-piles displace less soil laterally during driving, and so they are low-displacement piles. In contrast, bored piles are nondisplacement piles because their placement causes very little change in the state of stress in the soil.

13.5 *Load Transfer Mechanism*

The load transfer mechanism from a pile to the soil is complicated. To understand it, consider a pile of length L, as shown in Figure 13.7a. The load on the pile is

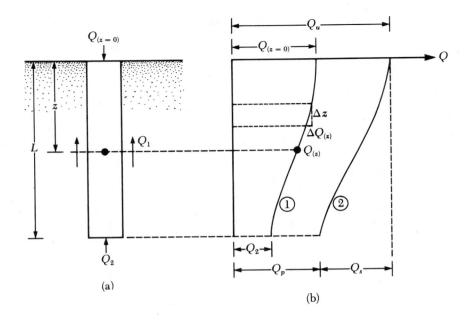

(a)

(b)

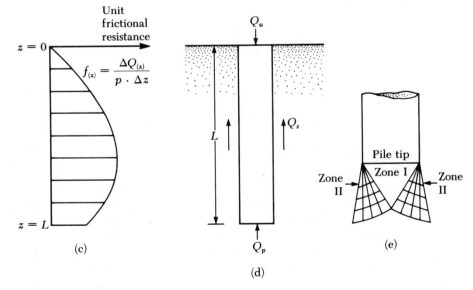

(c)

(d)

(e)

FIGURE 13.7   Load transfer mechanism for piles

gradually increased from 0 to $Q_{(z=0)}$ at the ground surface. Part of this load will be resisted by the side friction developed along the shaft, Q_1, and part by the soil below the tip of the pile, Q_2. Now, how are Q_1 and Q_2 related to the total load? If measurements are made to obtain the load carried by the pile shaft, $Q_{(z)}$, at any depth z, the nature of variation will be like curve 1 of Figure 13.7b. The *frictional resistance per unit area, $f_{(z)}$,* at any depth z may be determined as

$$f_{(z)} = \frac{\Delta Q_{(z)}}{(p)(\Delta z)} \tag{13.4}$$

where p = perimeter of the pile cross section. Figure 13.7c shows the variation of $f_{(z)}$ with depth.

If the load Q at the ground surface is gradually increased, maximum frictional resistance along the pile shaft will be fully mobilized when the relative displacement between the soil and the pile is about 5 to 10 mm, irrespective of pile size and length L. However, the maximum point resistance $Q_2 = Q_p$ will not be mobilized until the pile tip has moved about 10% to 25% of the pile width (or diameter). The lower limit applies to driven piles and the upper limit to bored piles. At ultimate load (Figure 13.7d and curve 2 in Figure 13.7b), $Q_{(z=0)} = Q_u$. Thus,

$$Q_1 = Q_s$$

and

$$Q_2 = Q_p$$

The preceding explanation indicates that Q_s (or the unit skin friction, f, along the pile shaft) is developed at a *much smaller pile displacement compared to the point resistance, Q_p.*

At ultimate load, the failure surface in the soil at the pile tip (bearing capacity failure caused by Q_p) is like that shown in Figure 13.7e. Note that pile foundations are deep foundations and that the soil fails mostly in a *punching mode,* as illustrated previously in Figures 11.2c and 11.3. That is, a *triangular zone,* I, is developed at the pile tip, which is pushed downward without producing any other visible slip surface. In dense sands and stiff clayey soils, a *radial shear zone,* II, may partially develop. Hence, the load displacement curves of piles will resemble those shown in Figure 11.2c.

13.6 *Equations for Estimation of Pile Capacity*

The ultimate load-carrying capacity of a pile, Q_u, is given by a simple equation as the load carried at the pile point plus the total frictional resistance (skin friction) derived from the soil–pile interface (Figure 13.8), or

$$Q_u = Q_p + Q_s \tag{13.5}$$

where Q_p = load-carrying capacity of the pile point
Q_s = frictional resistance

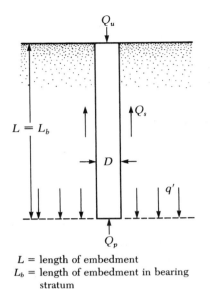

L = length of embedment
L_b = length of embedment in bearing
   stratum

FIGURE 13.8 Ultimate load-capacity of a pile

Numerous published studies cover the determination of the values of Q_p and Q_s. Excellent reviews of many of these investigations have been provided by Vesic (1977), Meyerhof (1976), and Coyle and Castello (1981). These studies provide insight into the problem of determining the ultimate pile capacity.

Load-Carrying Capacity of the Pile Point, Q_p

The ultimate bearing capacity of shallow foundations was discussed in Chapter 11. The general bearing capacity equation for shallow foundations was given in Chapter 11 (for vertical loading) as

$$q_u = cN_c F_{cs} F_{cd} + qN_q F_{qs} F_{qd} + \frac{1}{2} \gamma B N_\gamma F_{\gamma s} F_{\gamma d}$$

Hence, in general, the ultimate bearing capacity may be expressed as

$$q_u = cN_c^* + qN_q^* + \gamma B N_\gamma^* \tag{13.6}$$

where N_c^*, N_q^*, and N_γ^* are the bearing capacity factors that include the necessary shape and depth factors.

Pile foundations are deep. However, the ultimate resistance per unit area developed at the pile tip, q_p, may be expressed by an equation similar in form to Eq. 13.6, although the values of N_c^*, N_q^*, and N_γ^* will change. The notation used in this chapter for the width of the pile is D. Hence, substituting D for B in Eq. (13.6) gives

$$q_u = q_p = cN_c^* + qN_q^* + \gamma DN_\gamma^* \tag{13.7}$$

Because the width, D, of a pile is relatively small, the term γDN_γ^* may be dropped from the right side of the preceding equation without introducing a serious error, or

$$q_p = cN_c^* + q'N_q^* \tag{13.8}$$

Note that the term q has been replaced by q' in Eq. (13.8) to signify effective vertical stress. Hence, the load-carrying capacity of the pile point is

$$Q_p = A_p q_p = A_p(cN_c^* + q'N_q^*) \tag{13.9}$$

where A_p = area of the pile tip
c = cohesion of the soil supporting the pile tip
q_p = unit point resistance
q' = effective vertical stress at the level of the pile tip
N_c^*, N_q^* = bearing capacity factors

There are several methods for calculating the magnitude of q_p. In this text, the method suggested by Meyerhof (1976) will be used.

Load-Carrying Capacity of the Pile Point in Sand   In sand, the cohesion c is equal to 0. Thus, Eq. (13.9) takes the form

$$Q_p = A_p q_p = A_p q'N_q^* \tag{13.10}$$

The variation of N_q^* with the soil friction angle, ϕ, is shown in Figure 13.9. Meyerhof pointed out that the point bearing capacity, q_p, of a pile in sand generally increases with the depth of embedment in the bearing stratum and reaches a maximum value at an embedment ratio of $L_b/D = (L_b/D)_{cr}$. Note that in a homogeneous soil, L_b is equal to the actual embedment length of the pile, L (see Figure 13.8). However, in Figure 13.6b, where a pile has penetrated into a bearing stratum, $L_b < L$. Beyond the critical embedment ratio, $(L_b/D)_{cr}$, the value of q_p remains constant ($q_p = q_l$). That is, as shown in Figure 13.10 for the case of a homogeneous soil, $L = L_b$. Hence, Q_p should not exceed the limiting value, or $A_p q_l$, so

$$Q_p = A_p q'N_q^* \leq A_p q_l \tag{13.11}$$

The limiting point resistance is

$$q_l \, (\text{kN/m}^2) = 50N_q^* \tan \phi \tag{13.12}$$

where ϕ = soil friction angle in the bearing stratum.

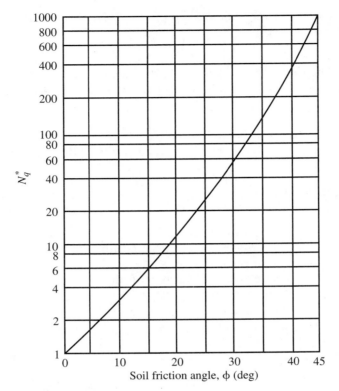

FIGURE 13.9 Meyerhof's bearing capacity factor, N_q^*

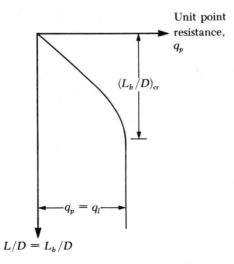

FIGURE 13.10 Variation of unit point resistance in a homogeneous sand

Based on field observations, Meyerhof (1976) also suggested that the ultimate point resistance, q_p, in a homogeneous granular soil ($L = L_b$) may be obtained from standard penetration numbers as

$$q_p \ (\text{kN/m}^2) = 40 N_{\text{cor}} \frac{L}{D} \leq 400 N_{\text{cor}} \tag{13.13}$$

where N_{cor} = average standard penetration number near the pile point (about $10D$ above and $4D$ below the pile point).

Load-Carrying Capacity of the Pile Point in Clay For piles in *saturated clays* in undrained conditions($\phi = 0$),

$$Q_p = N_c^* c_u A_p = 9 c_u A_p \tag{13.14}$$

where c_u = undrained cohesion of the soil below the pile tip.

Frictional Resistance, Q_s

The frictional or skin resistance of a pile may be written as

$$Q_s = \Sigma \, p \, \Delta L f \tag{13.15}$$

where p = perimeter of the pile section
ΔL = incremental pile length over which p and f are taken constant (Figure 13.11a)
f = unit friction resistance at any depth z

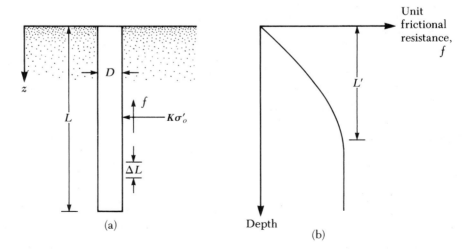

FIGURE 13.11 Unit frictional resistance for piles in sand

Frictional Resistance in Sand   The unit frictional resistance at any depth for a pile is

$$f = K\sigma'_o \tan \delta \qquad (13.16)$$

where K = earth pressure coefficient
σ'_o = effective vertical stress at the depth under consideration
δ = soil–pile friction angle

In reality, the magnitude of K varies with depth. It is approximately equal to the Rankine passive earth pressure coefficient, K_p, at the top of the pile and may be less than the at-rest earth pressure coefficient, K_o, at the pile tip. It also depends on the nature of the pile installation. Based on presently available results, the following average values of K are recommended for use in Eq. (13.16):

Pile type	K
Bored or jetted	$\approx K_o = 1 - \sin \phi$
Low-displacement driven	$\approx K_o = 1 - \sin \phi$ to $1.4K_o = 1.4(1 - \sin \phi)$
High-displacement driven	$\approx K_o = 1 - \sin \phi$ to $1.8K_o = 1.8(1 - \sin \phi)$

The effective vertical stress, σ'_o, for use in Eq. (13.16) increases with pile depth to a maximum limit at a depth of 15 to 20 pile diameters and remains constant thereafter, as shown in Figure 13.11b. This critical depth, L', depends on several factors, such as the soil friction angle and compressibility and relative density. A conservative estimate is to assume that

$$L' = 15D \qquad (13.17)$$

The values of δ from various investigations appear to be in the range of 0.5ϕ to 0.8ϕ. Judgment must be used in choosing the value of δ.

Meyerhof (1976) also indicated that the average unit frictional resistance, f_{av}, for high-displacement driven piles may be obtained from average standard penetration resistance values as

$$f_{av} \ (\text{kN/m}^2) = 2\overline{N}_{cor} \qquad (13.18)$$

where N_{cor} = average value of standard penetration resistance. For low-displacement driven piles,

$$f_{av} \ (\text{kN/m}^2) = \overline{N}_{cor} \qquad (13.19)$$

Thus,

$$Q_s = pLf_{av} \qquad (13.20)$$

Frictional (or Skin) Resistance in Clay Several methods are available for obtaining the unit frictional (or skin) resistance of piles in clay. Three of the presently accepted procedures are described briefly.

1. λ *Method:* This method was proposed by Vijayvergiya and Focht (1972). It is based on the assumption that the displacement of soil caused by pile driving results in a passive lateral pressure at any depth and that the average unit skin resistance is

$$f_{av} = \lambda(\overline{\sigma}'_o + 2c_u)$$

(13.21)

where $\overline{\sigma}'_o$ = mean effective vertical stress for the entire embedment length
c_u = mean undrained shear strength ($\phi = 0$ concept)

The value of λ changes with the depth of pile penetration (see Figure 13.12). Thus, the total frictional resistance may be calculated as

$$Q_s = pLf_{av}$$

Care should be taken in obtaining the values of $\overline{\sigma}'_o$ and c_u in layered soil. Figure 13.13 helps explain the reason. According to Figure 13.13b, the mean value of c_u is $(c_{u(1)}L_1 + c_{u(2)}L_2 + \cdots)/L$. Similarly, Figure 13.13c shows the plot of the variation of effective stress with depth. The mean effective stress is

$$\overline{\sigma}'_o = \frac{A_1 + A_2 + A_3 + \cdots}{L}$$

(13.22)

where $A_1, A_2, A_3, \ldots$ = areas of the vertical effective stress diagrams.
2. α *Method:* According to the α method, the unit skin resistance in clayey soils can be represented by the equation

$$f = \alpha c_u$$

(13.23)

where α = empirical adhesion factor. The approximate variation of the value of α is shown in Figure 13.14. Note that for normally consolidated clays with $c_u \leq$ about 50 kN/m², we have $\alpha = 1$. Thus,

$$Q_s = \Sigma fp \, \Delta L = \Sigma \alpha c_u p \, \Delta L$$

(13.24)

3. β *Method:* When piles are driven into saturated clays, the pore water pressure in the soil around the piles increases. This excess pore water pressure in normally consolidated clays may be 4 to 6 times c_u. However, within a month or so, this pressure gradually dissipates. Hence, the unit

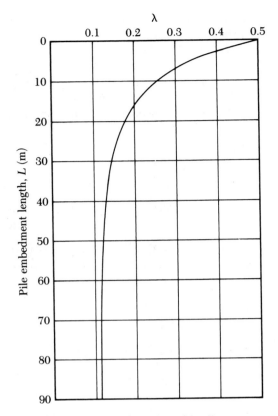

FIGURE 13.12 Variation of λ with pile embedment length (Redrawn after McClelland, 1974)

frictional resistance for the pile can be determined on the basis of the effective stress parameters of the clay in a remolded state ($c = 0$). Thus, at any depth,

$$f = \beta\sigma'_o \tag{13.25}$$

where σ'_o = vertical effective stress
$\quad\quad\quad \beta = K \tan \phi_R \tag{13.26}$
$\quad\quad\quad \phi_R$ = drained friction angle of remolded clay
$\quad\quad\quad K$ = earth pressure coefficient

Conservatively, we can calculate the magnitude of K as the earth pressure coefficient at rest, or

$$K = 1 - \sin \phi_R \quad \text{(for normally consolidated clays)} \tag{13.27}$$

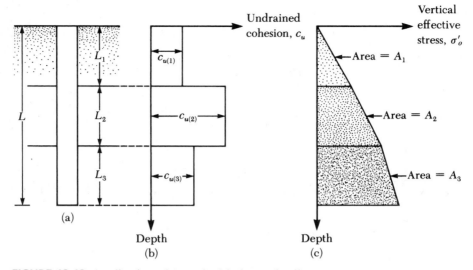

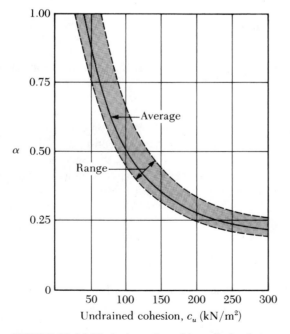

FIGURE 13.13 Application of λ method in layered soil

FIGURE 13.14 Variation of α with undrained cohesion of clay

and

$$K = (1 - \sin \phi_R)\sqrt{OCR} \qquad \text{(for overconsolidated clays)} \qquad (13.28)$$

where OCR = overconsolidation ratio.

Combining Eqs. (13.25), (13.26), (13.27), and (13.28) for normally consolidated clays yields

$$f = (1 - \sin \phi_R) \tan \phi_R \sigma_o' \qquad (13.29)$$

and for overconsolidated clays,

$$f = (1 - \sin \phi_R) \tan \phi_R \sqrt{OCR}\,\sigma_o' \qquad (13.30)$$

With the value of f determined, the total frictional resistance may be evaluated as

$$Q_s = \Sigma\, fp\, \Delta L$$

Allowable Pile Capacity

After the total ultimate load-carrying capacity of a pile has been determined by summing the point bearing capacity and the frictional (or skin) resistance, a reasonable factor of safety should be used to obtain the total allowable load for each pile, or

$$Q_{\text{all}} = \frac{Q_u}{FS} \qquad (13.31)$$

where Q_{all} = allowable load-carrying capacity for each pile
FS = factor of safety

The factor of safety generally used ranges from 2.5 to 4, depending on the uncertainties of the ultimate load calculation.

EXAMPLE 13.1

A fully embedded precast concrete pile 12 m long is driven into a homogeneous sand layer ($c = 0$). The pile is square in cross section with sides measuring 305 mm. The dry unit weight of sand, γ_d, is 16.8 kN/m³, the average soil friction angle is 35°, and the corrected standard penetration resistance near the vicinity of the pile tip is 16. Calculate the ultimate point load on the pile.

    a. Use Meyerhof's method with Eq. (13.11).
    b. Use Meyerhof's method with Eq. (13.13).

Solution

    a. This soil is homogeneous, so $L_b = L$. For $\phi = 35°$, $N_q^* \approx 120$. Thus,

$$q' = \gamma_d L = (16.8)(12) = 201.6 \text{ kN/m}^2$$

$$A_p = \frac{305 \times 305}{1000 \times 1000} = 0.0929 \text{ m}^2$$

$$Q_p = A_p q' N_q^* = (0.0929)(201.6)(120) = 2247.4 \text{ kN}$$

However, from Eq. (13.12), we have

$$q_l = 50 N_q^* \tan \phi = 50(120) \tan 35° = 4201.25 \text{ kN/m}^2$$

so

$$Q_p = A_p q_l = (0.0929)(4201.25) = 390.3 \text{ kN} < A_p q' N_q^*$$

and

$$Q_p \approx \textbf{390 kN}$$

b. The average standard penetration resistance near the pile tip is 16. So, from Eq. (13.13), we have

$$q_p = 40 N_{cor} \frac{L}{D} \leq 400 N_{cor}$$

$$\frac{L}{D} = \frac{12}{0.305} = 39.34$$

$$Q_p = A_p q_p = (0.0929)(40)(16)39.34 = 2339 \text{ kN}$$

However, the limiting value is

$$Q_p = A_p 400 N_{cor} = (0.0929)(400)(16) = 594.6 \text{ kN} \approx \textbf{595 kN} \qquad \blacksquare$$

EXAMPLE 13.2

Consider a precast concrete pile 12 m long in a homogeneous soil layer. The pile cross section = 305 mm × 305 mm, the unit weight of sand, $\gamma_d = 16.8 \text{ kN/m}^3$, and the soil friction angle, $\phi = 35°$. Determine the total frictional resistance. Use Eqs. (13.15), (13.16), and (13.17). Also use $K = 1.4$ and $\delta = 0.6\phi$.

Solution   The unit skin friction at any depth is given by Eq. (13.16) as

$$f = K\sigma_o' \tan \delta$$

Also from Eq. (13.17), we have

$$L' = 15D$$

So, for depth $z = 0\text{–}15D$, $\sigma_o' = \gamma z = 16.8z$ (kN/m²), and beyond $z \geq 15D$, $\sigma_o' = \gamma(15D) = (16.8)(15 \times 0.305) = 76.86 \text{ kN/m}^2$. This result is shown in Figure 13.15. The *frictional resistance from $z = 0$ to $15D$* is

$$Q_s = pL'f_{av} = [(4)(0.305)][15D]\left[\frac{(1.4)(76.86)\tan(0.6 \times 35)}{2}\right]$$

$$= (1.22)(4.575)(20.65) = 115.26 \text{ kN}$$

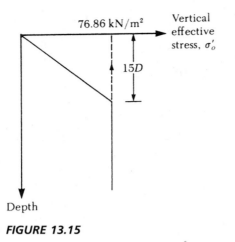

FIGURE 13.15

The *frictional resistance from* $z = 15D$ to 12 m is

$$Q_s = p(L - L')f_{z=15D} = [(4)(0.305)][12 - 4.575][(1.4)(76.86) \tan (0.6 \times 35)]$$

$$= (1.22)(7.425)(41.3) = 374.1 \text{ kN}$$

So, the total frictional resistance is $115.26 + 374.1 = 489.35 \text{ kN} \approx \mathbf{490 \text{ kN}}$ ∎

EXAMPLE 13.3

A concrete pile 458 mm $\times$ 458 mm in cross section is embedded in a saturated clay. The length of embedment is 16 m. The undrained cohesion, c_u, of clay is 60 kN/m^2, and the unit weight of clay is 18 kN/m^3. Use a factor of safety of 5 to determine the allowable load the pile can carry.

    a. Use the α method.
    b. Use the λ method.

Solution

    a. From Eq. (13.14),

$$Q_p = A_p q_p = A_p c_u N_c^* = (0.458 \times 0.458)(60)(9) = 113.3 \text{ kN}$$

From Eqs. (13.23) and (13.24),

$$Q_s = \alpha c_u p L$$

From the average plot of Figure 13.14 for $c_u = 60$ kN/m^2, $\alpha \approx 0.77$ and

$$Q_s = (0.77)(60)(4 \times 0.458)(16) = 1354 \text{ kN}$$

$$Q_{\text{all}} = \frac{Q_p + Q_s}{FS} = \frac{113.3 + 1354}{5} \approx \mathbf{294 \text{ kN}}$$

b. From Eq. (13.21),

$$f_{av} = \lambda(\overline{\sigma}'_o + 2c_u)$$

We are given $L = 16.0$ m. From Figure 13.12 for $L = 16$ m, $\lambda \approx 0.2$, so

$$f_{av} = 0.2\left[\left(\frac{18 \times 16}{2}\right) + 2(60)\right] = 52.8 \quad \text{kN/m}^2$$

$$Q_s = pLf_{av} = (4 \times 0.458)(16)(52.8) = 1548 \text{ kN}$$

As in part a, $Q_p = 113.3$, so

$$Q_{all} = \frac{Q_p + Q_s}{FS} = \frac{113.3 + 1548}{5} = \textbf{332 kN} \qquad \blacksquare$$

EXAMPLE 13.4

A driven pile in clay is shown in Figure 13.16a. The pile has a diameter of 406 mm.

a. Calculate the net point bearing capacity. Use Eq. (13.14).
b. Calculate the skin resistance (1) by using Eqs. (13.23) and (13.24) (α method), (2) by using Eq. (13.21) (λ method), and (3) by using Eq. (13.25) (β method). For all clay layers, $\phi_R = 30°$. The top 10 m of clay is normally consolidated. The bottom clay layer has an *OCR* of 2.
c. Estimate the net allowable pile capacity. Use $FS = 4$.

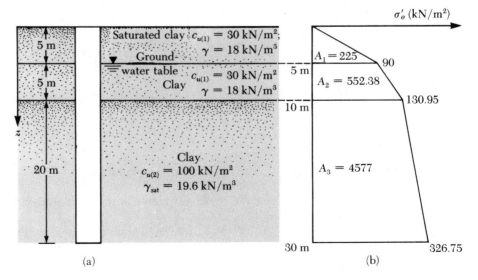

FIGURE 13.16

Solution   The area of the cross section of the pile is

$$A_p = \frac{\pi}{4} D^2 = \frac{\pi}{4} (0.406)^2 = 0.1295 \text{ m}^2$$

Part a: Calculation of Net Point Bearing Capacity   From Eq. (13.14), we have

$$Q_p = A_p q_p = A_p N_c^* c_{u(2)} = (0.1295)(9)(100) = \textbf{116.55 kN}$$

Part b: Calculation of Skin Resistance

(1) From Eq. (13.24),

$$Q_s = \Sigma \, \alpha c_u p \, \Delta L$$

For the top soil layer, $c_{u(1)} = 30$ kN/m². According to the average plot of Figure 13.14, $\alpha_1 = 1.0$. Similarly, for the bottom soil layer, $c_{u(2)} = 100$ kN/m²; $\alpha_2 = 0.5$. Thus,

$$\begin{aligned}
Q_s &= \alpha_1 c_{u(1)}[(\pi)(0.406)]10 + \alpha_2 c_{u(2)}[(\pi)(0.406)]20 \\
&= (1)(30)[(\pi)(0.406)]10 + (0.5)(100)[(\pi)(0.406)]20 \\
&= 382.7 + 1275.5 = \textbf{1658.2 kN}
\end{aligned}$$

(2) The average value of c_u is

$$\frac{c_{u(1)}(10) + c_{u(2)}(20)}{30} = \frac{(30)(10) + (100)(20)}{30} = 76.7 \text{ kN/m}^2$$

To obtain the average value of $\bar{\sigma}_o'$, the diagram for vertical effective stress variation with depth is plotted in Figure 13.16b. From Eq. (13.22),

$$\bar{\sigma}_o' = \frac{A_1 + A_2 + A_3}{L} = \frac{225 + 552.38 + 4577}{30} = 178.48 \text{ kN/m}^2$$

The magnitude of λ from Figure 13.12 is 0.14. So

$$f_{av} = 0.14[178.48 \times (2)(76.7)] = 46.46 \text{ kN/m}^2$$

Hence,

$$Q_s = pLf_{av} = \pi(0.406)(30)(46.46) = \textbf{1777.8 kN}$$

(3) The top clay layer (10 m) is normally consolidated and $\phi_R = 30°$.
    For $z = 0$–5 m [Eq. (13.29)],

$$\begin{aligned}
f_{av(1)} &= (1 - \sin \phi_R)\tan \phi_R \sigma_{o(av)}' \\
&= (1 - \sin 30°)(\tan 30°) \left(\frac{0 + 90}{2} \right) = 13.0 \text{ kN/m}^2
\end{aligned}$$

Similarly, for $z = 5$–10 m,

$$f_{av(2)} = (1 - \sin 30°)(\tan 30°)\left(\frac{90 + 130.95}{2}\right) = 31.9 \text{ kN/m}^2$$

For $z = 10$–30 m [Eq. (13.30)],

$$f_{av} = (1 - \sin \phi_R)\tan \phi_R \sqrt{OCR}\ \sigma'_{o(av)}$$

For $OCR = 2$,

$$f_{av(3)} = (1 - \sin 30°)(\tan 30°)\sqrt{2}\left(\frac{130.95 + 326.75}{2}\right) = 93.43 \text{ kN/m}^2$$

So

$$Q_s = p[f_{av(1)}(5) + f_{av(2)}(5) + f_{av(3)}(20)]$$
$$= (\pi)(0.406)[(13)(5) + (31.9)(5) + (93.43)(20)] = \mathbf{2669.7\ kN}$$

Part c: Calculation of Net Ultimate Capacity, Q_u Comparing the three values of Q_s shows that the α and λ methods give similar results. So we use

$$Q_s = \frac{1658.2 + 1777.8}{2} \approx 1718 \text{ kN}$$

Thus,

$$Q_u = Q_p + Q_s = 116.55 + 1718 = 1834.55 \text{ kN}$$

$$Q_{all} = \frac{Q_u}{FS} = \frac{1834.55}{4} = \mathbf{458.6\ kN} \quad\blacksquare$$

13.7 Load-Carrying Capacity of Pile Point Resting on Rock

Sometimes piles are driven to an underlying layer of rock. In such cases, the engineer must evaluate the bearing capacity of the rock. The ultimate unit point resistance in rock (Goodman, 1980) is approximately

$$q_p = q_{u\text{-}R}(N_\phi + 1) \tag{13.32}$$

where $N_\phi = \tan^2 (45 + \phi/2)$
   $q_{u\text{-}R} =$ unconfined compression strength of rock
   $\phi =$ drained angle of friction

The unconfined compression strength of rock can be determined by laboratory tests on rock specimens collected during field investigation. However, extreme caution

Table 13.8 Typical
unconfined compressive
strength of rocks

Rock type	$q_{u\text{-}R}$ (MN/m²)
Sandstone	70–140
Limestone	105–210
Shale	35–70
Granite	140–210
Marble	60–70

should be used in obtaining the proper value of $q_{u\text{-}R}$ because laboratory specimens are usually small in diameter. As the diameter of the specimen increases, the unconfined compression strength decreases, which is referred to as the *scale effect*. For specimens larger than about 1 m in diameter, the value of $q_{u\text{-}R}$ remains approximately constant. There appears to be a fourfold to fivefold reduction in the magnitude of $q_{u\text{-}R}$ in this process. The scale effect in rock is primarily caused by randomly distributed large and small fractures and also by progressive ruptures along the slip lines. Hence, we always recommend that

$$q_{u\text{-}R(\text{design})} = \frac{q_{u\text{-}R(\text{lab})}}{5}$$

(13.33)

Table 13.8 lists some representative values of (laboratory) unconfined compression strengths of rock. Representative values of the rock friction angle, ϕ, are given in Table 13.9.

A factor of safety of *at least* 3 should be used to determine the allowable load-carrying capacity of the pile point. Thus,

$$Q_{p(\text{all})} = \frac{[q_{u\text{-}R}(N_\phi + 1)]A_p}{FS}$$

(13.34)

Table 13.9 Typical values of angle of friction, ϕ, of rocks

Rock type	Angle of friction, ϕ (deg)
Sandstone	27–45
Limestone	30–40
Shale	10–20
Granite	40–50
Marble	25–30

EXAMPLE 13.5

An H-pile that has a length of embedment of 26 m is driven through a soft clay layer to rest on sandstone. The sandstone has a laboratory unconfined compression strength of 76 MN/m² and a friction angle of 28°. Use a factor of safety of 5 and estimate the allowable point bearing capacity, given $A_p = 15.9 \times 10^{-3}$ m².

Solution   From Eqs. (13.33) and (13.34),

$$Q_{p(\text{all})} = \frac{\left\{ \left[\dfrac{q_{u\text{-}R(\text{lab})}}{5} \right] \left[\tan^2 \left(45 + \dfrac{\phi}{2} \right) + 1 \right] \right\} A_p}{FS}$$

$$= \frac{\left\{ \left[\dfrac{76 \times 10^3 \text{ kN/m}^2}{5} \right] \left[\tan^2 \left(45 + \dfrac{28}{2} \right) + 1 \right] \right\} (15.9 \times 10^{-3} \text{ m}^2)}{5}$$

$$= \mathbf{182 \ kN} \qquad \blacksquare$$

13.8   *Settlement of Piles*

The elastic settlement of a pile under a vertical working load, Q_w, is determined by three factors:

$$S_e = S_{e(1)} + S_{e(2)} + S_{e(3)} \tag{13.35}$$

where S_e = total pile settlement
   $S_{e(1)}$ = settlement of pile shaft
   $S_{e(2)}$ = settlement of pile caused by the load at the pile point
   $S_{e(3)}$ = settlement of pile caused by the load transmitted along the pile shaft

Determination of $S_{e(1)}$

If the pile material is assumed to be elastic, the deformation of the pile shaft can be evaluated using the fundamental principles of mechanics of materials:

$$S_{e(1)} = \frac{(Q_{wp} + \xi Q_{ws})L}{A_p E_p} \tag{13.36}$$

where Q_{wp} = load carried at the pile point under working load condition
   Q_{ws} = load carried by frictional (skin) resistance under working load condition
   A_p = area of the pile cross section
   L = length of the pile
   E_p = modulus of elasticity of the pile material

The magnitude of ξ depends on the nature of the unit friction (skin) resistance distribution along the pile shaft. If the distribution of f is uniform or parabolic, as shown in Figures 13.17a and b, $\xi = 0.5$. However, for a triangular distribution of f (Figure 13.17c), the magnitude of ξ is about 0.67 (Vesic, 1977).

Determination of $S_{e(2)}$

The settlement of a pile caused by the load carried at the pile point may be expressed as

$$S_{e(2)} = \frac{q_{wp}D}{E_s}(1 - \mu_s^2)I_{wp} \tag{13.37}$$

where D = width or diameter of pile
q_{wp} = point load per unit area at the pile point = Q_{wp}/A_p
E_s = modulus of elasticity of soil at or below the pile point
μ_s = Poisson's ratio of soil
I_{wp} = influence factor ≈ 0.85

Vesic (1977) also proposed a semiempirical method to obtain the magnitude of the settlement, $S_{e(2)}$:

$$S_{e(2)} = \frac{Q_{wp}C_p}{Dq_p} \tag{13.38}$$

where q_p = ultimate point resistance of the pile
C_p = an empirical coefficient

Representative values of C_p for various soils are given in Table 13.10.

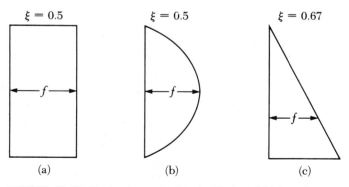

$\xi = 0.5$ $\xi = 0.5$ $\xi = 0.67$

(a)   (b)   (c)

FIGURE 13.17 Various types of unit friction (skin) resistance distribution along the pile shaft

Table 13.10 Typical values of C_p as recommended by Vesic (1977) [Eq. (13.38)]

Soil type	Driven pile	Bored pile
Sand (dense to loose)	0.02–0.04	0.09–0.18
Clay (stiff to soft)	0.02–0.03	0.03–0.06
Silt (dense to loose)	0.03–0.05	0.09–0.12

Determination of $S_{e(3)}$

The settlement of a pile caused by the load carried along the pile shaft is given by a relation similar to Eq. (13.37), or

$$S_{e(3)} = \left(\frac{Q_{ws}}{pL}\right)\frac{D}{E_s}(1 - \mu_s^2)I_{ws} \tag{13.39}$$

where p = perimeter of the pile
L = embedded length of the pile
I_{ws} = influence factor

Note that the term Q_{ws}/pL in Eq. (13.39) is the average value of f along the pile shaft. The influence factor, I_{ws}, has a simple empirical relation (Vesic, 1977):

$$I_{ws} = 2 + 0.35\sqrt{\frac{L}{D}} \tag{13.40}$$

Vesic (1977) also proposed a simple empirical relation similar to Eq. (13.38) for obtaining $S_{e(3)}$:

$$S_{e(3)} = \frac{Q_{ws}C_s}{Lq_p} \tag{13.41}$$

where C_s = an empirical constant = $(0.93 + 0.16\sqrt{L/D})C_p$. $\qquad$ (13.42)

The values of C_p for use in Eq. (13.41) may be estimated from Table 13.10.

EXAMPLE 13.6

A 12-m-long precast concrete pile is fully embedded in sand. The cross section of the pile measures 0.305 m × 0.305 m. The allowable working load for the pile is 337 kN, of which 240 kN is contributed by skin friction. Determine the elastic settlement of the pile for $E_p = 21 \times 10^6$ kN/m², $E_s = 30,000$ kN/m², and $\mu_s = 0.3$.

Solution   We will use Eq. (13.35):

$$S_e = S_{e(1)} + S_{e(2)} + S_{e(3)}$$

From Eq. (13.36),

$$S_{e(1)} = \frac{(Q_{wp} + \xi Q_{ws})L}{A_p E_p}$$

Let $\xi = 0.6$ and $E_p = 21 \times 10^6$ kN/m^2. Then

$$S_{e(1)} = \frac{[97 + (0.6)(240)]12}{(0.305)^2(21 \times 10^6)} = 0.00148 \text{ m} = 1.48 \text{ mm}$$

From Eq. (13.37),

$$S_{e(2)} = \frac{q_{wp}D}{E_s}(1 - \mu_s^2)I_{wp}$$

$$I_{wp} = 0.85$$

$$q_{wp} = \frac{Q_{wp}}{A_p} = \frac{97}{(0.305)^2} = 1042.7 \text{ kN/m}^2$$

So

$$S_{e(2)} = \left[\frac{(1042.7)(0.305)}{30,000}\right](1 - 0.3^2)(0.85) = 0.0082 \text{ m} = 8.2 \text{ mm}$$

Again, from Eq. (13.39),

$$S_{e(3)} = \left(\frac{Q_{ws}}{pL}\right)\frac{D}{E_s}(1 - \mu_s^2)I_{ws}$$

$$I_{ws} = 2 + 0.35\sqrt{\frac{L}{D}} = 2 + 0.35\sqrt{\frac{12}{0.305}} = 4.2$$

So

$$S_{e(3)} = \frac{240}{(\pi \times 0.305)(12)}\left(\frac{0.305}{30,000}\right)(1 - 0.3^2)(4.2) = 0.00081 \text{ m} = 0.81 \text{ mm}$$

Hence, the total settlement is

$$S_e = 1.48 + 8.2 + 0.81 = \textbf{10.49 mm} \qquad \blacksquare$$

13.9   *Pile-Driving Formulas*

To develop the desired load-carrying capacity, a point bearing pile must penetrate the dense soil layer sufficiently or have sufficient contact with a layer of rock. This requirement cannot always be satisfied by driving a pile to a predetermined depth because soil profiles vary. For that reason, several equations have been developed to calculate the ultimate capacity of a pile during driving. These dynamic equations are widely used in the field to determine whether the pile has reached a satisfactory bearing value at the predetermined depth. One of the earliest of these dynamic equations—commonly referred to as the *Engineering News Record (ENR) formula*—is derived from the work–energy theory; that is,

energy imparted by the hammer per blow
= (pile resistance)(penetration per hammer blow)

According to the ENR formula, the pile resistance is the ultimate load, Q_u, expressed as

$$Q_u = \frac{W_R h}{S + C} \tag{13.43}$$

where W_R = weight of the ram
      h = height of fall of the ram
      S = penetration of the pile per hammer blow
      C = a constant

The pile penetration, S, is usually based on the average value obtained from the last few driving blows. In the equation's original form, the following values of C were recommended:

For drop hammers: $C = 2.54$ cm (if the units of S and h are in centimeters)
For steam hammers: $C = 0.254$ m (if the units of S and h are in centimeters)

Also, a factor of safety of $FS = 6$ was recommended to estimate the allowable pile capacity. Note that, for single- and double-acting hammers, the term $W_R h$ can be replaced by EH_E (where E = hammer efficiency and H_E = rated energy of the hammer). Thus,

$$Q_u = \frac{EH_E}{S + C} \tag{13.44}$$

The ENR pile-driving formula has been revised several times over the years. A recent form—the *modified ENR formula*—is

$$Q_u = \frac{EW_R h}{S + C} \frac{W_R + n^2 W_p}{W_R + W_p} \tag{13.45}$$

where E = hammer efficiency
      C = 0.254 cm if the units of S and h are in centimeters
     W_p = weight of the pile
      n = coefficient of restitution between the ram and the pile cap

The efficiencies of various pile-driving hammers, E, are in the following ranges:

Hammer type	Efficiency, E
Single- and double-acting hammers	0.7–0.85
Diesel hammers	0.8–0.9
Drop hammers	0.7–0.9

Representative values of the coefficient of restitution, n, follow:

Pile material	Coefficient of restitution, n
Cast iron hammer and concrete piles (without cap)	0.4–0.5
Wood cushion on steel piles	0.3–0.4
Wooden piles	0.25–0.3

A factor of safety of 4 to 6 may be used in Eq. (13.45) to obtain the allowable load-bearing capacity of a pile.

The Michigan State Highway Commission (1965) undertook a study to obtain a rational pile-driving equation. At three diverse sites, a total of 88 piles were driven. Based on these tests, Michigan adopted a modified ENR formula:

$$Q_u = \frac{1.25 H_E}{S + C} \frac{W_R + n^2 W_p}{W_R + W_p} \tag{13.46}$$

where H_E = manufacturer's maximum rated hammer energy (kN-m)
$\quad\quad C = 2.54 \times 10^{-3}$ m

The unit of S is meters in Eq. (13.46). A factor of safety of 6 is recommended.

Another equation referred to as the *Danish formula* also yields results as reliable as any other equation's:

$$Q_u = \frac{EH_E}{S + \sqrt{\dfrac{EH_E L}{2A_p E_p}}} \tag{13.47}$$

where E = hammer efficiency
$\quad\quad H_E$ = rated hammer energy
$\quad\quad E_p$ = modulus of elasticity of the pile material
$\quad\quad L$ = length of the pile
$\quad\quad A_p$ = area of the pile cross section

Consistent units must be used in Eq. (13.47). A factor of safety varying from 3 to 6 is recommended to estimate the allowable load-bearing capacity of piles.

Other frequently used equations for pile driving are those given by the Pacific Coast Uniform Building Code (International Conference of Building Officials, 1982) and by Janbu (1953).

Pacific Coast Uniform Building Code Formula

$$Q_u = \frac{(EH_E)\left(\dfrac{W_R + nW_p}{W_R + W_p}\right)}{S + \dfrac{Q_u L}{AE}} \tag{13.48}$$

The value of n in Eq. (13.48) should be 0.25 for steel piles and 0.1 for all other piles. A factor of safety of 4 is generally recommended.

Janbu's Formula

$$Q_u = \frac{EH_E}{K'_u S} \tag{13.49}$$

where $K'_u = C_d(1 + \sqrt{1 + \lambda/C_d})$ \hfill (13.50)

$C_d = 0.75 + 0.15(W_p/W_R)$ \hfill (13.51)

$\lambda = (EH_E L/A_p E_p S^2)$ \hfill (13.52)

A factor of safety of 4 to 5 is generally recommended.

EXAMPLE 13.7

A precast concrete pile 305 mm $\times$ 305 mm in cross section is driven by a Vulcan hammer (Model No. 08). We have these values:

maximum rated hammer energy = 35 kN-m (Table 13.6)

weight of ram = 36 kN (Table 13.6)

total length of pile = 20 m

hammer efficiency = 0.8

coefficient of restitution = 0.45

weight of pile cap = 3.2 kN

number of blows for last 25.4 mm of penetration = 5

Estimate the allowable pile capacity by using each of these equations:

a. Eq. (13.44) (use $FS = 6$)
b. Eq. (13.45) (use $FS = 5$)
c. Eq. (13.47) (use $FS = 4$)

Solution

a. Eq. (13.44) is

$$Q_u = \frac{EH_E}{S + C}$$

We have $E = 0.8$, $H_E = 35$ kN-m, and

$$S = \frac{25.4}{5} = 5.08 \text{ mm} = 0.508 \text{ cm}$$

So

$$Q_u = \frac{(0.8)(35)(100)}{0.508 + 0.254} = 3674 \text{ kN}$$

Hence,

$$Q_{all} = \frac{Q_u}{FS} = \frac{3674}{6} = \textbf{612 kN}$$

b. Eq. (13.45) is

$$Q_u = \frac{EW_R h}{S + C} \frac{W_R + n^2 W_p}{W_R + W_p}$$

Weight of pile $= LA_p \gamma_c = (20)(0.305)^2(23.58) = 43.87$ kN and

$$W_p = \text{weight of pile} + \text{weight of cap} = 43.87 + 3.2 = 47.07 \text{ kN}$$

So

$$Q_u = \left[\frac{(0.8)(35)(100)}{0.508 + 0.254}\right]\left[\frac{36 + (0.45)^2(47.07)}{36 + 47.07}\right]$$

$$= (3674)(0.548) \approx 2013 \text{ kN}$$

$$Q_{all} = \frac{Q_u}{FS} = \frac{2013}{5} = 402.6 \text{ kN} \approx \textbf{403 kN}$$

c. Eq. (13.47) is

$$Q_u = \frac{EH_E}{S + \sqrt{\dfrac{EH_E L}{2A_p E_p}}}$$

We have $E_p \approx 20.7 \times 10^6$ kN/m². So

$$\sqrt{\frac{EH_EL}{2A_pE_p}} = \sqrt{\frac{(0.8)(35)(20)}{(2)(0.305)^2(20.7 \times 10^6)}} = 0.0121 \text{ m} = 1.21 \text{ cm}$$

Hence,

$$Q_u = \frac{(0.8)(35)(100)}{0.508 + 1.21} = 1630 \text{ kN}$$

$$Q_{\text{all}} = \frac{Q_u}{FS} = \frac{1630}{4} = \textbf{407.5 kN} \qquad \blacksquare$$

13.10 *Stress on Piles During Pile Driving*

The maximum stress developed on a pile during the driving operation can be estimated from the pile-driving formulas presented in the preceding section. To illustrate, we use the modified ENR formula given in Eq. (13.45)

$$Q_u = \frac{EW_Rh}{S + C} \frac{W_R + n^2W_p}{W_R + W_p}$$

In this equation, S equals the average penetration per hammer blow, which can also be expressed as

$$S = \frac{2.54}{N} \qquad (13.53)$$

where S is in centimeters
N = number of hammer blows per inch of penetration

Thus,

$$Q_u = \frac{EW_Rh}{\dfrac{2.54}{N} + 0.254 \text{ (cm)}} \frac{W_R + n^2W_p}{W_R + W_p} \qquad (13.54)$$

Different values of N may be assumed for a given hammer and pile and Q_u calculated. The driving stress can then be calculated for each value of N and Q_u/A_p. This procedure can be demonstrated with a set of numerical values. Assume that a prestressed concrete pile 25 m in length has to be driven by an 11B3 (MKT) hammer. The pile sides measure 254 mm. From Table 13.3 for this pile, we have

$$A_p = 645 \text{ cm}^2$$

The weight of the pile is

$$W_p = (A_pL)\gamma_c = \left(\frac{645}{10,000}\right)(25)(23.58) \approx 38 \text{ kN}$$

Let the weight of the cap be 3 kN. So, $W_p = 38 + 3 = 41$ kN. Again, from Table 13.6, for an 11B3 hammer,

$$\text{rated energy} = 26 \text{ kN-m} = 26 \times 100 \text{ kN-cm} = H_E = W_R h$$

The weight of the ram is 22 kN. Assume that the hammer efficiency, $E = 0.85$, and $n = 0.35$. Substituting these values into Eq. (13.54), we get

$$Q_u = \left[\frac{0.85(26 \times 100)}{\dfrac{2.54}{N} + 0.254} \right] \left[\frac{22 + 0.35^2(41)}{22 + 41} \right] = \frac{948}{\dfrac{2.54}{N} + 0.254}$$

We can substitute various values of N into the preceding relationship and calculate Q_u. Figure 13.18 shows a plot of Q_u/A_p (pile-driving stress) versus N. From such a curve, the number of blows per 2.54 cm of pile penetration corresponding to the allowable pile-driving stress can be determined easily.

In practice, the driving stresses in wooden piles are limited to about $0.7 f_u$. Similarly, for concrete and steel piles, the driving stresses are limited to about $0.6 f_c'$ and $0.85 f_y$, respectively.

In most cases, wooden piles are driven with a hammer energy of less than 60 kN-m. The driving resistances are mostly limited to 4 to 5 blows per 2.54 cm of pile penetration. For concrete and steel piles, the usual N values adopted are 6–8 and 12–14, respectively.

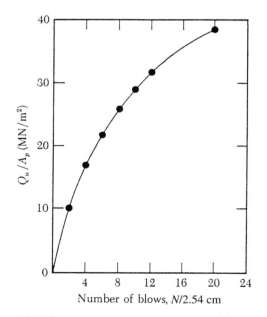

FIGURE 13.18 Plot of pile-driving stress versus number of blow counts per 2.54 cm

13.11   Pile Load Tests

In most large projects, a specific number of load tests must be conducted on piles. The primary reason is the unreliability of prediction methods. The vertical and lateral load-bearing capacity of a pile can be tested in the field. Figure 13.19a shows a schematic diagram of the pile load test arrangement for testing in *axial compression* in the field. The load is applied to the pile by a hydraulic jack. Step loads are

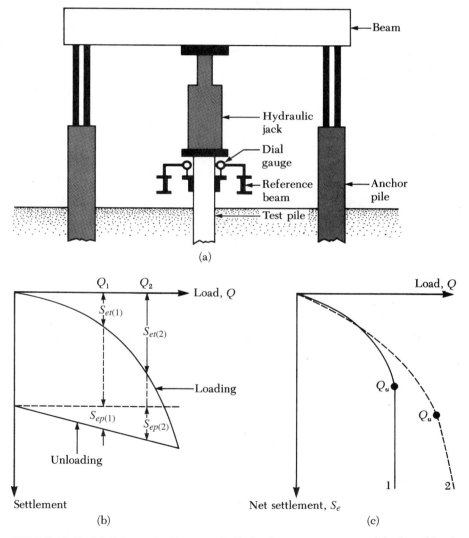

(a)

(b)

(c)

FIGURE 13.19 (a) Schematic diagram of pile load test arrangement; (b) plot of load against total settlement; (c) plot of load against net settlement

applied to the pile, and sufficient time is allowed to elapse after each load so that a small amount of settlement occurs. The settlement of the pile is measured by dial gauges. The amount of load to be applied for each step will vary, depending on local building codes. Most building codes require that each step load be about one-fourth of the proposed working load. The load test should be carried out to a total load of at least two times the proposed working load. After reaching the desired pile load, the pile is gradually unloaded.

Load tests on piles in sand can be carried out immediately after the piles are driven. However, care should be taken in deciding the time lapse between driving and starting the load test when piles are embedded in clay. This time lapse can range from 30 to 60 days or more because the soil requires some time to gain its *thixotropic strength.*

Figure 13.19b shows a load settlement diagram obtained from field loading and unloading. For any load, Q, the net pile settlement can be calculated as follows: When $Q = Q_1$,

$$\text{net settlement, } S_{e(1)} = S_{et(1)} - S_{ep(1)}$$

When $Q = Q_2$,

$$\text{net settlement, } S_{e(2)} = S_{et(2)} - S_{ep(2)}$$

and so on,

where S_e = net settlement
$\quad\quad\quad S_{ep}$ = elastic settlement of the pile itself
$\quad\quad\quad S_{et}$ = total settlement

These values of Q can be plotted on a graph against the corresponding net settlement, S_e, as shown in Figure 13.19c. The ultimate load of the pile can be determined from this graph. Pile settlement may increase with load to a certain point, beyond which the load–settlement curve becomes vertical. The load corresponding to the point where the Q–S_e curve becomes vertical is the ultimate load, Q_u, for the pile; it is shown by curve 1 in Figure 13.19c. In many cases, the latter stage of the load–settlement curve is almost linear, showing a large degree of settlement for a small increment of load; it is shown by curve 2 in Figure 13.19c. The ultimate load, Q_u, for such a case is determined from the point of the Q–S_e curve where this steep linear portion starts.

The load test procedure just described requires the application of step loads on the piles and measurement of settlement and is called a *load-controlled* test. Another technique used for the pile load test is the *constant-rate-of-penetration* test. In it, the load on the pile is continuously increased to maintain a constant rate of penetration, which can vary from 0.25 to 2.5 mm/min. This test gives a load–settlement plot similar to that obtained from the load-controlled test. Another type of pile load test is *cyclic loading,* in which an incremental load is repeatedly applied and removed.

Negative Skin Friction

Negative skin friction is a downward drag force exerted on the pile by the soil surrounding it. This action can occur under conditions such as the following:

1. If a fill of clay soil is placed over a granular soil layer into which a pile is driven, the fill will gradually consolidate. This consolidation process will exert a downward drag force on the pile (Figure 13.20a) during the period of consolidation.
2. If a fill of granular soil is placed over a layer of soft clay, as shown in Figure 13.20b, it will induce the process of consolidation in the clay layer and thus exert a downward drag on the pile.
3. Lowering of the water table will increase the vertical effective stress on the soil at any depth, which will induce consolidation settlement in clay. If a pile is located in the clay layer, it will be subjected to a downward drag force.

In some cases, the downward drag force may be excessive and cause foundation failure. This section outlines two tentative methods for calculating negative skin friction.

Clay Fill over Granular Soil (Figure 13.20a)

Similar to the β method presented in Section 13.6, the negative (downward) skin stress on the pile is

$$f_n = K' \sigma'_o \tan \delta \tag{13.55}$$

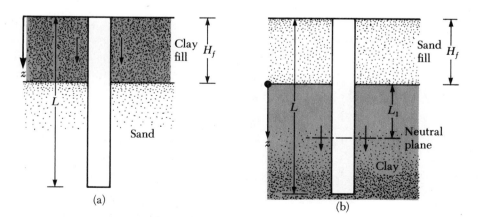

FIGURE 13.20 Negative skin friction

where K' = earth pressure coefficient = $K_o = 1 - \sin\phi$
σ'_o = vertical effective stress at any depth $z = \gamma'_f z$
γ'_f = effective unit weight of fill ϕ
δ = soil–pile friction angle $\approx 0.5\phi$–0.7ϕ

Hence, the total downward drag force, Q_n, on a pile is

$$Q_n = \int_0^{H_f} (pK'\gamma'_f \tan\delta)z\, dz = \frac{pK'\gamma'_f \tan\delta}{2} \tag{13.56}$$

where H_f = height of the fill. If the fill is above the water table, the effective unit weight, γ'_f, should be replaced by the moist unit weight.

Granular Soil Fill over Clay (Figure 13.20b)

In this case, the evidence indicates that the negative skin stress on the pile may exist from $z = 0$ to $z = L_1$, which is referred to as the *neutral depth* (see Vesic, 1977, pp. 25–26, for discussion). The neutral depth may be given as (Bowles, 1982)

$$L_1 = \frac{L - H_f}{L_1}\left(\frac{L - H_f}{2} + \frac{\gamma'_f H_f}{\gamma'}\right) - \frac{2\gamma'_f H_f}{\gamma'} \tag{13.57}$$

where γ'_f and γ' = effective unit weights of the fill and the underlying clay layer, respectively.

Once the value of L_1 is determined, the downward drag force is obtained in the following manner: The unit negative skin friction at any depth from $z = 0$ to $z = L_1$ is

$$f_n = K'\sigma'_o \tan\delta \tag{13.58}$$

where $K' = K_o = 1 - \sin\phi$
$\sigma'_o = \gamma'_f H_f + \gamma'z$
$\delta = 0.5\phi$–0.7ϕ

Hence, the total drag force is

$$Q_n = \int_0^{L_1} pf_n\, dz = \int_0^{L_1} pK'(\gamma'_f H_f + \gamma'z)\tan\delta\, dz$$
$$= (pK'\gamma'_f H_f \tan\delta)L_1 + \frac{L_1^2 pK'\gamma' \tan\delta}{2} \tag{13.59}$$

If the soil and the fill are above the water table, the effective unit weights should be replaced by moist unit weights. In some cases, the piles can be coated with bitumen in the down-drag zone to avoid this problem.

EXAMPLE 13.8

Refer to Figure 13.20a; $H_f = 2$ m. The pile is circular in cross section with a diameter of 0.305 m. For the fill that is above the water table, $\gamma_f = 16$ kN/m^3 and $\phi = 32°$. Determine the total drag force.

Solution   From Eq. (13.56), we have

$$Q_n = \frac{pK'\gamma_f' H_f^2 \tan \delta}{2}$$

$$p = \pi(0.305) = 0.958 \text{ m}$$

$$K' = 1 - \sin \phi = 1 - \sin 32° = 0.47$$

$$\delta = (0.6)(32) = 19.2°$$

$$Q_n = \frac{(0.958)(0.47)(16)(2)^2 \tan 19.2°}{2} = \textbf{5.02 kN} \qquad \blacksquare$$

EXAMPLE 13.9

Refer to Figure 13.20b. Here, $H_f = 2$ m, pile diameter $= 0.305$ m, $\gamma_f = 16.5$ kN/m^3, $\phi_{clay} = 34°$, $\gamma_{sat(clay)} = 17.2$ kN/m^3, and $L = 20$ m. The water table coincides with the top of the clay layer. Determine the downward drag force.

Solution   The depth of the neutral plane is given in Eq. (13.57) as

$$L_1 = \frac{L - H_f}{L_1}\left(\frac{L - H_f}{2} + \frac{\gamma_f H_f}{\gamma'}\right) - \frac{2\gamma_f H_f}{\gamma'}$$

Note that γ_f' in Eq. (13.57) has been replaced by γ_f because the fill is above the water table. So

$$L_1 = \frac{20 - 2}{L_1}\left[\frac{(20 - 2)}{2} + \frac{(16.5)(2)}{(17.2 - 9.81)}\right] - \frac{(2)(16.5)(2)}{(17.2 - 9.81)}$$

$$= \frac{242.4}{L_1} - 8.93 = 11.75 \text{ m}$$

Now, referring to Eq. (13.59), we have

$$Q_n = (pK'\gamma_f H_f \tan \delta)L_1 + \frac{L_1^2 pK'\gamma' \tan \delta}{2}$$

$$p = \pi(0.305) = 0.958 \text{ m}$$

$$K' = 1 - \sin 34° = 0.44$$

$$Q_n = (0.958)(0.44)(16.5)(2)[\tan(0.6 \times 34)](11.75)$$

$$+ \frac{(11.75)^2(0.958)(0.44)(17.2 - 9.81)[\tan(0.6 \times 34)]}{2}$$

$$= 60.78 + 79.97 = \mathbf{140.75\ kN} \qquad\blacksquare$$

13.13 *Group Piles—Efficiency*

In most cases, piles are used in groups to transmit the structural load to the soil (Figure 13.21). A *pile cap* is constructed over *group piles*. The pile cap can be in contact with the ground, as in most cases (Figure 13.21a), or well above the ground, as in offshore platforms (Figure 13.21b).

Determination of the load-bearing capacity of group piles is extremely complicated and has not yet been fully resolved. When the piles are placed close to each other, a reasonable assumption is that the stresses transmitted by the piles to the soil will overlap (Figure 13.21c), thus reducing the load-bearing capacity of the piles. Ideally, the piles in a group should be spaced so that the load-bearing capacity of the group should be no less than the sum of the bearing capacity of the individual piles. In practice, the minimum center-to-center pile spacing, d, is $2.5D$ and in ordinary situations is actually about $3D$ to $3.5D$.

The efficiency of the load-bearing capacity of a group pile may be defined as

$$\eta = \frac{Q_{g(u)}}{\Sigma\, Q_u} \tag{13.60}$$

where η = group efficiency
$Q_{g(u)}$ = ultimate load-bearing capacity of the group pile
Q_u = ultimate load-bearing capacity of each pile without the group effect

Many structural engineers use a simplified analysis to obtain the group efficiency for friction piles, particularly in sand. This type of analysis can be explained with the aid of Figure 13.21a. Depending on their spacing within the group, the piles may act in one of two ways: (1) as a *block* with dimensions $L_g \times B_g \times L$ or (2) as *individual piles*. If the piles act as a block, the frictional capacity is $f_{av}p_g L \approx Q_{g(u)}$. [*Note:* p_g = perimeter of the cross section of block = $2(n_1 + n_2 - 2)d + 4D$, and f_{av} = average unit frictional resistance.] Similarly, for each pile acting individually, $Q_u \approx pLf_{av}$. (*Note:* p = perimeter of the cross section of each pile.) Thus,

$$\eta = \frac{Q_{g(u)}}{\Sigma\, Q_u} = \frac{f_{av}[2(n_1 + n_2 - 2)d + 4D]L}{n_1 n_2 p L f_{av}}$$

$$= \frac{2(n_1 + n_2 - 2)d + 4D}{p n_1 n_2} \tag{13.61}$$

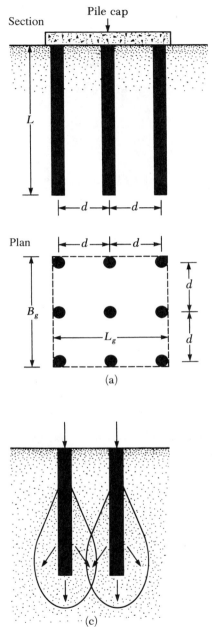

Section

Pile cap

L

$\leftarrow d \rightarrow \leftarrow d \rightarrow$

Plan

$\leftarrow d \rightarrow \leftarrow d \rightarrow$

B_g

d

d

L_g

(a)

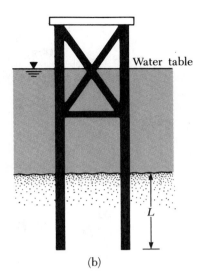

Water table

L

(b)

Number of piles in group $= n_1 \times n_2$
Note: $L_g \geq B_g$
$L_g = (n_1 - 1)d + 2(D/2)$
$B_g = (n_2 - 1)d + 2(D/2)$

(c)

FIGURE 13.21 Pile groups

Hence,

$$Q_{g(u)} = \left[\frac{2(n_1 + n_2 - 2)d + 4D}{pn_1n_2} \right] \Sigma Q_u \tag{13.62}$$

From Eq. (13.62), if the center-to-center spacing, d, is large enough, then $\eta > 1$. In that case, the piles will behave as individual piles. Thus, in practice, if $\eta < 1$,

$$Q_{g(u)} = \eta \Sigma Q_u$$

and, if $\eta \geq 1$,

$$Q_{g(u)} = \Sigma Q_u$$

Another equation often referred to by design engineers is the *Converse–Labarre equation:*

$$\eta = 1 - \left[\frac{(n_1 - 1)n_2 + (n_2 - 1)n_1}{90n_1n_2} \right] \theta \tag{13.63}$$

where θ (deg) $= \tan^{-1} (D/d)$. $\tag{13.64}$

Piles in Sand

Model test results on group piles in sand have shown that group efficiency can be greater than 1 because soil compaction zones are created around the piles during driving. Based on the experimental observations of the behavior of group piles in sand to date, two general conclusions may be drawn:

1. For *driven* group piles in *sand* with $d \geq 3D$, $Q_{g(u)}$ may be taken to be ΣQ_u, which includes the frictional and the point bearing capacities of individual piles.
2. For *bored* group piles in *sand* at conventional spacings ($d \approx 3D$), $Q_{g(u)}$ may be taken to be $\frac{2}{3}$ to $\frac{3}{4}$ times ΣQ_u (frictional and point bearing capacities of individual piles).

Piles in Clay

The ultimate load-bearing capacity of group piles in clay may be estimated with the following procedure:

1. Determine $\Sigma Q_u = n_1n_2(Q_p + Q_s)$. From Eq. (13.14),

$$Q_p = A_p[9c_{u(p)}]$$

where $c_{u(p)}$ = undrained cohesion of the clay at the pile tip. Also, from Eq. (13.24),

$$Q_s = \Sigma \alpha p c_u \Delta L$$

So

$$\Sigma Q_u = n_1 n_2 [9 A_p c_{u(p)} + \Sigma \alpha p c_u \, \Delta L] \tag{13.65}$$

2. Determine the ultimate capacity by assuming that the piles in the group act as a block with dimensions of $L_g \times B_g \times L$. The skin resistance of the block is

$$\Sigma p_g c_u \, \Delta L = \Sigma 2(L_g + B_g) c_u \, \Delta L$$

Calculate the point bearing capacity from

$$A_p q_p = A_p c_{u(p)} N_c^* = (L_g B_g) c_{u(p)} N_c^*$$

The variation of N_c^* with L/B_g and L_g/B_g is illustrated in Figure 13.22. Thus, the ultimate load is

$$\Sigma Q_u = L_g B_g c_{u(p)} N_c^* + \Sigma 2(L_g + B_g) c_u \, \Delta L \tag{13.66}$$

3. Compare the values obtained from Eqs. (13.65) and (13.66). The *lower* of the two values is $Q_{g(u)}$.

Piles in Rock

For point bearing piles resting on rock, most building codes specify that $Q_{g(u)} = \Sigma Q_u$, provided that the minimum center-to-center spacing of piles is $D + 300$ mm. For H-piles and piles with square cross sections, the magnitude of D is equal to the diagonal dimension of the pile cross section.

General Comments

A pile cap resting on soil, as shown in Figure 13.21a, will contribute to the load-bearing capacity of a pile group. However, this contribution may be neglected for

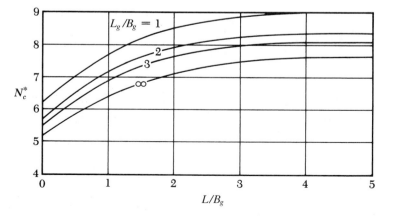

FIGURE 13.22 Variation of N_c^* with L_g/B_g and L/B_g

design purposes because the support may be lost as a result of soil erosion or excavation during the life of the project.

EXAMPLE 13.10

Refer to Figure 13.21a and 13.23. We are given $n_1 = 4$, $n_2 = 3$, $D = 305$ mm, and $d = 2.5D$. The piles are square in cross section and are embedded in sand. Use Eqs. (13.61) and (13.63) to obtain the group efficiency.

Solution From Eq. (13.61),

$$\eta = \frac{2(n_1 + n_2 - 2)d + 4D}{pn_1n_2}$$

$$d = 2.5D = (2.5)(305) = 762.5 \text{ mm}$$

$$p = 4D = (4)(305) = 1220 \text{ mm}$$

So

$$\eta = \frac{2(4 + 3 - 2)762.5 + 1220}{(1220)(4)(3)} = 0.604 = \mathbf{60.4\%}$$

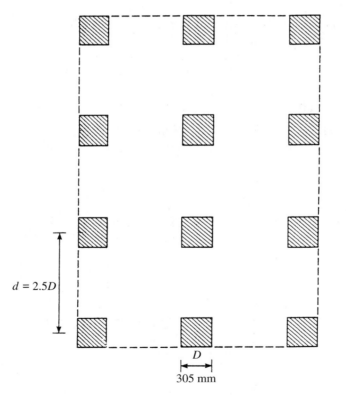

FIGURE 13.23

According to Eq. (13.63),

$$\eta = 1 - \left[\frac{(n_1 - 1)n_2 + (n_2 - 1)n_1}{90n_1n_2}\right] \tan^{-1}\left(\frac{D}{d}\right)$$

However,

$$\tan^{-1}\left(\frac{D}{d}\right) = \tan^{-1}\left(\frac{1}{2.5}\right) = 21.8°$$

So

$$\eta = 1 - \left[\frac{(3)(3) + (2)(4)}{(90)(3)(4)}\right](21.8°) = 0.657 = \textbf{65.7\%}$$ ∎

EXAMPLE 13.11

Refer to Figure 13.21a. For this group pile, $n_1 = 4$, $n_2 = 3$, $D = 305$ mm, $d = 1220$ mm, and $L = 15$ m. The piles are square in cross section and are embedded in a homogeneous clay with $c_u = 70$ kN/m². Use $FS = 4$ and determine the allowable load-bearing capacity of the group pile.

Solution   From Eq. (13.65),

$$\Sigma Q_u = n_1n_2[9A_pc_{u(p)} + \Sigma \alpha pc_u \, \Delta L]$$
$$A_p = (0.305)(0.305) = 0.093 \text{ m}^2$$
$$p = (4)(0.305) = 1.22 \text{ m}$$

From Figure 13.14 for $c_u = 70$ kN/m², $\alpha = 0.63$. So

$$\Sigma Q_u = (4)(3)[(9)(0.093)(70) + (0.63)(1.22)(70)(15)]$$
$$= 12(58.59 + 807.03) \approx 10{,}387 \text{ kN}$$

Again, from Eq. (13.66), the ultimate block capacity is

$$\Sigma Q_u = L_gB_gc_{u(p)}N_c^* + \Sigma 2(L_g + B_g)c_u \, \Delta L$$

So

$$L_g = (n_1 - 1)d + 2\left(\frac{D}{2}\right) = (4 - 1)(1.22) + 0.305 = 3.965 \text{ m}$$

$$B_g = (n_2 - 1)d + 2\left(\frac{D}{2}\right) = (3 - 1)(1.22) + 0.305 = 2.745 \text{ m}$$

$$\frac{L}{B_g} = \frac{15}{2.745} = 5.46$$

$$\frac{L_g}{B_g} = \frac{3.965}{2.745} = 1.44$$

From Figure 13.22, $N_c^* \approx 8.6$. Thus,

$$\Sigma Q_u = (3.965)(2.745)(70)(8.6) + 2(3.965 + 2.745)(70)(15)$$

$$= 6552 + 14{,}091 = 20{,}643 \text{ kN}$$

So

$$Q_{g(u)} = 10{,}387 \text{ kN} < 20{,}643 \text{ kN}$$

$$Q_{g(\text{all})} = \frac{Q_{g(u)}}{FS} = \frac{10{,}387}{4} \approx \mathbf{2597 \text{ kN}} \qquad \blacksquare$$

13.14   Elastic Settlement of Group Piles

Several investigations relating to the settlement of group piles with widely varying results have been reported in the literature. The simplest relation for the settlement of group piles was given by Vesic (1969) as

$$S_{g(e)} = \sqrt{\frac{B_g}{D}} \, S_e \qquad (13.67)$$

where $S_{g(e)}$ = elastic settlement of group piles

B_g = width of pile group section (see Figure 13.21a)

D = width or diameter of each pile in the group

S_e = elastic settlement of each pile at comparable working load (see Section 13.8)

For pile groups in sand and gravel, Meyerhof (1976) suggested the following empirical relation for elastic settlement:

$$S_{g(e)} \text{ (mm)} = \frac{0.92q\sqrt{B_g}\,I}{N_{\text{cor}}} \qquad (13.68)$$

where q (kN/m^2) = $Q_g/(L_g B_g)$ $\qquad$ (13.69)

L_g and B_g = length and width of the pile group section, respectively (m)

N_{cor} = average corrected standard penetration number within seat of settlement ($\approx B_g$ deep below the tip of the piles)

$$I = \text{influence factor} = 1 - L/8B_g \geq 0.5 \qquad (13.70)$$

L = length of embedment of piles (m)

Similarly, the pile group settlement is related to the cone penetration resistance as

$$S_{g(e)} = \frac{qB_g I}{2q_c}$$ (13.71)

where q_c = average cone penetration resistance within the seat of settlement. In Eq. (13.71), all symbols are in consistent units.

13.15 *Consolidation Settlement of Group Piles*

The consolidation settlement of a pile group can be estimated by assuming an approximate distribution method that is commonly referred to as the 2:1 method. The calculation procedure involves the following steps (Figure 13.24):

1. Let the depth of embedment of the piles be L. The group is subjected to a total load of Q_g. If the pile cap is below the original ground surface, Q_g equals the total load of the superstructure on the piles minus the effective weight of soil above the pile group removed by excavation.
2. Assume that the load Q_g is transmitted to the soil beginning at a depth of $2L/3$ from the top of the pile, as shown in Figure 13.24 ($z = 0$). The load Q_g spreads out along 2 vertical:1 horizontal lines from this depth. Lines aa' and bb' are the two 2:1 lines.
3. Calculate the effective stress increase caused at the middle of each soil layer by the load Q_g:

$$\Delta\sigma'_i = \frac{Q_g}{(B_g + z_i)(L_g + z_i)}$$ (13.72)

where $\Delta\sigma'_i$ = stress increase at the middle of layer i
L_g, B_g = length and width of the plan of pile group, respectively
z_i = distance from $z = 0$ to the middle of the clay layer, i

For example, in Figure 13.24 for layer 2, $z_i = L_1/2$; for layer 3, $z_i = L_1 + L_2/2$; and for layer 4, $z_i = L_1 + L_2 + L_3/2$. Note, however, that there will be no stress increase in clay layer 1 because it is above the horizontal plane ($z = 0$) from which the stress distribution to the soil starts.
4. Calculate the settlement of each layer caused by the increased stress:

$$\Delta S_{c(i)} = \left[\frac{\Delta e_{(i)}}{1 + e_{0(i)}}\right] H_i$$ (13.73)

where $\Delta S_{c(i)}$ = consolidation settlement of layer i
$\Delta e_{(i)}$ = change of void ratio caused by the stress increase in layer i

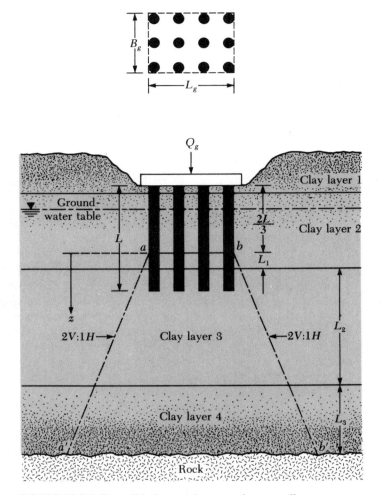

FIGURE 13.24 Consolidation settlement of group piles

$e_{o(i)}$ = initial void ratio of layer i (before construction)

H_i = thickness of layer i (*Note:* In Figure 13.24, for layer 2, $H_i = L_1$; for layer 3, $H_i = L_2$; and for layer 4, $H_i = L_3$.)

The relations for $\Delta e_{(i)}$ are given in Chapter 6.

5. Calculate the total consolidation settlement of the pile group by

$$\Delta S_{c(g)} = \Sigma \, \Delta S_{c(i)} \tag{13.74}$$

Note that the consolidation settlement of piles may be initiated by fills placed nearby, adjacent floor loads, and lowering of water tables.

**EXAMPLE
13.12**

A group pile in clay is shown in Figure 13.25. Determine the consolidation settlement of the pile groups. All clays are normally consolidated.

Solution   Because the lengths of the piles are 15 m each, the stress distribution starts at a depth of 10 m below the top of the pile. We have $Q_g = 2000$ kN.

Calculation of Settlement of Clay Layer 1   For normally consolidated clays,

$$\Delta S_{c(1)} = \left[\frac{C_{c(1)}H_1}{1 + e_{0(1)}} \right] \log \left[\frac{\sigma'_{o(1)} + \Delta\sigma'_{(1)}}{\sigma'_{o(1)}} \right]$$

$$\Delta\sigma'_{(1)} = \frac{Q_g}{(L_g + z_1)(B_g + z_1)} = \frac{2000}{(3.3 + 3.5)(2.2 + 3.5)} = 51.6 \text{ kN/m}^2$$

$$\sigma'_{o(1)} = 2(16.2) + 12.5(18.0 - 9.81) = 134.8 \text{ kN/m}^2$$

So

$$\Delta S_{c(1)} = \left[\frac{(0.3)(7)}{1 + 0.82} \right] \log \left[\frac{134.8 + 51.6}{134.8} \right] = 0.1624 \text{ m} = \textbf{162.4 mm}$$

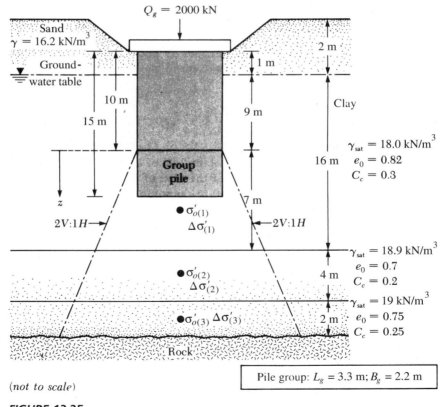

(not to scale)

FIGURE 13.25

Settlement of Layer 2

$$\Delta S_{c(2)} = \left[\frac{C_{c(2)} H_2}{1 + e_{o(2)}} \log \left[\frac{\sigma'_{o(2)} + \Delta \sigma'_{(2)}}{\sigma'_{o(2)}} \right] \right]$$

$$\sigma'_{o(2)} = 2(16.2) + 16(18.0 - 9.81) + 2(18.9 - 9.81) = 181.62 \text{ kN/m}^2$$

$$\Delta \sigma'_{(2)} = \frac{2000}{(3.3 + 9)(2.2 + 9)} = 14.52 \text{ kN/m}^2$$

Hence,

$$\Delta S_{c(2)} = \left[\frac{(0.2)(4)}{1 + 0.7} \right] \log \left[\frac{181.62 + 14.52}{181.62} \right] = 0.0157 \text{ m} = \mathbf{15.7 \ mm}$$

Settlement of Layer 3

$$\sigma'_{o(3)} = 181.62 + 2(18.9 - 9.81) + 1(19 - 9.81) = 208.99 \text{ kN/m}^2$$

$$\Delta \sigma'_{(3)} = \frac{2000}{(3.3 + 12)(2.2 + 12)} = 9.2 \text{ kN/m}^2$$

$$\Delta S_{c(3)} = \left[\frac{(0.25)(2)}{1 + 0.75} \right] \log \left[\frac{208.99 + 9.2}{208.99} \right] = 0.0054 \text{ m} = \mathbf{5.4 \ mm}$$

Hence, the total settlement is

$$\Delta S_{c(g)} = 162.4 + 15.7 + 5.4 = \mathbf{183.5 \ mm} \quad\blacksquare$$

DRILLED SHAFTS

As mentioned in the introduction of this chapter, drilled shafts are cast-in-place piles that generally have a diameter of about 750 mm or more. The use of drilled-shaft foundations has many advantages:

1. A single drilled shaft may be used instead of a group of piles and the pile cap.
2. Constructing drilled shafts in deposits of dense sand and gravel is easier than driving piles.
3. Drilled shafts may be constructed before grading operations are completed.
4. When piles are driven by a hammer, the ground vibration may cause damage to nearby structures, which the use of drilled shafts avoids.
5. Piles driven into clay soils may produce ground heaving and cause previously driven piles to move laterally. This does not occur during construction of drilled shafts.
6. There is no hammer noise during the construction of drilled shafts, as there is during pile driving.
7. Because the base of a drilled shaft can be enlarged, it provides great resistance to the uplifting load.

8. The surface over which the base of the drilled shaft is constructed can be visually inspected.

9. Construction of drilled shafts generally utilizes mobile equipment, which, under proper soil conditions, may prove to be more economical than methods of constructing pile foundations.

10. Drilled shafts have high resistance to lateral loads.

There are also several drawbacks to the use of drilled-shaft construction. The concreting operation may be delayed by bad weather and always needs close supervision. Also, as in the case of braced cuts, deep excavations for drilled shafts may cause substantial ground loss and damage to nearby structures.

13.16 *Types of Drilled Shafts*

Drilled shafts are classified according to the ways in which they are designed to transfer the structural load to the substratum. Figure 13.26a shows a drilled shaft that has a *straight shaft*. It extends through the upper layer(s) of poor soil, and its tip rests on a strong load-bearing soil layer or rock. The shaft can be cased with steel shell or pipe when required (as in the case of cased, cast-in-place concrete piles). For such shafts, the resistance to the applied load may develop from end bearing and also from side friction at the shaft perimeter and soil interface.

A *drilled shaft with bell* (Figures 13.26b and c) consists of a straight shaft with a bell at the bottom, which rests on good bearing soil. The bell can be constructed in the shape of a dome (Figure 13.26b), or it can be angled (Figure 13.26c). For

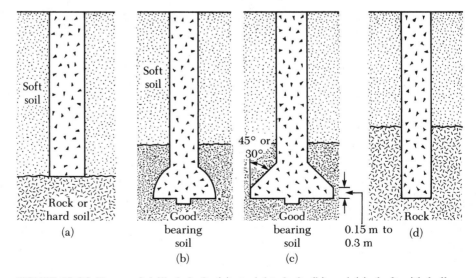

FIGURE 13.26 Types of drilled shaft: (a) straight shaft; (b) and (c) shaft with bell; (d) straight shafts socketed into rock

angled bells, the underreaming tools commercially available can make 30° to 45° angles with the vertical. For the majority of drilled shafts constructed in the United States, the entire load-carrying capacity is assigned to the end bearing only. However, under certain circumstances, the end-bearing capacity and the side friction are taken into account. In Europe, both the side frictional resistance and the end-bearing capacity are always taken into account.

Straight shafts can also be extended into an underlying rock layer (Figure 13.26d). In calculating the load-bearing capacity of such drilled shafts, engineers take into account the end bearing and the shear stress developed along the shaft perimeter and rock interface.

13.17   *Construction Procedures*

Most shaft excavations are now done mechanically. Open helix augers (flight augers) are common excavation tools. These augers have cutting edges or cutting teeth. Those with cutting edges are used mostly for drilling in soft, homogeneous soil; those with cutting teeth are used for drilling in hard soil and hard pan. The auger is attached to a square shaft referred to as the *Kelly* and pushed into the soil and rotated. When the flights are filled with soil, the auger is raised above the ground surface, and the soil is dumped into a pile by rotating the auger at high speed. These augers are available in various diameters; sometimes they may be as large as 3 m or more.

When the excavation is extended to the level of the load-bearing stratum, the auger is replaced by underreaming tools to shape the bell, if required. An underreamer essentially consists of a cylinder with two cutting blades hinged to the top of the cylinder (Figure 13.27). When the underreamer is lowered into the hole, the cutting blades stay folded inside the cylinder. When the bottom of the hole is reached, the blades are spread outward, and the underreamer is rotated. The loose soil falls inside the cylinder, which is raised periodically and emptied until the bell is completed. Most underreamers can cut bells with diameters as large as three times the diameter of the shaft.

Another common shaft-drilling device is the *bucket type drill.* It is essentially

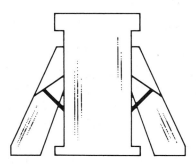

FIGURE 13.27 Underreamer

a bucket with an opening and cutting edges at the bottom. The bucket is attached to the Kelly and rotated. The loose soil is collected in the bucket, which is periodically raised and emptied. Holes as large as 5 to 6 m in diameter can be drilled with this type of equipment.

When rock is encountered during drilling, *core barrels* with *tungsten carbide teeth* attached to the bottom of the barrels are used. *Shot barrels* are also used for drilling into very hard rock. The principle of rock coring by a shot barrel is shown in Figure 13.28. The drill stem is attached to the shot barrel's plate. The barrel has some feeder slots through which chilled steel shots are supplied to the bottom of the bore hole. The steel shots cut the rock when the barrel is rotated. Water is supplied to the drill hole through the drill stem. Fine rock and steel particles (produced by the grinding of the steel shots) are washed upward, and they settle on the upper portion of the barrel.

The *Benoto machine* is another type of drilling equipment that is generally used when drilling conditions are difficult and the soil contains many boulders. It essentially consists of a steel tube that can be oscillated and pushed into the soil. A tool usually referred to as the *hammer grab*, which is fitted with cutting blades and jaws, is used to break up the soil and rock inside the tube and remove them.

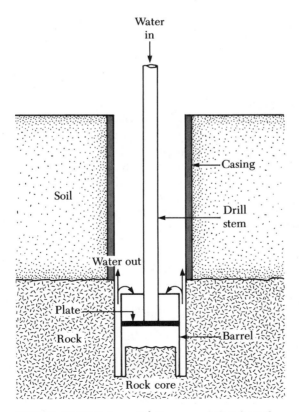

FIGURE 13.28 Schematic diagram of shot barrel

Use of Casings and Drilling Mud

When holes for drilled shafts are made in soft clays, the soil tends to squeeze in and close the hole. In such situations, casings may be used to keep the hole open and may have to be driven before excavation begins. Holes made in gravelly and sandy soils also tend to cave in. Excavation of holes for drilled shafts in these soils can be continued either by casing as the hole progresses or by using *drilling mud*.

Inspection of the Bottom of the Hole

In many instances, the bottom of the hole must be inspected to ensure that the load-bearing stratum is what was anticipated and that the bell is properly done. For these reasons, an inspector must descend to the bottom of the hole. Several safety precautions must be observed during this procedure:

1. If a casing is not already in the hole, one should be lowered by crane into it to prevent the hole and the bell from collapsing.
2. The hole should be tested for the presence of poisonous or explosive gases. The testing can be done by using a miner's safety lamp.
3. The inspector should wear a safety harness.
4. The inspector should also carry a safety lamp and an air tank.

Other Considerations

For the design of ordinary drilled shafts without casings, a minimum amount of vertical steel reinforcement is always desirable. Minimum reinforcement is 1% of the gross cross-sectional area of the shaft. In California, a reinforcing cage with a length of about 4 m is used in the top part of the drilled shaft, and no reinforcement is provided at the bottom. This procedure helps in the construction process because the cage is placed after most of the concreting is complete.

For drilled shafts with nominal reinforcement, most building codes suggest using a design concrete strength, f_c, on the order of $f'_c/4$. Thus, the minimum shaft diameter becomes

$$f_c = 0.25 f'_c = \frac{Q_w}{A_{gs}} = \frac{Q_w}{\frac{\pi}{4} D_s^2}$$

or

$$D_s = \sqrt{\frac{Q_w}{\left(\frac{\pi}{4}\right)(0.25)f'_c}} = 2.257 \sqrt{\frac{Q_w}{f'_c}}$$

where D_s = diameter of the shaft
f'_c = 28-day concrete strength
Q_w = working load
A_{gs} = gross cross-sectional area of the shaft

13.18   *Estimation of Load-Bearing Capacity*

The ultimate load of a drilled shaft (Figure 13.29) is

$$Q_u = Q_p + Q_s \qquad (13.75)$$

where Q_u = ultimate load
Q_p = ultimate load-carrying capacity at the base
Q_s = frictional (skin) resistance

The equation for the ultimate base load is similar to that for shallow foundations:

$$Q_p = A_p(cN_c^* + q'N_q^* + 0.3\gamma D_b N_\gamma^*) \qquad (13.76)$$

where N_c^*, N_q^*, N_γ^* = the bearing capacity factors
q' = vertical effective stress at the level of the bottom of the drilled shaft

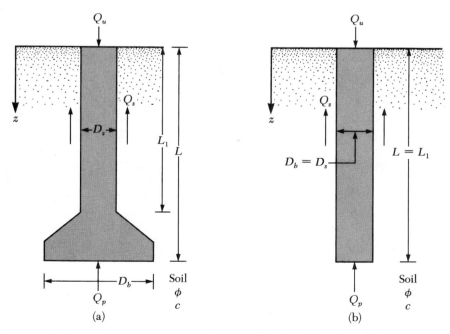

FIGURE 13.29 Ultimate bearing capacity of drilled shafts: (a) with bell; (b) straight shaft

$$D_b = \text{diameter of the base (see Figures 13.29a and b)}$$
$$A_p = \text{area of the base} = \pi/4 D_b^2$$

In most cases, the last term (containing N_γ^*) is neglected except for relatively short shafts, so

$$Q_p = A_p(cN_c^* + q'N_q^*) \tag{13.77}$$

The net load-carrying capacity at the base (that is, the gross load minus the weight of the drilled shaft) may be approximated as

$$Q_{p(net)} = A_p(cN_c^* + q'N_q^* - q') = A_p[cN_c^* + q'(N_q^* - 1)] \tag{13.78}$$

The expression for the frictional, or skin, resistance, Q_s, is similar to that for piles:

$$Q_s = \int_0^{L_1} pf\,dz \tag{13.79}$$

where p = shaft perimeter = πD_s
     f = unit frictional (or skin) resistance

Drilled Shafts in Sand

For shafts in sand, $c = 0$ and hence Eq. (13.78) simplifies to

$$Q_{p(net)} = A_p q'(N_q^* - 1) \tag{13.80}$$

Determination of N_q^* is always a problem for deep foundations, as in the case of piles. Note, however, that all shafts are *drilled,* unlike the majority of piles, which are *driven.* For similar initial soil conditions, the actual value of N_q^* may be substantially lower for objects drilled and placed *in situ* than for objects that are driven. Vesic (1967) compared the theoretical results obtained by several investigators relating to the variation of N_q^* with the soil friction angle. These investigators include DeBeer, Meyerhof, Hansen, Vesic, and Terzaghi. The values of N_q^* given by Vesic (1963) are approximately the lower bound and hence are used in this text (see Figure 13.30).

The frictional resistance at ultimate load, Q_s, developed in a drilled shaft may be calculated from the relation given in Eq. (13.79), in which

$$p = \text{shaft perimeter} = \pi D_s$$

$$f = \text{unit frictional (or skin) resistance} = K\sigma_o' \tan \delta \tag{13.81}$$

where K = earth pressure coefficient $\approx K_o = 1 - \sin \phi$
     σ_o' = effective vertical stress at any depth z

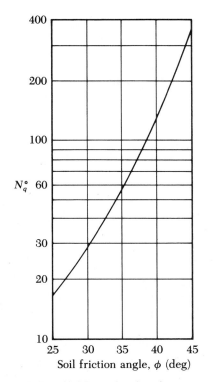

FIGURE 13.30 Vesic's bearing capacity factor, N_q^*, for deep foundations

Thus,

$$Q_s = \int_0^{L_1} pf\, dz = \pi D_s(1 - \sin \phi) \int_0^{L_1} \sigma_o' \tan \delta\, dz \qquad (13.82)$$

The value of σ_o' will increase to a depth of about $15D_s$ and will remain constant thereafter, as shown in Figure 13.11.

An appropriate factor of safety should be applied to the ultimate load to obtain the net allowable load, or

$$Q_{u(\text{net})} = \frac{Q_{p(\text{net})} + Q_s}{FS} \qquad (13.83)$$

A reliable estimate of the soil friction angle, ϕ, must be made in obtaining the net base resistance, $Q_{p(\text{net})}$. Figure 13.31 shows a conservative correlation between the soil friction angle and the corresponding corrected standard penetration resistance numbers in granular soils. However, these friction angles are valid only for low confining pressures. At higher confining pressures, which occur in the case of deep foundations, ϕ can decrease substantially for medium to dense sands. This

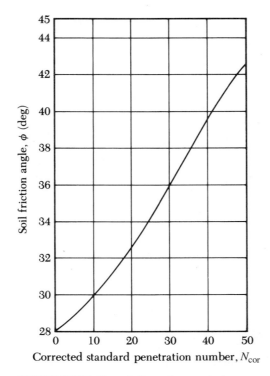

FIGURE 13.31 Correlation of corrected standard penetration number with the soil friction angle

decrease affects the value of N_q^* to be used for estimating Q_p. For example, Vesic (1977) showed that, for Chattahoochee River sand at a relative density of about 80%, the triaxial angle of friction is about 45° at a confining pressure of 70 kN/m². However, at a confining pressure of 10.5 MN/m², the friction angle is about 32.5°, which will ultimately result in a tenfold decrease of N_q^*. Thus, for general working conditions of drilled shafts, the estimated friction angle determined from Figure 13.31 should be reduced by about 10% to 15%. In general, the existing experimental values show the range of N_q^* for standard drilled shafts given in Table 13.11.

Drilled Shafts in Clay

From Eq. (13.78), for saturated clays with $\phi = 0$, $N_q^* = 1$; hence, the net base resistance becomes

$$Q_{p(\text{net})} = A_p c_u N_c^*$$ (13.84)

where c_u = undrained cohesion. The bearing capacity factor, N_c^*, is usually taken

Table 13.11 Range of N_q^* for standard drilled shafts

Sand type	Relative density of sand	Range of N_q^*
Loose	40 or less	10–20
Medium	40–60	25–40
Dense	60–80	30–50
Very dense	>80	75–90

to be 9. Figure 13.22 indicates that, when the L/D_b ratio (*note: L_g and B_g in Figure 13.22 are equal to D_b*) is 4 or more, $N_c^* = 9$, which is the condition for most drilled shafts. Experiments by Whitaker and Cooke (1966) showed that, for belled shafts, the full value of $N_c^* = 9$ is realized with a base movement of about 10% to 15% of D_b. Similarly, for those with straight shafts ($D_b = D_s$), the full value of $N_c^* = 9$ is obtained with a base movement of about 20% of D_b.

The expression for the skin resistance of drilled shafts in clay is similar to Eq. (13.24), or

$$Q_s = \sum_{L=0}^{L=L_1} \alpha^* c_u\, p\, \Delta L \tag{13.85}$$

where p = perimeter of the shaft cross section. The value of α^* that can be used in Eq. (13.85) has not been fully established. However, the field test results available at this time indicate that α^* may vary between 1.0 and 0.3.

Kulhawy and Jackson (1989) reported the field test results of 106 drilled shafts without bell: 65 in uplift and 41 in compression. The best correlation for the magnitude of α^* obtained from these results is

$$\alpha^* = 0.21 + 0.25 \left(\frac{p_a}{c_u}\right) \leq 1 \tag{13.86}$$

where p_a = atmospheric pressure = 101.3 kN/m² and c_u is in kN/m². So, conservatively, we may assume that

$$\alpha^* = 0.4 \tag{13.87}$$

EXAMPLE 13.13

A soil profile is shown in Figure 13.32. A point bearing drilled shaft with a bell is to be placed in the dense sand and gravel layer. The working load, Q_w, is 2000 kN.

  a. Determine the shaft diameter for $f_c' = 21{,}000$ kN/m².
  b. Use Eq. (13.80) and a factor of safety of 4 to determine the bell diameter, D_b. Ignore the frictional resistance of the shaft.

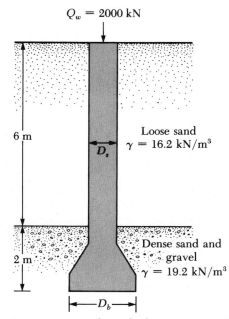

$$Q_w = 2000 \text{ kN}$$

Loose sand
$\gamma = 16.2 \text{ kN/m}^3$

6 m

D_s

Dense sand and gravel
$\gamma = 19.2 \text{ kN/m}^3$

2 m

$|\!\!\leftarrow\!\!-D_b\!-\!\!\rightarrow\!\!|$

Average corrected standard penetration number $= 40 = N_{\text{cor}}$

FIGURE 13.32

Solution

    a. From Section 13.17, we know

$$D_s = 2.257 \sqrt{\frac{Q_w}{f_c'}}$$

For $Q_w = 2000$ kN and $f_c' = 21{,}000$ kN/m²,

$$D_s = 2.257 \sqrt{\frac{2000}{21{,}000}} = 0.697 \text{ m}$$

The shaft diameter $D_s = \mathbf{1 \ m}$.

    b. We use Eq. (13.80):

$$Q_{p(\text{net})} = A_p q'(N_q^* - 1)$$

For $N_{\text{cor}} = 40$, Figure 13.31 indicates that $\phi \approx 39.5°$. To be conservative, we use a reduction of about 10%, or $\phi = 35.6$. From Figure 13.30, $N_q^* \approx 60$. So

$$q' = 6(16.2) + 2(19.2) = 135.6 \text{ kN/m}^2$$

$$Q_{p(\text{net})} = (Q_w)(FS) = (2000)(4) = 8000 \text{ kN}$$

$$8000 = (A_p)(135.6)(60 - 1)$$

$$A_p = 1.0 \text{ m}^2$$

$$D_b = \sqrt{\frac{1.0}{\pi/4}} = \mathbf{1.13\ m}$$

**EXAMPLE
13.14**

A drilled shaft is shown in Figure 13.33. Assume that $Q_w = 2800$ kN.

a. For $f'_c = 28,000$ kN/m², determine whether the proposed diameter of the shaft is adequate.
b. Determine the net ultimate point load-carrying capacity.
c. Determine the ultimate skin resistance.
d. Calculate the factor of safety with respect to the working load, Q_w.

Solution

a. From Section 13.18, we know

$$D_s = 2.257 \sqrt{\frac{Q_w}{f'_c}} = 2.257 \sqrt{\frac{2800}{28,000}} = 0.714 \text{ m}$$

The proposed shaft diameter $D_s = \mathbf{1\ m} > 0.714$ m—OK

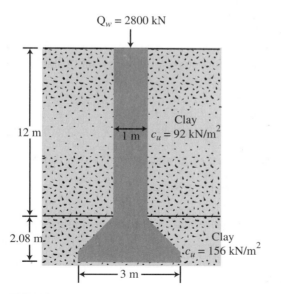

FIGURE 13.33

b. We find

$$Q_{p(\text{net})} = c_u N_c^* A_p = (156)(9)\left[\left(\frac{\pi}{4}\right)(3)^2\right] = \mathbf{9924.4\,kN}$$

c. According to Eq. (13.85)

$$Q_s = \pi D_s L_1 c_u \alpha^*$$

Assuming that $\alpha^* = 0.4$ [Eq. (13.87)] yields

$$Q_s = (\pi)(1)(12)(92)(0.4) = \mathbf{1387.3\,kN}$$

d. The factor of safety is

$$\frac{Q_u}{Q_w} = \frac{9924.4 + 1387.3}{2800} = \mathbf{4.04} \qquad \blacksquare$$

13.19   Settlement of Drilled Shafts at Working Load

The settlement of drilled shafts at working load is calculated in a manner similar to the one outlined in Section 13.8. In many cases, the load carried by shaft resistance is small compared to the load carried at the base. In such cases, the contribution of $S_{e(3)}$ may be ignored. Note that, in Eqs. (13.37) and (13.38), the term D should be replaced by D_b for shafts.

13.20   Reese and O'Neill's Method for Calculating Load-Bearing Capacity

Based on a data base of 41 loading tests, Reese and O'Neill (1989) proposed a method to calculate the load-bearing capacity of drilled shafts. The method is applicable to the following ranges:

1. Shaft diameter: $D_s = 0.52$ to 1.2 m
2. Bell depth: $L = 4.7$ to 30.5 m
3. $c_u = 29$ to 287 kN/m²
4. Standard field penetration resistance: $N_F = 5$ to 60
5. Overconsolidation ratio: 2 to 15
6. Concrete slump: 100 to 225 mm

Reese and O'Neill's procedure, with reference to Figure 13.34, gives

$$Q_u = \sum_{i=1}^{N} f_i p\, \Delta L_i + q_p A_p \qquad (13.88)$$

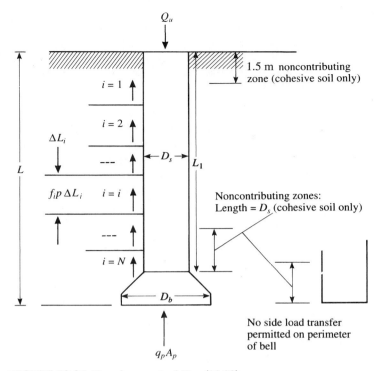

FIGURE 13.34 Development of Eq. (13.88)

where f_i = ultimate unit shearing resistance in layer i
  p = perimeter of the shaft = πD_s
  q_p = unit point resistance
  A_p = area of the base = $(\pi/4)D_b^2$

Following are the relationships for determining Q_u in cohesive and granular soils.

Cohesive Soil

Based on Eq. (13.88), we have

$$f_i = \alpha_i^* \, c_{u(i)} \tag{13.89}$$

The following values are recommended for α_i^*:

  $\alpha_i^* = 0$ for the top 1.5 m and bottom 1 diameter, D_s, of the drilled shaft. (*Note:* If $D_b > D_s$, then $\alpha^* = 0$ for 1 diameter above the top of the bell and for the peripheral area of the bell itself.)
  $\alpha_i^* = 0.55$ elsewhere

and

$$q_p \text{ (kN/m}^2) = 6c_{ub}\left(1 + 0.2\,\frac{L}{D_b}\right) \le 9c_{ub} \le 3.83 \text{ MN/m}^2 \qquad (13.90)$$

where c_{ub} = average undrained cohesion within $2D_b$ below the base (kN/m²).

If D_b is large, excessive settlement will occur at the ultimate load per unit area, q_p, as given by Eq. (13.90). Thus, for $D_b > 1.9$ m, q_p may be replaced by q_{pr}, or

$$q_{pr} = F_r q_p \qquad (13.91)$$

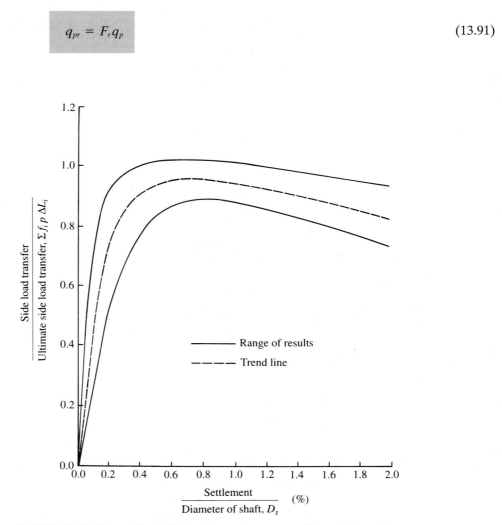

FIGURE 13.35 Normalized side load transfer versus settlement for cohesive soil (After Reese and O'Neill, 1989)

where $F_r = \dfrac{2.5}{0.0254\,\psi_1 D_b\,(\text{m}) + \psi_2} \leq 1$ (13.92)

$$\psi_1 = 0.0071 + 0.0021\left(\frac{L}{D_b}\right) \leq 0.015 \qquad\qquad (13.93)$$

$$\psi_2 = 7.787(c_{ub})^{0.5} \qquad (0.5 \leq \psi_2 \leq 1.5) \qquad\qquad (13.94)$$
$$\uparrow$$
$$\text{kN/m}^2$$

Figures 13.35 and 13.36 may be used to evaluate short-term settlements. (Note that the ultimate bearing capacity in Figure 13.36 is q_p, not q_{pr}.) To do so, follow this procedure:

1. Select a value of settlement, S_e.
2. Calculate $\sum\limits_{i=1}^{N} f_i p\,\Delta L_i$ and $q_p A_p$, as given in Eq. (13.88).

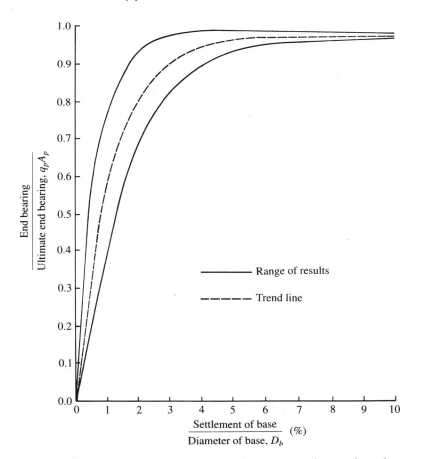

FIGURE 13.36 Normalized base load transfer versus settlement for cohesive soil (After Reese and O'Neill, 1989)

3. Using Figures 13.35 and 13.36 and the calculated values in step 2, determine the *side load* and the *end bearing load*.
4. The sum of the side load and the end bearing load is the total applied load.

Cohesionless Soil

Based on Eq. (13.88), we have

$$f_i = \beta \sigma'_{ozi} \tag{13.95}$$

where σ'_{ozi} = vertical effective stress at the middle of layer i

$$\beta = 1.5 - 0.244 z_i^{0.5} \qquad (0.25 \le \beta \le 1.2) \tag{13.96}$$

z_i = depth of the middle of layer i (m)

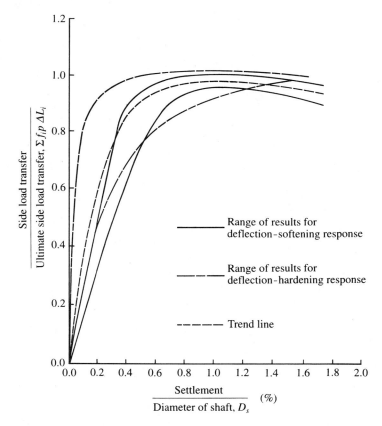

FIGURE 13.37 Normalized side load transfer versus settlement for cohesionless soil (After Reese and O'Neill, 1989)

The point bearing capacity is

$$q_p \ (\text{kN/m}^2) = 57.5 \ N_F \leq 4.3 \ \text{MN/m}^2 \tag{13.97}$$

where N_F = mean *uncorrected* standard penetration number within a distance of $2D_b$ below the base of the drilled shaft.

As in Eq. (13.91), to control excessive settlement, the magnitude of q_p may be modified as follows:

$$q_{pr} = \frac{1.27}{D_b \ (\text{m})} \ q_p \qquad (\text{for } D_b \geq 1.27 \ \text{m}) \tag{13.98}$$

Figures 13.37 and 13.38 may be used to calculate short-term settlements. They are similar to Figures 13.35 and 13.36 for clay.

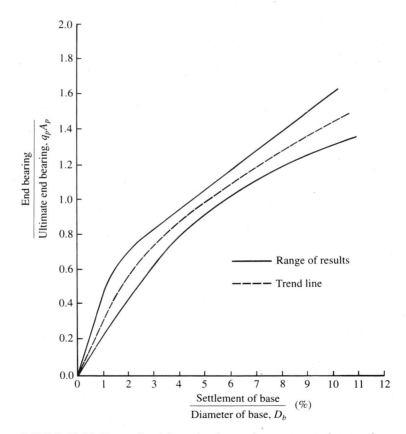

FIGURE 13.38 Normalized base load transfer versus settlement for cohesionless soil (After Reese and O'Neill, 1989)

**EXAMPLE
13.15**

A drilled shaft in a cohesive soil is shown in Figure 13.39. Use the procedure outlined in this section to determine these values:

a. The ultimate load-carrying capacity
b. The load-carrying capacity for an allowable settlement of 12.7 mm

Solution

Part a From Eq. (13.89), we have

$$f_i = \alpha_i^* c_{u(i)}$$

From Figure 13.39,

$$\Delta L_1 = 3.66 - 1.5 = 2.16 \text{ m}$$

$$\Delta L_2 = (6.1 - 3.66) - D_s = 2.44 - 0.76 = 1.68 \text{ m}$$

$$c_{u(1)} = 38 \text{ kN/m}^2$$

$$c_{u(2)} = 57.5 \text{ kN/m}^2$$

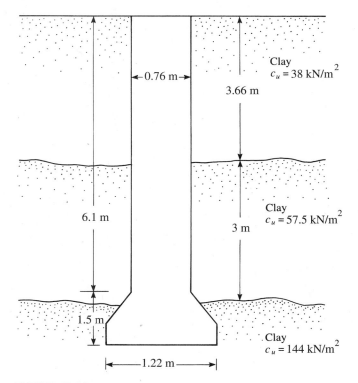

FIGURE 13.39

Hence,

$$\Sigma f_i p \, \Delta L_i = \Sigma \alpha_i^* c_{u(i)} p \, \Delta L_i$$

$$= (0.55)(38)(\pi \times 0.76)(2.16) + (0.55)(57.5)(\pi \times 0.76)(1.68)$$

$$= 234.6 \text{ kN}$$

From Eq. (13.90), we have

$$q_p = 6c_{ub}\left(1 + 0.2\frac{L}{D_b}\right) = (6)(144)\left[1 + 0.2\left(\frac{6.1 + 1.5}{1.22}\right)\right] = 1940 \text{ kN/m}^2$$

Check:

$$q_p = 9c_{ub} = (9)(144) = 1296 \text{ kN/m}^2 < 1940 \text{ kN/m}^2$$

So, we use $q_p = 1296 \text{ kN/m}^2$:

$$q_p A_p = q_p\left(\frac{\pi}{4}D_b^2\right) = (1296)\left[\left(\frac{\pi}{4}\right)(1.22)^2\right] \approx 1515 \text{ kN}$$

Hence,

$$Q_u = \Sigma \alpha_i^* c_{u(i)} p \, \Delta L_i + q_p A_p = 234.6 + 1515 = \textbf{1749.6 kN}$$

Part b   We have

$$\frac{\text{allowable settlement}}{D_s} = \frac{12.7}{(0.76)(1000)} = 0.0167 = 1.67\%$$

The trend line shown in Figure 13.35 indicates that, for a normalized settlement of 1.67%, the normalized side load is about 0.89. Thus, the side load is

$$(0.89)(\Sigma f_i p \, \Delta L_i) = (0.89)(234.6) = 208.8 \text{ kN}$$

Again,

$$\frac{\text{allowable settlement}}{D_b} = \frac{12.7}{(1.22)(1000)} = 0.0104 = 1.04\%$$

The trend line shown in Figure 13.36 indicates that, for a normalized settlement of 1.04%, the normalized end bearing is about 0.57. So the base load is

$$(0.57)(q_p A_p) = (0.57)(1515) = 863.6 \text{ kN}$$

Thus, the total load is

$$Q = 208.8 + 863.6 = \textbf{1072.4 kN} \qquad \blacksquare$$

EXAMPLE 13.16

A drilled shaft is shown in Figure 13.40. The uncorrected average standard penetration number within a distance of $2D_b$ below the base of the shaft is about 35.

    a. Determine the ultimate load-carrying capacity.
    b. What is the load-carrying capacity for a settlement of 12.7 mm?

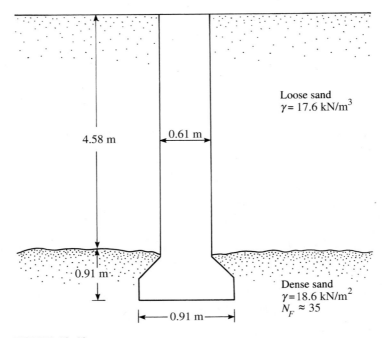

FIGURE 13.40

Solution

Part a   From Eqs. (13.95) and (13.96), we have

$$f_i = \beta \sigma'_{ozi}$$

$$\beta = 1.5 - 0.244 z_i^{0.5}$$

For this problem, $z_i = 4.58/2 = 2.29$ m, so

$$\beta = 1.5 - (0.244)(2.29)^{0.5} = 1.13$$

$$\sigma'_{ozi} = \gamma z_i = (17.6)(2.29) = 40.3 \text{ kN/m}^2$$

Hence,

$$f_i = (1.13)(40.3) = 45.5 \text{ kN/m}^2$$

From Eq. (13.88),

$$\Sigma f_i p \, \Delta L_i = (45.5)(\pi \times 0.61)(4.58) = 399.4 \text{ kN}$$

Again, from Eq. (13.97),

$$q_p = 57.5 N_F = (57.5)(35) = 2012 \text{ kN/m}^2$$

From Eq. (13.88),

$$q_p A_p = (2012) \left[\left(\frac{\pi}{4} \right) (0.91)^2 \right] = 1308.6 \text{ kN}$$

Thus,

$$Q_u = \Sigma f_i p \, \Delta L_i + q_p A_p = 399.4 + 1308.6 = \mathbf{1708 \text{ kN}}$$

Part b   We have

$$\frac{\text{allowable settlement}}{D_s} = \frac{12.7}{(0.61)(1000)} = 0.021 = 2.1\%$$

The trend line shown in Figure 13.37 indicates that, for a normalized settlement of 2.1%, the normalized side load is about 0.9. Thus, the side load transfer is

$$(0.9)(399.4) \approx 359.5 \text{ kN}$$

Similarly,

$$\frac{\text{allowable settlement}}{D_b} = \frac{12.7}{(0.91)(1000)} = 0.014 = 1.4\%$$

The trend line shown in Figure 13.38 indicates that, for a normalized settlement of 1.4%, the normalized base load is 0.312. So the base load is

$$(0.312)(1308.6) = 408.3 \text{ kN}$$

Hence, the total load is

$$Q = 359.5 + 408.3 \approx \mathbf{767.8 \text{ kN}} \qquad \blacksquare$$

Problems

13.1   A concrete pile is 15 m long and 406 mm × 406 mm in cross section. The pile is fully embedded in sand, for which $\gamma = 17.3$ kN/m³ and $\phi = 30°$.
a. Calculate the ultimate point load, Q_p.
b. Determine the total frictional resistance for $K = 1.3$ and $\delta = 0.8\phi$.

13.2   Redo Problem 13.1 for $\gamma = 18.4$ kN/m³ and $\phi = 37°$.

13.3   A driven closed-ended pipe pile is shown in Figure 13.41.
a. Find the ultimate point load.
b. Determine the ultimate frictional resistance, Q_s; use $K = 1.4$ and $\delta = 0.6\phi$.
c. Calculate the allowable load of the pile; use $FS = 4$.

13.4   A concrete pile 20 m long with a cross section of 381 mm × 381 mm is fully embedded in a saturated clay layer. For the clay, $\gamma_{sat} = 18.5$ kN/m³, $\phi = 0$, and $c_u = 70$ kN/m². Assume that the water table lies below the tip of the pile. Determine the allowable load that the pile can carry ($FS = 3$). Use the α method to estimate the skin resistance.

13.5   Redo Problem 13.4 using the λ method for estimating the skin resistance.

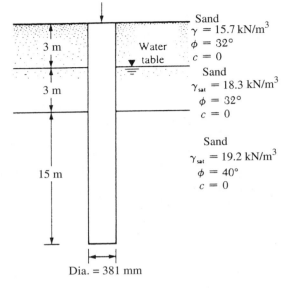

Sand
$\gamma = 15.7 \text{ kN/m}^3$
$\phi = 32°$
$c = 0$

Sand
$\gamma_{sat} = 18.3 \text{ kN/m}^3$
$\phi = 32°$
$c = 0$

Sand
$\gamma_{sat} = 19.2 \text{ kN/m}^3$
$\phi = 40°$
$c = 0$

3 m

Water
table

3 m

15 m

Dia. = 381 mm

FIGURE 13.41

13.6   A concrete pile 405 mm × 405 mm in cross section is shown in Figure 13.42. Calculate the ultimate skin resistance by using each method:
a. α method
b. λ method
c. β method
Use $\phi_R = 25°$ for all clays, which are normally consolidated.

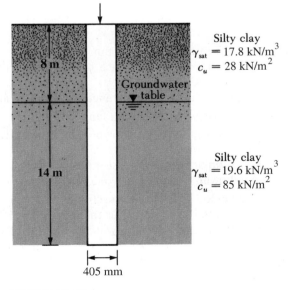

Silty clay
$\gamma_{sat} = 17.8 \text{ kN/m}^3$
$c_u = 28 \text{ kN/m}^2$

Groundwater table

8 m

Silty clay
$\gamma_{sat} = 19.6 \text{ kN/m}^3$
$c_u = 85 \text{ kN/m}^2$

14 m

405 mm

FIGURE 13.42

13.7 A concrete pile is 20 m long and has a cross section of 381 mm × 381 mm. The pile is embedded in sand having $\gamma = 18.9$ kN/m³ and $\phi = 38°$. The allowable working load is 760 kN. If 450 kN are contributed by the frictional resistance and 310 kN are from the point load, determine the elastic settlement of the pile. Here, $E_p = 21 \times 10^6$ kN/m², $E_s = 35 \times 10^3$ kN/m², $\mu_s = 0.35$, and $\zeta = 0.62$.

13.8 A steel H-pile (Section: HP 360 × 1.707, see Table 13.1) is driven by an MKT S-20 hammer (see Table 13.6). The length of the pile is 24 m, the coefficient of restitution is 0.35, the weight of the pile cap is 4 kN, hammer efficiency is 0.84, and the number of blows for the last 25.4 mm of penetration is 10. Estimate the ultimate pile capacity using Eq. (13.44). For the pile, $E_p = 210 \times 10^6$ kN/m².

13.9 Redo Problem 13.8 using Eq. (13.45).

13.10 Redo Problem 13.8 using Eq. (13.47).

13.11 The plan of a group pile (friction pile) in sand is shown in Figure 13.43. The piles are circular in cross section and have an outside diameter of 460 mm. The center-to-center spacing of the piles, d, is 920 mm. Use Eq. (13.61) and find the efficiency of the pile group.

13.12 Solve Problem 13.11 using the Converse–Labarre equation.

13.13 The plan of a group pile is shown in Figure 13.43. Assume that the piles are embedded in a saturated homogeneous clay having $c_u = 80$ kN/m². For the piles, $D = 406$ mm, center-to-center spacing = 850 mm, and $L = 20$ m. Find the allowable load-carrying capacity of the pile group. Use $FS = 3$.

13.14 Redo Problem 13.13 for $d = 762$ mm, $L = 15$ m, $D = 381$ mm, and $c_u = 50$ kN/m².

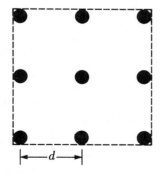

FIGURE 13.43

13.15 The section of a 3×4 group pile in a layered saturation clay is shown in Figure 13.44. The piles are square in cross section (356 mm × 356 mm). The center-to-center spacing, d, of the piles is 890 mm. Determine the allowable load-bearing capacity of the pile group. Use $FS = 4$.

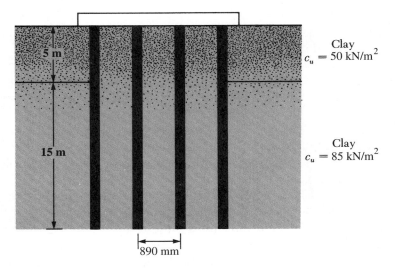

Clay
$c_u = 50$ kN/m^2

Clay
$c_u = 85$ kN/m^2

5 m

15 m

890 mm

FIGURE 13.44

13.16 Figure 13.45 shows a group pile in clay. Determine the consolidation settlement of the group.

13.17 Figure 13.20a shows a pile. Let $L = 15$ m, $D = 457$ mm, $H_f = 3.5$ m, $\gamma_f = 17.6$ kN/m^3, $\phi_{fill} = 28°$. Determine the total downward drag force on the pile. Assume that the fill is located above the water table and that $\delta = 0.6\phi_{fill}$.

13.18 Redo Problem 13.17 assuming that the water table coincides with the top of the fill and that $\gamma_{sat(fill)} = 19.6$ kN/m^3. If the other quantities remain the same, what would be the downward drag force on the pile? Assume that $\delta = 0.6\phi_{fill}$.

13.19 Refer to Figure 13.20b. Let $L = 19$ m, $\gamma_{fill} = 15.2$ kN/m^3, $\gamma_{sat(clay)} = 19.5$ kN/m^3, $\phi_{clay} = 30°$, $H_f = 3.2$ m, and $D = 0.46$ m. The water table coincides with the top of the clay layer. Determine the total downward drag on the pile. Assume that $\delta = 0.5\phi_{clay}$.

13.20 A drilled shaft is shown in Figure 13.46. For the shaft, $L_1 = 4$ m, $L_2 = 2.5$ m, $D_s = 1$ m, and $D_b = 2$ m. For the soil, $\gamma_c = 16.8$ kN/m^3, $c_u = 26$ kN/m^2, $\gamma_s = 18.6$ kN/m^3, and $\phi = 37.5°$. Determine the net allowable point bearing capacity (factor of safety = 4). Do not reduce the friction angle of sand, ϕ (use Figure 13.30).

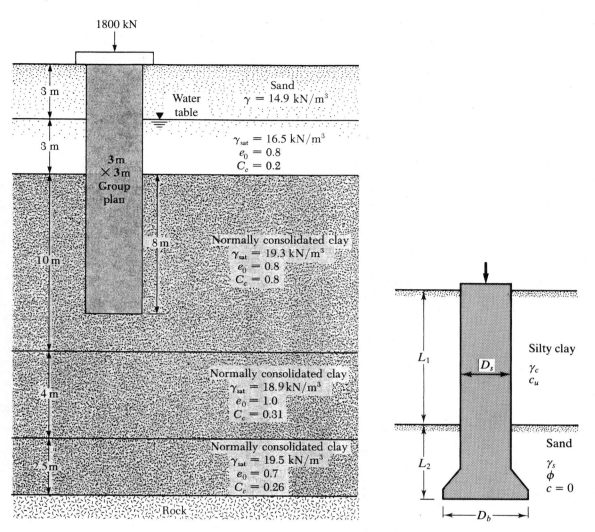

FIGURE 13.45

FIGURE 13.46

13.21 For the drilled shaft described in Problem 13.20, what skin resistance would develop for the top 4 m, which is in clay?

13.22 Figure 13.47 shows a drilled shaft without a bell. Here, $L_1 = 9$ m, $L_2 = 2.8$ m, $D_s = 1.1$ m, $c_{u(1)} = 50$ kN/m², and $c_{u(2)} = 105$ kN/m². Find these values:
a. The net ultimate point bearing capacity
b. The ultimate skin resistance
c. The working load, Q_w (*FS* = 3)

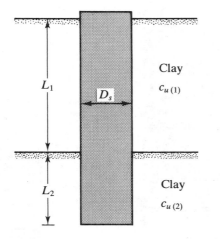

Clay
$c_{u\,(1)}$

L_1

D_s

Clay
$c_{u\,(2)}$

L_2

FIGURE 13.47

13.23 For the drilled shaft described in Problem 13.22, estimate the total elastic settlement at working load. Use Eqs. (13.36), (13.38), and (13.39). Assume that $E_p = 21 \times 10^6$ kN/m², $\mu_s = 0.3$, $E_s = 14 \times 10^3$ kN/m², $\xi = 0.65$. Assume 50% mobilization of skin resistance at working load.

13.24 For the drilled shaft described in Problem 13.22, determine these values:
a. The ultimate load-carrying capacity.
b. The load-carrying capacity for a settlement of 12.7 mm
Use the procedure outlined in Section 13.20.

13.25 Refer to Figure 13.48, for which $L = 6$ m, $L_1 = 5$ m, $D_s = 1.2$ m, $D_b = 1.7$ m, $\gamma = 15.7$ kN/m³, and $\phi = 33°$. The average uncorrected standard penetration number within $2D_b$ below the base is 32. Determine these values:
a. The ultimate load-carrying capacity
b. The load-carrying capacity for a settlement of 12.7 mm
Use the procedure outlined in Section 13.20.

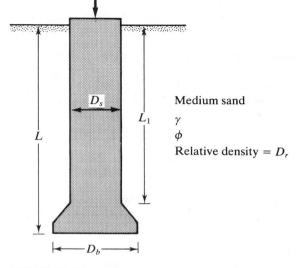

Medium sand
γ
ϕ
Relative density = D_r

FIGURE 13.48

References

American Society of Civil Engineers (1959). "Timber Piles and Construction Timbers," *Manual of Practice,* No. 17, American Society of Civil Engineers, New York.

Bowles, J. E. (1982). *Foundation Design and Analysis,* McGraw-Hill, New York.

Coyle, H. M., and Castello, R. R. (1981). "New Design Correlations for Piles in Sand," *Journal of the Geotechnical Engineering Division,* American Society of Civil Engineers, Vol. 107, No. GT7, 965–986.

Goodman, R. E. (1980). *Introduction to Rock Mechanics,* Wiley, New York.

International Conference of Building Officials (1982). "Uniform Building Code," Whittier, CA.

Janbu, N. (1953). *An Energy Analysis of Pile Driving with the Use of Dimensionless Parameters,* Norwegian Geotechnical Institute, Oslo, Publication No. 3.

Kulhawy, F. H., and Jackson, C. S. (1989). "Some Observations on Undrained Side Resistance of Drilled Shafts," *Proceedings,* Foundation Engineering: Current Principles and Practices, American Society of Civil Engineers, Vol. 2, 1011–1025.

McClelland, B. (1974). "Design of Deep Penetration Piles for Ocean Structures," *Journal of the Geotechnical Engineering Division,* American Society of Civil Engineers, Vol. 100, No. GT7, 709–747.

Meyerhof, G. G. (1976). "Bearing Capacity and Settlement of Pile Foundations," *Journal of the Geotechnical Engineering Division,* American Society of Civil Engineers, Vol. 102, No. GT3, 197–228.

Michigan State Highway Commission (1965). *A Performance Investigation of Pile Driving Hammers and Piles,* Lansing, 338 pp.

Reese, L. C., and O'Neill, M. W. (1989). "New Design Method for Drilled Shafts from Common Soil and Rock Tests," *Proceedings,* Foundation Engineering: Current Principles and Practices, American Society of Civil Engineers, Vol. 2, 1026–1039.

Vesic, A. S. (1963). "Bearing Capacity of Deep Foundations in Sand," *Highway Research Record,* No. 39, Highway Research Board, National Academy of Science, Washington, D.C., 112–153.

Vesic, A. S. (1967). "Ultimate Load and Settlement of Deep Foundations in Sand," *Proceedings,* Symposium on Bearing Capacity and Settlement of Foundations, Duke University, Durham, NC, p. 53.

Vesic, A. S. (1969). "Experiments with Instrumented Pile Groups in Sand," American Society for Testing and Materials; Special Technical Publication, No. 444, 177–222.

Vesic, A. S. (1977). *Design of Pile Foundations,* National Cooperative Highway Research Program Synthesis of Practice No. 42, Transportation Research Board, Washington, D.C.

Vijayvergiya, V. N., and Focht, J. A., Jr. (1972). *A New Way to Predict Capacity of Piles in Clay,* Offshore Technology Conference Paper 1718, Fourth Offshore Technology Conference, Houston.

Whitaker, T., and Cooke, R. W. (1966). "An Investigation of the Shaft and Base Resistance of Large Bored Piles in London Clay," *Proceedings,* Conference on Large Bored Piles, Institute of Civil Engineers, London, 7–49.

Supplementary References for Further Study

Briaud, J. L., Moore, B. H., and Mitchell, G. B. (1989). "Analysis of Pile Load Test at Lock and Dam 26," *Proceedings,* Foundation Engineering: Current Principles and Practices, American Society of Civil Engineers, Vol. 2, 925–942.

Briaud, J. L., and Tucker, L. M. (1984). "Coefficient of Variation of *In-Situ* Tests in Sand," *Proceedings,* Symposium on Probabilistic Characterization of Soil Properties, Atlanta, 119–139.

Broms, B. B. (1965). "Design of Laterally Loaded Piles," *Journal of the Soil Mechanics and Foundations Division,* American Society of Civil Engineers, Vol. 91, No. SM3, 79–99.

Davisson, M. T., and Gill, H. L. (1963). "Laterally Loaded Piles in a Layered Soil System," *Journal of the Soil Mechanics and Foundations Division,* American Society of Civil Engineers, Vol. 89, No. SM3, 63–94.

Kishida, H., and Meyerhof, G. G. (1965). "Bearing Capacity of Pile Groups Under Eccentric Loads in Sand," *Proceedings,* Sixth International Conference on Soil Mechanics and Foundation Engineering, Montreal, Vol. 2, 270–274.

Liu, J. L., Yuan, Z. L., and Zhang, K. P. (1985). "Cap-Pile-Soil Interaction of Bored Pile Groups," *Proceedings,* Eleventh International Conference on Soil Mechanics and Foundation Engineering, San Francisco, Vol. 3, 1433–1436.

Matlock, H., and Reese, L. C. (1960). "Generalized Solution for Laterally Loaded Piles," *Journal of the Soil Mechanics and Foundations Division,* American Society of Civil Engineers, Vol. 86, No. SM5, Part I, 63–91.

Meyerhof, G. G. (1995). "Behavior of Pile Foundations Under Special Loading Conditions: 1994 R. M. Hardy Keynote Address," *Canadian Geotechnical Journal,* Vol. 32, No. 2, 204–222.

Answers to Selected Problems

1.1 **a.** _____

U.S. sieve no.	Percent finer
4	100
10	95.2
20	84.2
40	61.4
60	41.6
100	20.4
200	7

b. $D_{10} = 0.09$ mm, $D_{30} = 0.185$ mm, $D_{60} = 0.41$ mm
c. $C_u = 4.56$
d. $C_z = 0.93$

1.3 $C_u = 6.2$, $C_z = 2.0$

1.5 **a.** _____

U.S. sieve no.	Percent finer
4	100
6	100
10	100
20	98.2
40	48.3
60	12.3
100	7.8
200	4.7

b. $D_{10} = 0.23$ mm, $D_{30} = 0.33$ mm, $D_{60} = 0.48$ mm
c. $C_u = 2.09$
d. $C_z = 0.99$

1.7 Gravel: 0%; sand: 46%; silt: 31%; clay: 23%

1.9 Gravel: 0%; sand: 70%; silt: 16%; clay: 14%

1.11 Gravel: 0%; sand: 66%; silt: 20%; clay: 14%

CHAPTER 2

2.1 **a.** 19.19 kN/m³
b. 17.13 kN/m³
c. 0.56 **d.** 0.36
e. 58.3% **f.** 0.593×10^{-3} m³

2.3 **a.** 17.49 kN/m³ **b.** 0.509
c. 0.337 **d.** 51.8%

2.5 **a.** 1678 kg/m³
b. 1465 kg/m³
c. 240 kg/m³

2.7 **a.** 16.2 kN/m³ **b.** 0.66
c. 0.4 **d.** 24.1%

2.9 **a.** 18.5 kN/m³ **b.** 0.487
c. 2.32 **d.** 16.9 kN/m³

2.11 0.61

2.13 19.4%

2.15

Soil no.	Group symbol	Group name
1	ML	Sandy silt
2	CH	Fat clay with sand
3	CL	Sandy lean clay
4	SC	Clayey sand with gravel
5	ML	Sandy silt
6	SC	Clayey sand

2.17

Soil	Group symbol	Group name
A	SP	Poorly graded sand
B	SW-SM	Well-graded sand with silt
C	MH	Elastic silt with sand
D	CH	Fat clay
E	SC	Clayey sand

2.19

Soil	Group symbol	Group name
A	SC	Clayey sand
B	GM-GC	Silty clayey gravel with sand
C	CH	Fat clay with sand
D	ML	Sandy silt
E	SM	Silty sand with gravel

CHAPTER 3

3.1

w (%)	γ_{zav} (kN/m³)
5	23.2
8	21.7
10	20.7
12	19.9
15	18.8

3.3

a. $\gamma_d = \dfrac{S\gamma_w}{w + \dfrac{S}{G_s}}$

b.

w (%)	γ_{zav} (kN/m³)
5	22.3
10	19.8
15	17.8
20	16.2

3.5 $\gamma_{d(max)} = 16.8$ kN/m³, $w_{(opt)} = 14.4\%$, w for 95% compaction = 11%

3.7 **a.** 86.7%   **b.** 50.3%

3.9 $\gamma_{d(field)} = 15.8$ kN/m³, $R = 95.8\%$

3.11 **a.** 18.6 kN/m³   **b.** 97.9%

CHAPTER 4

4.1 0.97×10^{-4} m³/m/sec

4.3 0.057 cm/sec

4.5 $h = 340$ mm, $v = 0.019$ cm/sec

4.7 **a.** 2.27×10^{-3} cm/sec
b. 376.4 mm

4.9 Eq. (4.20): 0.018 cm/sec
Eq. (4.21): 0.015 cm/sec

4.11 5.67×10^{-2} cm/sec

4.13 0.709×10^{-6} cm/sec

4.15 17.06×10^{-6} m³/m/sec

4.17 2.42×10^{-5} m³/m/sec

CHAPTER 5

5.1

Point	σ (kN/m²)	u (kN/m²)	σ' (kN/m²)
A	0	0	0
B	69.2	0	69.2
C	163.7	49.05	114.65
D	281.9	107.91	173.99

5.3

Point	σ (kN/m²)	u (kN/m²)	σ' (kN/m²)
A	0	0	0
B	55.08	0	55.08
C	135.52	39.24	96.28
D	173.9	58.86	115.04

5.5 **a.**

Point	σ (kN/m²)	u (kN/m²)	σ' (kN/m²)
A	0	0	0
B	64.8	0	64.8
C	169.2	49.05	120.15

b. 2.65 m

5.7 6.9m

5.9 0.06 kN/m²

5.11 0.4 m

5.13 0.68 kN/m²

5.15 245 kN/m²

5.17 184.4 kN/m²

5.19 **a.** 17 kN/m²   **b.** 40 kN/m²
        c. 5.5 kN/m²

CHAPTER 6

6.1 218.7 mm

6.3 196 mm

6.5 45 mm

6.7 **b.** 310 kN/m²   **c.** 0.53

6.9 0.87

6.11 **a.** 5.08×10^{-4} m²/kN
        b. 1.146×10^{-7} cm/sec

6.13 80.6 days

6.15 2.94×10^{-3} cm²/sec

6.17 **a.** 31.3%
         b. 190 days
         c. 47.5 days

6.19

Point	$\Delta\sigma$ (kN/m²)
A	110
B	11
C	44.6

6.21 **a.** 159 mm
         b. 12.05 months
         c. 98 kN/m²

6.23 **a.** 153 mm
         b. 12.65 months
         c. 84 kN/m²

6.25 64.9%

CHAPTER 7

7.1 $\phi = 34°$, shear force = 142 N

7.3 0.164 kN

7.5 23.5°

7.7 **a.** 61.55°
        b. $\sigma' = 294.5$ kN/m², $\tau = 109.4$ kN/m²

7.9 **a.** 65.5°
        b. $\sigma' = 236.76$ kN/m², $\tau = 188.17$ kN/m²

7.11 105.2 kN/m²

7.13 **a.** 414 kN/m²
         b. $\tau = 138$ kN/m², $\tau_f = 146.2$ kN/m², $\tau_f > \tau$

7.15 94 kN/m²

7.17 $\phi_{cu} = 15°$, $\phi = 23.3°$

7.19 185.8 kN/m²

7.21 91 kN/m²

7.23 -83 kN/m²

CHAPTER 8

8.1 **a.** 13.9%
        b. 48.44 mm

8.3 50.4 kN/m²

8.5   30°

8.7   36°

8.9   5.06 m

8.11   **a.** 51.4 kN/m²
      b. 39.7 kN/m²
      c. 40.6 kN/m²

8.13   4.27

8.15   **a.** 30 kN/m²
      b. 1.84

8.17   **a.** 0.65
      b. 1.37
      c. 2131 kN/m²

CHAPTER 9

9.1   **a.** $P_o = 87.3$ kN/m, $\bar{z} = 1.67$ m
      b. $P_o = 56.87$ kN/m, $\bar{z} = 1.33$ m

9.3   **a.** $P_p = 169.6$ kN/m, $\sigma'_p = 138.5$ kN/m²
      b. $P_p = 593.3$ kN/m, $\sigma'_p = 296.8$ kN/m²

9.5   **a.** $P_p = 623.38$ kN/m, $\bar{z} = 1.73$ m
      b. $P_p = 1259.54$ kN/m, $\bar{z} = 2.77$ m

9.7   **a.** At $z = 0$, $\sigma_a = -24$ kN/m²; at $z = 6$ m, $\sigma_a = 90$ kN/m²
      b. 1.26 m
      c. 198 kN/m
      d. 213.3 kN/m

9.9   12 kN/m

9.11   **a.** $P_a = 80$ kN/m, 0.888 m from bottom of wall
      b. $P_a = 56.3$ kN/m, 1.1 m from bottom of wall

9.13   **a.** 1261 kN/m
      b. 1787 kN/m
      c. 2735 kN/m

CHAPTER 10

10.1   9.9m

10.3

β (deg)	H_{cr} (m)
20	∞
25	9.9
30	5.2
35	3.69
40	2.98

10.5   0.98

10.7   5.76

10.9   39.4 m

10.11   1.8

10.13   **a.** 8.21 m   **b.** 14.1 m
      c. 6.98 m

10.15   4.4 m

10.17   1.27

10.19   29°

10.21   **a.** 2   **b.** 1.9

10.23   1.83

CHAPTER 11

11.1   311 kN/m²

11.3   1743 kN

11.5   8834 kN

11.7   2 m

11.9   19,789 kN

11.11   1851 kN

11.13   14.55 mm

11.15   27 mm

11.17   586 kN/m²

11.19   284.5 kN/m²

11.21   11.7 m

11.23   4.2 m

CHAPTER 12

12.1 $FS_{(overturning)} = 3.41$, $FS_{(sliding)} = 1.5$, $FS_{(bearing\ capacity)} = 5.5$

12.3 $FS_{(overturning)} = 2.81$, $FS_{(sliding)} = 1.56$, $FS_{(bearing\ capacity)} = 3.22$

12.5 $FS_{(overturning)} = 6.21$, $FS_{(sliding)} = 2.35$

12.7 **a.** 3.68×10^{-4} m³/m
b. 1.53×10^{-3} m³/m

12.9 3.34×10^{-4} m³/m

12.11 **a.** $c_{av} = 84.5$ kN/m², $\gamma_{av} = 17.07$ kN/m³
b. $\sigma = 35.85$ kN/m² (Use Figure 12.17.)

12.13 22.94×10^{-5} m³/m

12.15 1.75

CHAPTER 13

13.1 **a.** 261.7 kN **b.** 1184 kN

13.3 **a.** 1435 kN **b.** 976 kN
c. 603 kN

13.5 609 kN

13.7 21.84 mm

13.9 9565 kN

13.11 71%

13.13 3831 kN

13.15 4478 kN

13.17 24.8 kN

13.19 42.9 kN

13.21 130.7 kN

13.23 8 mm

13.25 **a.** 5000 kN **b.** 1802 kN

Index